Statistical Analysis of Stochastic Processes in Time

Many observed phenomena, from the changing health of a patient to values on the stock market, are characterised by quantities that vary over time: stochastic processes are designed to study them. Much theoretical work has been done but virtually no modern books are available to show how the results can be applied. This book fills that gap by introducing practical methods of applying stochastic processes to an audience knowledgeable only in the basics of statistics. It covers almost all aspects of the subject and presents the theory in an easily accessible form that is highlighted by application to many examples. These examples arise from dozens of areas, from sociology through medicine to engineering. Complementing these are exercise sets making the book suited for introductory courses in stochastic processes.

Software is provided within the freely available R system for the reader to be able to apply all the models presented.

J. K. LINDSEY is Professor of Quantitative Methodology, University of Liège. He is the author of 14 books and more than 120 scientific papers.

CAMBRIDGE SERIES IN STATISTICAL AND PROBABILISTIC MATHEMATICS

This series of high-quality upper-division textbooks and expository monographs covers all aspects of stochastic applicable mathematics. The topics range from pure and applied statistics to probability theory, operations research, optimization, and mathematical programming. The books contain clear presentations of new developments in the field and also of the state of the art in classical methods. While emphasizing rigorous treatment of theoretical methods, the books also contain applications and discussions of new techniques made possible by advances in computational practice.

Already published

1. *Bootstrap Methods and Their Application*, by A. C. Davison and D. V. Hinkley
2. *Markov Chains*, by J. Norris
3. *Asymptotic Statistics*, by A. W. van der Vaart
4. *Wavelet Methods for Time Series Analysis*, by Donald B. Percival and Andrew T. Walden
5. *Bayesian Methods*, by Thomas Leonard and John S. J. Hsu
6. *Empirical Processes in M-Estimation*, by Sara van de Geer
7. *Numerical Methods of Statistics*, by John F. Monahan
8. *A User's Guide to Measure Theoretic Probability*, by David Pollard
9. *The Estimation and Tracking of Frequency*, by B. G. Quinn and E. J. Hannan
10. *Data Analysis and Graphics using R*, by John Maindonald and John Braun
11. *Statistical Models*, by A. C. Davison
12. *Semiparametric Regression*, by David Ruppert, M. P. Wand and R. J. Carroll
13. *Exercises in Probability*, by L. Chaumont and M. Yor

Statistical Analysis of Stochastic Processes in Time

J. K. Lindsey
University of Liège

CAMBRIDGE UNIVERSITY PRESS
Cambridge, New York, Melbourne, Madrid, Cape Town,
Singapore, São Paulo, Delhi, Mexico City

Cambridge University Press
The Edinburgh Building, Cambridge CB2 8RU, UK

Published in the United States of America by Cambridge University Press, New York

www.cambridge.org
Information on this title: www.cambridge.org/9781107405325

First published 2004
First paperback edition 2011

A catalogue record for this publication is available from the British Library

Library of Congress Cataloguing in Publication Data

ISBN 978-0-521-83741-5 Hardback
ISBN 978-1-107-40532-5 Paperback

Contents

Preface

Throughout their history, human beings have been fascinated by time. Indeed, what is history but an interpretation of time? Each civilisation has had its own special conception of time. Our present anti-civilisation only knows 'time is money'! No one can deny that the study of time is important. This text attempts to make more widely available some of the tools useful in such studies.

Thus, my aim in writing this text is to introduce research workers and students to ways of modelling a wide variety of phenomena that occur over time. My goal is explicitly to show the broadness of the field and the many inter-relations within it. The material covered should enable mathematically literate scientists to find appropriate ways to handle the analysis of their own specific research problems. It should also be suitable for an introductory course on the applications of stochastic processes. It will allow the instructor to demonstrate the unity of a wide variety of procedures in statistics, including connections to other courses. If time is limited, it will be possible to select only certain chapters for presentation.

No previous knowledge of stochastic processes is required. However, an introductory course on statistical modelling, at the level of Lindsey (2004), is a necessary prerequisite. Although not indispensable, it may be helpful to have more extensive knowledge of several areas of statistics, such as generalised linear and categorical response models. Familiarity with classical introductory statistics courses based on point estimation, hypothesis testing, confidence intervals, least squares methods, personal probabilities, ... will be a definite handicap.

Many different types of stochastic processes have been proposed in the literature. Some involve very complex and intractable distributional assumptions. Here, I shall restrict attention to a selection of the simpler processes, those for which explicit probability models, and hence likelihood functions, can be specified and which are most useful in statistical applications modelling empirical data. More complex models, including those requiring special estimation techniques such as Monte Carlo Markov Chain, are beyond the scope of this text. Only parametric models are covered, although descriptive 'nonparametric' procedures, such as the Kaplan–Meier estimates, are used for examining model fit.

The availability of explicit probability models is important for at least two reasons:

(i) Probability statements can be made about observable data, including the observed data:

    (a) A likelihood is available for making inferences.
    (b) Predictions can be made.

(ii) If the likelihood can be calculated, models can be compared to see which best fit the data, instead of making empty claims about wonderful models with no empirical basis, as is most often done in the statistical literature.

Isolated from a probability model basis, parameter estimates, with their standard errors, are of little scientific value.

Many standard models, such as those for survival, point processes, Markov chains, and time series, are presented. However, because of the book's wide scope, it naturally cannot cover them in as great a depth as a book dedicated to only one of them. In addition, certain areas, such as survival analysis and time series, occupy vast literatures to which complete justice cannot be made here. Thus, in order to provide a reasonably equitable coverage, these two topics are explored especially briefly; the reader can consult a good introductory text on either of these topics for additional details.

Many basic theoretical results are presented without proof. The interested reader can pursue these in further detail by following up the 'Further reading' list at the end of each chapter. On the other hand, for the readers primarily interested in developing appropriate stochastic models to apply to their data, the sections labelled 'Theory' can generally be skimmed or skipped and simply used as a reference source when required for deeper understanding of their applications.

Stochastic processes usually are classified by the type of recording made, that is, whether they are discrete events or continuous measurements, and by the time spacing of recording, that is, whether time is discrete or continuous. Applied statisticians and research workers usually are particularly interested in the type of response so that I have chosen the major division of the book in this way, distinguishing between categorical events and continuous measurements. Certain models, such as Markov chains using logistic or log linear models, are limited to discrete time, but most of the models can be applied in either discrete or continuous time.

Classically, statistics has distinguished between linear and nonlinear models, primarily for practical reasons linked with numerical methods and with inference procedures. With modern computing power, such a distinction is no longer necessary and will be ignored here. The main remaining practical difference is that nonlinear models generally require initial values of parameters to be supplied in the estimation procedure, whereas linear models do not.

It is surprisingly difficult to find material on fitting stochastic models to data. Most of the literature concentrates either on the behaviour of stochastic models under specific restrictive conditions, with illustrative applications rarely involving real data, or on the estimation of some asymptotic statistics, such as means or variances. Unavoidably, most of the references for further reading given at the ends of the chapters are of much more difficult level than the present text.

My final year undergraduate social science students have helped greatly in developing this course over the past 25 years. The early versions of the course were based on Bartholomew (1973), but, at that time, it was very difficult or impossible actually to analyse data in class using those methods. However, this rapidly evolved, eventually to yield Lindsey (1992). The present text reflects primarily the more powerful software now available. Here, I have supplemented the contents of my current course with extra theoretical explanations of the stochastic processes and with examples drawn from a wide variety of areas besides the social sciences.

Thus, I provide the analysis of examples from many areas, including botany (leaf growth), criminology (recidivism), demography (migration, human mortality), economics and finance (capital formation, share returns), education (university enrolment), engineering (degradation tests, road traffic), epidemiology (AIDS cases, respiratory mortality, spermarche), industry (mining accidents), medicine (blood pressure, leukæmia, bladder and breast cancer), meteorology (precipitation), pharmacokinetics (drug efficacy, radioactive tracers), political science (voting behaviour), psychology (animal learning), sociology (divorces, social mobility), veterinary science (cow hormones, sheep infections), and zoology (locust activity, nematode control). Still further areas of application are covered in the exercises.

The data for the examples and exercises, as well as the R code for all of the examples, can be found at `popgen0146uns50.unimaas.nl/~jlindsey`, along with the required R libraries. With this material, the reader can see exactly how I performed the analyses described in the text and adapt the code to his or her particular problems.

This text is not addressed to probabilists and academic statisticians, who will find the definitions unrigorous and the proofs missing. Rather, it is aimed at the scientist, looking for realistic statistical models to help in understanding and explaining the specific conditions of his or her empirical data. As mentioned above, the reader primarily interested in applying stochastic processes can omit reading the theory sections and concentrate on the examples. When necessary, reference can then be made to the appropriate parts of theory.

I thank Bruno Genicot, Patrick Lindsey, and Pablo Verde who provided useful comments on earlier versions of certain chapters of this text.

Notation and symbols

Notation is generally explained when it is first introduced. However, for reference, some of the more frequently used symbols are listed below.

Vectors are bold lower case and matrices bold upper case Greek or Roman letters. $\top$ denotes the transpose of a vector or matrix.

h, i, j, k, l	arbitrary indices
Y, y	random response variable and its observed value
T, t	time
h	lag
S	sum of random variables
$x, \mathbf{X}$	explanatory variables
N, n	number of events
$\Delta N, \Delta n$	change in number of events
Δt	interval width
$\mathcal{F}$	previous history
μ	location parameter
σ^2	variance
π	probability (usually binary)
τ	change point parameter
ψ, ξ	random parameters
$\alpha, \beta, \delta, \epsilon, \zeta, \phi, \nu, \upsilon, \kappa, \theta$	arbitrary parameters
ρ	(auto)correlation or other dependence parameter
M	order of a Markov process
R	length of a series
$\Pr(y)$	probability of response
$f(y)$	probability density function
$F(y)$	cumulative distribution function
$S(y)$	survival function
$p(\cdot)$	probability of a random parameter
$h(\cdot)$	arbitrary regression function
$g(\cdot)$	link function
Λ	integrated intensity (function)
λ	intensity (function)

$\mathrm{E}(\cdot)$	expected value
γ	(auto)covariance (function)
$\boldsymbol{\Sigma}$	covariance matrix
$\mathrm{I}(\cdot)$	indicator function
$B(\cdot,\cdot)$	beta function
$\Gamma(\cdot)$	gamma function
$\mathrm{L}(\cdot)$	likelihood function
$\mathbf{T}$	transition (probability) matrix
$\boldsymbol{\Lambda}$	transition intensity matrix
$\boldsymbol{\pi}$	marginal or conditional probability distribution
$\boldsymbol{\nu}$	first passage distribution
$\mathbf{F}$	diagonal matrix of probabilities
$\mathbf{D}$	diagonal matrix of eigenvalues
$\mathbf{W}$	matrix of eigenvectors
$\mathbf{b}$	vector of deterministic input
$\boldsymbol{\epsilon}$	vector of random input

Part I

Basic principles

1

What is a stochastic process?

Intuitively, a stochastic process describes some phenomenon that evolves over time (a *process*) and that involves a random (a *stochastic*) component. Empirically, we observe such a process by recording values of an appropriate response variable at various points in time. In interesting cases, the phenomenon under study will usually depend on covariates. Some of these may also vary over time, whereas others will define the differing (static) conditions under which various 'copies' of the process occur.

1.1 Definition

Thus, a stochastic process involves some response variable, say Y_t, that takes values varying randomly in some way over time t (or space, although that will not be considered here). Y_t may be a scalar or a vector, but I shall concentrate primarily on scalar responses in this text (see, however, Section 8.3). Generally, in the study of such processes, the term 'random' is replaced by '*stochastic*'; hence, the name.

An observed value or realisation y_t of the response variable Y_t is called the *state* of the process at time t. We might call an observation of the state of a process an event. However, I shall restrict the meaning of *event* to the occurrence of a *change* of state. Thus, the number of possible different events will depend, among other things, on the number of distinct states.

More generally, the probability of the process being in some given state at some point in time may depend on some function of previous events and of covariates. Usually, the probabilities of possible events will be conditional on the state of the process. Such relationships will thus be determined by the type of model being fitted.

The main properties distinguishing among observed stochastic processes are:

(i) The frequency or periodicity with which observations are made over time.
(ii) The set of all of its possible observable values, that is, of all possible responses or states of the series, called the *state space*.
(iii) The sources and forms of randomness present, including the nature of the dependence among the values in a series of realisations of the random variable Y_t.

(iv) The number of 'copies' of the process available (only one or several), which will determine how adequate information can be obtained for modelling.

Let us look more closely at each of these aspects.

1.1.1 Time

Observations of a stochastic process (at least of the kinds that interest us here) are made over *time*. If these observations are at equally spaced intervals, time is said to be *discrete*. Otherwise, time is *continuous*. Notice, however, that a process can never really be observed continuously because that would imply an infinite number of observations, even in a small interval of time. Hence, the distinction is primarily used to determine what kind of model will be applied. Continuous-time models can be used for any data but may be more complex, and may not be appropriate if changes can only occur at discrete time points. Discrete-time models require equally-spaced observations, unless some simple mechanism for missingness can be introduced.

Attention may centre on

(i) the *states* at given points in time,
(ii) the *events*, that is, on what change of state occurs at each particular time point, or
(iii) the *times* of events.

Consider simple examples:

- When an economist measures monthly unemployment, time is only an equally-spaced indicator of when observations are made. The number of unemployed may change at each monthly recording. Either the level of employment (the state) or the amount of change (the event) may be of central interest. Time itself is not of direct concern except for ordering the observations.
- In contrast, when a doctor records the times of a patient's repeated infections, the state might be defined to be the total number of such infections so far suffered by that patient. Each observation (event) is the same, an infection, and the time between these events is essential. With substantial loss of information, this could be reduced to discrete time by recording only the numbers of infections in equally-spaced time intervals.

A response may only be recorded when some specific event of interest occurs. However, in order to determine the timing of that event, fairly continual observation of the process usually is necessary, that is, a series of intermediate, implicit recordings of no event. If observation begins at some natural time point, such as birth, at which no response value occurs, the mechanism determining the time to the first event will usually be different from that between subsequent events.

1.1.2 State space

At any given point in time, a process will be in some *state*. This usually is observed by recording the value of a response variable y_t at that point t. As always in statistical modelling, the set of possible states is a scientific construct that should be defined in a way appropriate to answer the questions at hand. Although, in principle, the state is what is being observed, certain models also assume a second process of unobservable *hidden* states (Chapters 7 and 11).

The set of all possible observable states is called the *state space*. This may be finite, yielding a categorical response variable, or infinite, giving either a discrete or a continuous response variable. Generally, a *minimal* state space is chosen, one in which the probability (density) of every state is nonzero.

If an observed response variable were truly continuous, every recording would be an event because no two would be the same. However, this is empirically impossible so that any observable process could possibly stay in the same state (given the limit of precision of the recording instrument) over two or more consecutive observation points.

A categorical response usually refers to a finite set of possible different states that may be observed. However, when only one type of response is of particular interest and it is fairly rare, the response might be recorded as binary, indicating presence (coded 1) or absence (coded 0) of that response. This particular response, with a binary state space, is often called a *point event* or a *recurrent event* (Chapter 4); thus, here the term 'event' refers both to one of the states and to the change of state. Above, I gave an example of repeated infections, where the cumulative number of infections was important. If, instead, repeated epileptic fits were recorded, the states might more appropriately be defined as having a fit or not on each particular day instead of the total number of fits so far.

In certain situations, the state may be defined by more than one response value, that is, $\mathbf{y}_t$ may be a vector containing quite distinct types of values. Thus, there may be a categorical value, such as a binary indicator that it rains or not, accompanied by a (usually quantitative) value, the *mark*, for example, how much rain fell (Section 8.3).

A vector $\mathbf{y}_t$ also is usually necessary when there are *endogenous* time-varying covariates. These are variables, other than the response of direct interest, that are influenced by the previous states of that response. Suppose, for example, that the condition of a patient, measured in some appropriate way, is the response of direct interest. If the dose of a medication that a patient receives depends upon his or her previous condition, then dose and condition will generally have to be handled simultaneously. They must together define the state and be allowed to vary stochastically interdependently, if a reasonable model is to be constructed.

A process may also involve time-varying *exogenous* covariates, that is, variables not influenced by the previous states of the process. The stochastic variability of such covariates is not usually of interest, so that the probability of the state y_t can be taken to be conditional on their observed values, as in classical regression models.

Practically, we see from this section and the preceding one that models for

Table 1.1. *Chapters in which different types of stochastic processes are covered.*

	State space	
Time	Categorical	Continuous
Discrete	5, 8	9
Continuous	3, 4, 6, 7, 8	9, 10, 11, 12, 13, 14

stochastic processes can be classified by the type of variable observed, that is, the state space, and by the frequency or regularity with which it is observed over time. The structure of this text according to these criteria is summarised in Table 1.1. Recall, however, that most models in continuous time also can be applied to observations in discrete time.

1.1.3 Randomness

In a *deterministic* process, one can predict the exact sequence of outcomes (states) from the initial conditions, although some, especially chaotic, systems are extremely sensitive to these conditions. In contrast, a *stochastic* process has an inherent component of unpredictability or randomness. In empirical observation, randomness is closely associated with unknownness or incomplete information. To handle such situations, the theory of probability is used. Thus, predictions involving stochastic processes will not indicate specific outcomes, but only the probabilities of occurrence of the different possible outcomes.

A stochastic process may involve many forms of randomness. A number will arise from the imperfections of the observation and modelling procedures. However, in scientific modelling, the most important should be inherent in the process under study.

The types of randomness that need to be allowed for in modelling a stochastic process can have several sources including the following:

(i) Unmeasurable variability is built into many scientific systems. This is true of quantum mechanics, but also of most biological and social processes.
(ii) Almost invariably, all of the measurable factors determining the process cannot be take into account. For example, unrecorded environmental conditions may change over the period in which the process is acting.
(iii) The initial conditions may be very difficult to determine precisely.

Unfortunately, an inappropriate model can also generate additional spurious randomness and dependencies.

Traditionally, essentially for mathematical simplicity, the Poisson, binomial, and normal distributions have been used in modelling the randomness of stochastic processes. However, we shall see as we proceed that a wide variety of other distributions may be more suitable in specific circumstances.

1.1.4 Stationarity, equilibrium, and ergodicity

The series of responses y_t of a stochastic process usually will not be independent. Thus, procedures must be available to introduce appropriate dependencies. Because complex models should generally be avoided in science, relatively simple methods for introducing such dependencies are desirable.

Multivariate distributions

Multivariate probability distributions provide the general framework in which to specify the ways in which responses are interdependent. Unfortunately, in the context of stochastic processes, these may be difficult to use, especially if long series of observations are available. Thus, with a series of R observations, a model involving a multivariate normal distribution will require the manipulation of an $R \times R$ covariance matrix. When R is large, this will often not be practical or efficient, at least for technical reasons.

Fortunately, for an ordered series of R responses, such as those that interest us, the multivariate distribution always can be decomposed into an ordered sequence of independent *conditional* univariate distributions:

$$f(y_0, y_1, \ldots, y_R) = f_0(y_0) f_1(y_1|y_0) \cdots f_R(y_R|y_0, \ldots, y_{R-1}) \tag{1.1}$$

where

$$f_t(y_t|y_0, \ldots, y_{t-1}) = \frac{f(y_0, \ldots, y_t)}{f(y_0, \ldots, y_{t-1})} \tag{1.2}$$

Notice that each conditional distribution $f_t(y_t|y_0, \ldots, y_{t-1})$ may be completely different from all the others, even though the multivariate distributions $f(\cdot)$ for different lengths of series will usually have the same form.

Generally, it will be easier to work with the conditional distributions than the multivariate one. For the multivariate normal distribution, this will be a series of univariate normal distributions, not involving directly the covariance matrix. I shall elaborate on this approach in Section 1.2 below.

In contrast to the multivariate distribution, and its univariate conditional decomposition, the series of univariate *marginal* distributions, although often of interest, cannot by itself specify a unique stochastic process (unless all successive states are independent). It is generally of limited direct use in constructing a realistic model and will, most often, rather be a byproduct of the construction. On the other hand, this sequence of univariate marginal distributions of a stochastic process does provide valuable information indicating how the process is evolving over time: its 'underlying' profile or trend.

In the decomposition in Equation (1.1), no restrictive assumptions have been made (except ordering). However, in order for such a multivariate model to be tractable, even fitted conditionally, rather strong assumptions often have to be made. The problem with the general specification just given is that every response has a different conditional distribution, and each new response will have yet another one. The situation is changing faster than information can be accumulated about it! Thus, we require some reasonable simplifying assumptions.

Stationarity

A stochastic process is said to be *strictly stationary* if all sequences of consecutive responses of equal length in time have identical multivariate distributions

$$f(Y_1 = y_1, \cdots, Y_R = y_R) = f(Y_{t+1} = y_1, \cdots, Y_{t+R} = y_R) \tag{1.3}$$

for all t and R. In other words, shifting a fixed-width time observation window along a strictly stationary series always yields the same multivariate distribution. Such an assumption reduces enormously the amount of empirical information necessary in order to model a stochastic process.

A less restrictive assumption, reducing even further the amount of information required, is that a process is *second-order stationary*. This is defined only by the mean, variance, and covariances:

$$\begin{aligned} \mathrm{E}(Y_t) &= \mu \\ \mathrm{E}[(Y_t - \mu)(Y_{t+h} - \mu)] &= \gamma(h) \end{aligned} \tag{1.4}$$

for all t and h. Because a multivariate normal distribution is completely defined by its first two moments, if the process is normal and second-order stationary, it is strictly stationary. This is not generally true of other distributions. In this text, stationarity will always be strict.

Stationarity is a characteristic of multivariate distributions. Thus, from Equation (1.1), it cannot be determined solely by the conditional distributions, but requires also that the initial marginal distribution $f_0(y_0)$ be specified. Of course, this, in turn, implies that the univariate marginal distributions at all other time points will also be known.

Stationarity can be an appropriate assumption if the stochastic process has no inherent time origin. However, in experimental situations, for example, where treatments are applied at a specific time point, this will not be true. The need for greater information due to lack of stationarity can often be compensated by studying replications of the process (Section 1.1.5).

Equilibrium

Although a process may not be stationary when it starts, it may reach an *equilibrium* after a sufficiently long time, independent of the initial conditions. In other words, if an equilibrium has been reached, the probability that the process is in each given state, or the proportion of time spent in each state, has converged to a constant that does not depend on the initial conditions. This generally implies that eventually the process approaches closely to a stationary situation in the sense that, if it initially had the equilibrium distribution of states, it would be stationary. (See Cox and Miller, 1965, pp. 9, 272.)

Ergodicity

The concept of *ergodicity* is closely related to that of equilibrium, although the former has various meanings in the literature on stochastic processes. Ergodic theorems provide identities between probability averages, such as an expected value, and long-run averages over a single realisation of the process. Thus, if the equilibrium probability of being in a given state equals the proportion of a long time period spent in that state, this is called an ergodic property of the process. In a similar way, the law of large numbers can be generalised to stochastic processes. (See Cox and Miller, 1965, p. 292.)

Regeneration points

Another important concept for some stochastic processes is that of a *regeneration point*. This is a time instant at which the process returns to a specific state such that future evolution of the process does not depend on how that state was reached. In other words, whenever such a process arrives at a regeneration point, all of its previous history is forgotten.

A well known case is the renewal process (Section 4.1.4) describing times between recurrent events which, as its name suggests, starts over again at each such event. (See Daley and Vere-Jones, 1988, p. 13.)

I shall look at further general procedures for simplifying models of stochastic processes in Section 1.2 below.

1.1.5 Replications

When studying a stochastic process, two approaches to obtaining adequate information can be envisaged. One can either observe

(i) one series for a long enough period, if it is reasonably stable, or
(ii) several short 'replications' of the process, if they are reasonably similar.

In certain situations, one has no choice but to use replications. Thus, for example, with survival data (Chapter 3), one single event, say death, terminates the process so that the only way to proceed is by collecting information on a large number of individuals and by assuming that the process is identical for all of them.

Both approaches can create problems. The phenomenon under study may not be stable enough to be observed over a very long time, say due to problems of lack of stationarity as discussed above. It may only be possible to assume that shorter segments of a series are from the same stochastic process. On the other hand, with replications, one must be able to assume that the different series recorded do, in fact, represent the same stochastic process. In certain situations, such as survival data, it may not even be possible to check this strong assumption empirically.

When replications of a stochastic process are modelled, extreme care must be taken with the time scale. If it is not chronological time, problems may arise. For example, in experiments with biological organisms, time may be measured either from birth or from start of a treatment. If all births do not occur simultaneously and treatment is not started at the same time for all subjects, events that occur at similar times after beginning treatment may occur neither closely together chronologically nor at similar ages. This can create difficult problems of interpretation due to confounding.

In the examples that I shall analyse in this text, I shall use either one long series or a set of several short ones, depending both on the type of problem and on the kind of data available. Generally, in the second case, when replications are present, I shall assume, for simplicity, that they come from the same process, perhaps with any differences explainable by time-constant (interprocess) covariates. Only in Section 7.3 and in Chapter 14 shall I look at some standard ways of modelling the differences among a set of series not described by observed covariates.

1.2 Dependence among states

As we have seen, dependencies among successive responses of a stochastic process can be modelled by multivariate distributions or equivalently by the corresponding product of conditional distributions. Certain general procedures are available that I shall review briefly here. Some of them arise from time series analysis (Chapter 9) but are of much wider applicability.

1.2.1 Constructing multivariate distributions

In a model for a stochastic process, some specific stochastic mechanism is assumed to generate the states. We often can expect that states of a series observed more closely together in time will be more similar, that is, more closely related. In other words, the state at a given time point will generally be related to those recently produced: the probability of a given state will be conditional, in some way, on the values of the process previously generated.

In certain situations, an adequate model without such dependence may be constructed. It usually will require the availability of appropriate time-varying covariates. If these have been recorded, and perhaps an appropriate time trend specified, the present state, conditional on the covariates, should be independent of previous states. However, in many cases, this will not be possible, because of lack of information, or will not be desirable, perhaps because of the complexity of the model required or its lack of generality.

In order to model time dependencies among the successive states of a stochastic process of interest, we may choose a given form either for the conditional distributions, on the right-hand side of Equation (1.1), or for the multivariate distribution, on the left-hand side. In general, the conditional distribution will be different from the multivariate and marginal distributions, because the ratio of two multivariate distributions does not yield a conditional distribution of the same form except in very special circumstances such as the normal distribution. The one or the other will most often be intractable.

Because a limited number of useful non-normal multivariate distributions is available, suitable models often can only be obtained by direct construction of the conditional distribution. Thus, usually, we shall need to set up some hierarchical series of conditional distributions, as in Equation (1.1). In this way, by means of the univariate conditional probabilities, we can construct an appropriate multivariate distribution for the states of the series.

The hierarchical relationship among the ordered states of a series implies, by recursion, that the conditional probabilities are independent, as in Equation (1.1). In this way, univariate analysis may be used and the model will be composed of a product of terms. Thus, multivariate distributions with known conditional form are usually much easier to handle than those with known marginal form.

However, as we have seen in Section 1.1.4, the general formulation of Equation (1.1) highlights some potential difficulties. Each state depends on a different number of previous states so that the conditional distribution $f_t(y_t|y_0, \cdots, y_{t-1})$ is different at each time point; this may or may not be reasonable. Usually, additional assumptions must be introduced. As well, the unconditional distribution of the first state is required. Its choice will depend upon the initial conditions for the process.

The ways in which the conditionality in the distributions is specified will depend on the type of state space. I shall, first, look at general procedures for any kind of state (Sections 1.2.2 and 1.2.3) and, then, at specific ones for continuous states (or perhaps counts, although this is unusual for the true state of a stochastic process) and for categorical states.

1.2.2 Markov processes

An important class of simple processes makes strong assumptions about dependence over time, in this way reducing the amount of empirical information required in order to model them.

A series of responses in discrete time such that

$$f(y_t|y_0, \cdots, y_{t-1}) = f(y_t|y_{t-1}) \tag{1.5}$$

so that each state only depends on the immediately preceding one, is known as a *Markov process*. In other words, the state at time t, given that at $t-1$, is independent of all the preceding ones. Notice that, in contrast to Equation (1.1), in the simplest situation here, the form of the conditional distribution $f(y_t|y_{t-1})$ does not change over time.

This definition can be extended both to dependence further back in time and to continuous time (Section 6.1). Recall, however, that such a conditional specification is not sufficient to imply stationarity. This will also depend on the initial conditions and/or on the length of time that the process has been functioning.

Equation (1.5) specifies a Markov process of order one. More generally, if the dependence extends only for a short, fixed distance back in time, so that the present state only depends on the M preceding states, it is said to be a Markov process of order M. The random variables Y_t and Y_{t-k} are conditionally independent for $k > M$, given the intermediate states.

It is often necessary to assume that a stochastic process has some finite order M (considerably less than the number of observations available over time), for otherwise it is nonstationary and its multivariate distribution continues to change with each additional observation, as we saw above.

If the response variable for a Markov process can only take discrete values or states, it is known as a *Markov chain* (Chapter 5 and Section 6.1.3). Usually, there will be a finite number of possible states (categories), and observations will be made at equally-spaced discrete intervals. Interest often centres on the conditional *transition probabilities* of changes between states. When time is continuous, we have to work with *transition rates* or *intensities* (Section 3.1.3) between states, instead of probabilities. When the response variable for a Markov process is continuous, we have a *diffusion process* (Chapter 10).

1.2.3 State dependence

One simple way to introduce Markov dependence is to construct a regression function for some parameter of the conditional probability distribution such that it incorporates previously generated states directly, usually in addition to the other covariates. The states may be either continuous or categorical. Thus, the location parameter μ (often the mean) of the conditional distribution of the series could be dependent in the following way:

$$\mu_t = \rho y_{t-1} + h(\boldsymbol{\beta}, \mathbf{x}_t) \tag{1.6}$$

for a process of order one, where $\mathbf{x}_t$ is a vector of possibly time-varying covariates in some regression function $h(\cdot)$, possibly nonlinear in the parameters $\boldsymbol{\beta}$. If there are more than two categorical states, μ_t usually will have to be replaced by a vector (often of conditional probabilities) and y_{t-1} will need to be modelled as a factor variable.

Here, the present location parameter, that is, the prediction of the mean present state, depends directly on the previous state of the process, as given by the previous observed response. Thus, this can be called a *state dependence* model.

An easy way to introduce state dependence is by creating *lagged* variables from the response and using them as covariates. For a first-order Markov process ($M = 1$), the values in the vector of successive observed states will be displaced by one position to create the lagged covariate so that the current state in the response vector corresponds to the previous state in the lagged vector. However, this means that we cannot use the first observed state, without additional assumptions, because we do not know its preceding state. If a higher-order model ($M > 1$) is used, the displacement will be greater and more than one observation will be involved in this way.

One case of this type of model for categorical states is the Markov chain mentioned above (see Chapter 5 and and Section 6.1.3). A situation in which the state space is continuous is the autoregression of time series, at least when there are no time-varying covariates (Chapter 9).

1.2.4 Serial dependence

For a continuous state space, a quite different possibility also exists, used widely in classical time series analysis (Chapter 9). This is to allow some parameter to depend, not directly on the previous observed state, but on the difference between that previous state and its prediction at that time. For a location parameter, this might be

$$\mu_t = \rho[y_{t-1} - h(\boldsymbol{\beta}, \mathbf{x}_{t-1})] + h(\boldsymbol{\beta}, \mathbf{x}_t) \qquad (1.7)$$

Notice that this will be identical to the state dependence model of Equation (1.6) if there are no covariates, and also that it generally will not work for a categorical state space because the subtraction has no meaning.

Dependence among states is now restricted to a more purely stochastic component, the difference between the previous observed state and its location regression function, called the *recursive residual* or *innovation*. This may be seen more clearly by rewriting Equation (1.7) in terms of these differences:

$$\mu_t - h(\boldsymbol{\beta}, \mathbf{x}_t) = \rho[y_{t-1} - h(\boldsymbol{\beta}, \mathbf{x}_{t-1})] \qquad (1.8)$$

The new predicted difference is related to the previous observed difference by ρ. As in Equation (1.6), the previous observed value y_{t-1} is being used to predict the new expected value μ_t, but here corrected by its previous prediction.

I shall call this a *serial dependence* model. The present location parameter depends on how far the previous state was from its prediction given by the corresponding location parameter, the previous residual or innovation.

Thus, both state and serial dependence yield conditional models: the response has some specified *conditional* distribution given the covariates and the previous state. In contrast, for some reason, most models constructed by probabilists, as generalisations of normal distribution serial dependence time series to other distributions, require the *marginal* distribution to have some required form (see, among others, Lawrance and Lewis, 1980). These are complex to construct mathematically and difficult to interpret scientifically.

1.2.5 Birth processes

If the state space has a small finite number of states, that is, it is categorical, different methods may need to be used. One possibility may be to condition on the number of previous recurrent events or, more generally, on the number of times that the process was previously in one or more of the states.

The classical situation arises when the two possible states are present or absence of a recurrent event. Then, a *birth process* counts the number of previous such events:

$$\pi_t = \rho n_{t-1} + h(\boldsymbol{\beta}, \mathbf{x}_t) \qquad (1.9)$$

where π_t is the conditional probability of the event at time t and n_{t-1} is the number of previous such events. Even without time-varying covariates, this process is clearly nonstationary.

Generally, some function of π_t, such as a logit transformation, will instead be used to ensure that it cannot take impossible values. More often, the process will be defined in terms of the (log) rate or intensity of events (Section 4.1) instead of the probability. I shall describe more general types of dependence for finite state spaces in Section 4.1.1, once I have introduced this concept of intensity.

Only three basic types of dependency have been presented here. Other more specific ones may often be necessary in special circumstances. Some of these will be the subject of the chapters to follow.

1.3 Selecting models

As always in statistical work, the ideal situation occurs when the phenomenon under study is well enough known so that scientific theory can tell us how to construct an appropriate model of the stochastic process. However, theory never arises in a vacuum, but must depend on empirical observations as well as on human power of abstraction. Thus, we must begin somewhere! The answers to a series of questions may help in constructing useful models.

1.3.1 Preliminary questions

When confronting the modelling of observations from some stochastic process, one may ask various fundamental questions:

- How was the point in time to begin observation chosen?
  - Is there a clear time origin for the process?
  - What role do the initial conditions play?
- Are observations made systematically or irregularly over time?
  - If observations are irregularly spaced, are these time points fixed in advance, random, or dependent on the previous history of the process itself?
  - Is the process changing continuously over time or only at specific time points?
  - Are all changes in the process recorded or only those when an observation happens to be made?
  - Does a record at a given time point indicate a new value in the series then or only that it changed some time since the previous observation?
- Is the process stationary?
  - Is the process increasing or decreasing systematically over time, such as a growth curve?
  - Is there periodic (daily, seasonal) variation?
  - Does the process change abruptly at some time point(s)?
  - Are there long-term changes?
- Is more than one (type of) response recorded at each time point?
  - Can several events occur simultaneously?
  - Do some quantitative measurements accompany the occurrence of an event?
- Does what is presently occurring depend on the previous history of the process?
  - Is it sufficient to take into account what happened immediately previously (the Markov assumption)?

  - Is there a cumulative effect over the history of the process, such as a birth effect?
- Is the process influenced by external phenomena?
  - Can these be described by time-varying covariates?
  - Is some other unrecorded random process, such as the weather, affecting the one of interest?
- If there is more than one series, do the differences among them arise solely from the randomness of the process?
  - Are the differences simply the result of varying initial conditions?
  - Do the series differ because of their dependence on their individual histories?
  - Can part of the difference among the series be explained by time-constant covariates with values specific to each series?
  - Are there static random differences among the series?
  - Does each series depend on a different realisation of one or more time-varying covariates (different weather recorded in different locations)?
  - Are there unrecorded random processes external to the series influencing each of them in a different way (different weather in different locations, but never recorded)?

Possible answers to some of these questions may be indicated by appropriate plots of the series under study. However, most require close collaboration with the scientists undertaking the study in order to develop a fruitful interaction between empirical observation and theory.

1.3.2 Inference

Some objective empirical procedure must be available in order to be able to select among models under consideration as possible descriptions of an observed stochastic process. With the exception of preliminary descriptive examination of the data, all analyses of such processes in this book will be based on the construction of probabilistic models. This means that the probability of the actually observed process(es) always can be calculated for any given values of the unknown parameters. This is called the *likelihood function*, a function of the parameters for fixed observed data. Here, all inferences will be based on this. Thus, the basic assumption is the Fisherian one that a model is more plausible or likely if it makes the observed data more probable.

A set of probability-based models that one is entertaining as having possibly generated the observed data defines the likelihood function. If this function is so complex as to be intractable, then there is a good chance that it cannot provide useful and interpretable information about the stochastic process.

However, the probability of the data for fixed parameter values, a likelihood value, does not, by itself, take into account the complexity of the model. More complex models generally will make the observed data more probable, but simpler models are more scientifically desirable. To allow for this, minus the logarithm of the maximised likelihood can be penalised by adding to it some function of the number of parameters estimated. Here, I shall simply add the number of parameters to this negative log likelihood, a form of the Akaike (1973) information criterion (AIC). Smaller values will indicate models fitting relatively better to the data, given the constraint on the degree of complexity.

Further reading

Jones and Smith (2001) give an elementary introduction to stochastic processes. Grimmett and Stirzaker (1992), Karlin and Taylor (1975; 1981), Karr (1991), and Ross (1989) provide more advanced standard general introductions. The reader also may like to consult some of the classical works such as Bailey (1964), Bartlett (1955), Chiang (1968), Cox and Miller (1965), Doob (1953), and Feller (1950).

Important recent theoretical works include Grandell (1997), Guttorp (1995), Küchler and Sørensen (1997), and MacDonald and Zucchini (1997). More applied texts include Snyder and Miller (1991), Thompson (1988), and the excellent introduction to the uses of stochastic processes in molecular biology, Ewens and Grant (2001).

An important book on multivariate dependencies is Joe (1997).

For inferences using the likelihood function and the AIC, see Burnham and Anderson (1998) and Lindsey (2004).

Exercises

1.1 Describe several stochastic processes that you can encounter while reading a daily newspaper.
   (a) What is the state space of each?
   (b) What are the possible events?
   (c) Is time discrete or continuous?
   (d) What covariates are available?
   (e) Will interest centre primarily on durations between events or on the states themselves?
   (f) What types of dependencies might be occurring over time?

1.2 Consider the following series, each over the past ten years:
   (a) the monthly unemployment figures in your country,
   (b) the daily precipitation in the region where you live, and
   (c) the times between your visits to a doctor.

   For each series:
   (a) Is the state space categorical or continuous?
   (b) Is time discrete or continuous?
   (c) What types of errors might be introduced in recording the observations?
   (d) Is there a clear time origin for the process?
   (e) Is it plausible to assume stationarity?

(f) Can you expect there to be dependence among the responses?
(g) Can you find appropriate covariates upon which the process might depend?

2

Basics of statistical modelling

In this chapter, I shall review some of the elementary principles of statistical modelling, not necessarily specifically related to stochastic processes. In this way, readers may perhaps more readily understand how models of stochastic processes relate to other areas of statistics with which they are more familiar. At the same time, I shall illustrate how many of these standard procedures are not generally applicable to stochastic processes using, as an example, a study of the duration of marriages before divorce. As in subsequent chapters, I shall entertain a wide variety of distributional assumptions for the response variable and use both linear and nonlinear regression functions to incorporate covariates into the models.

2.1 Descriptive statistics

Let us first examine the data that we shall explore in this chapter.

Divorces Marriage may be conceptualised as some kind of stochastic process describing the relationships within a couple, varying over time, that may eventually lead to rupture. In this light, the process ends at divorce and the duration of the marriage is the centre of interest.

In order to elucidate these ideas, a study was conducted in 1984 of all people divorcing in the city of Liège, Belgium, in that year, a total of 1727 couples. (For the data, see Lindsey, 1992, pp. 268–280). Here, I shall examine how the length of marriage before divorce may vary with certain covariates: the ex-spouses' ages and the person applying for the divorce (husband, wife, or mutual agreement).

Only divorced people were recorded, so that all durations are complete. However, this greatly restricts the conclusions that can be drawn. Thus, the design of this study makes these data rather difficult to model.

- The design was retrospective, looking back in time to see how long people were married when they divorced in 1984. Thus, all divorces occurred within the relatively short period of one year.
- On the other hand, the couples married at quite different periods in time. This could have an influence on the occurrence of divorce not captured by age.
- The study included only those couples who did divorce so that it can tell us

nothing about the probability of divorce. To be complete, such a study would somehow have to include a 'representative' group of people who were still married. These incompletely observed marriages would be censored (Section 3.1.3).

The reader should keep these problems in mind while reading the following analyses.

2.1.1 Summary statistics

Before beginning modelling, it is always useful first to look at some simple descriptive statistics.

Divorces In the divorce study, the mean length of marriage is 13.9 years, with mean ages 38.5 and 36.1, respectively, for the husband and the wife. Because length of marriage will be the response variable, we also should look at its variability; the variance is 75.9, or the standard deviation, 8.7. Thus, an interval of, say, two standard deviations around the mean length of marriage contains negative values; such an interval is meaningless. Symmetric intervals around the mean are not appropriate indicators of variability when the distribution is asymmetric, the typical case for many responses arising from stochastic processes.

A more useful measure would be intervals (contours) of equal probability about the mode, which has highest probability. For this, graphical methods are often appropriate.

2.1.2 Graphics

Visual methods often are especially appropriate for discovering simple relationships among variables. Two of the most useful in the context of modelling are histograms and scatterplots.

Divorces The histogram for duration of marriage is plotted in the upper left hand graph of Figure 2.1; its shape is typical of duration data. We see indeed that it is skewed, not having the form of a normal distribution. From this, intervals of equal probability can be visualised.

However, this histogram indicates the form of the distribution for all of the couples together. Models based on covariates generally make assumptions about the *conditional* distribution for each value of the covariates. Thus, a linear regression model carries the assumption that the conditional distribution is normal with constant variance for all sets of covariate values. This histogram does not provide information about such conditional distributions. Consider, as an example then, the explanatory variable, applicant. We can examine the histograms separately for each of the three types of applicant. These are also given in Figure 2.1. We can see that the form of the histogram differs quite substantially among these three groups.

The above procedure is especially suitable when an explanatory variable has only a few categories. If a quantitative variable, like age, is involved, another approach may be more appropriate. Let us, then, see how to examine graphically

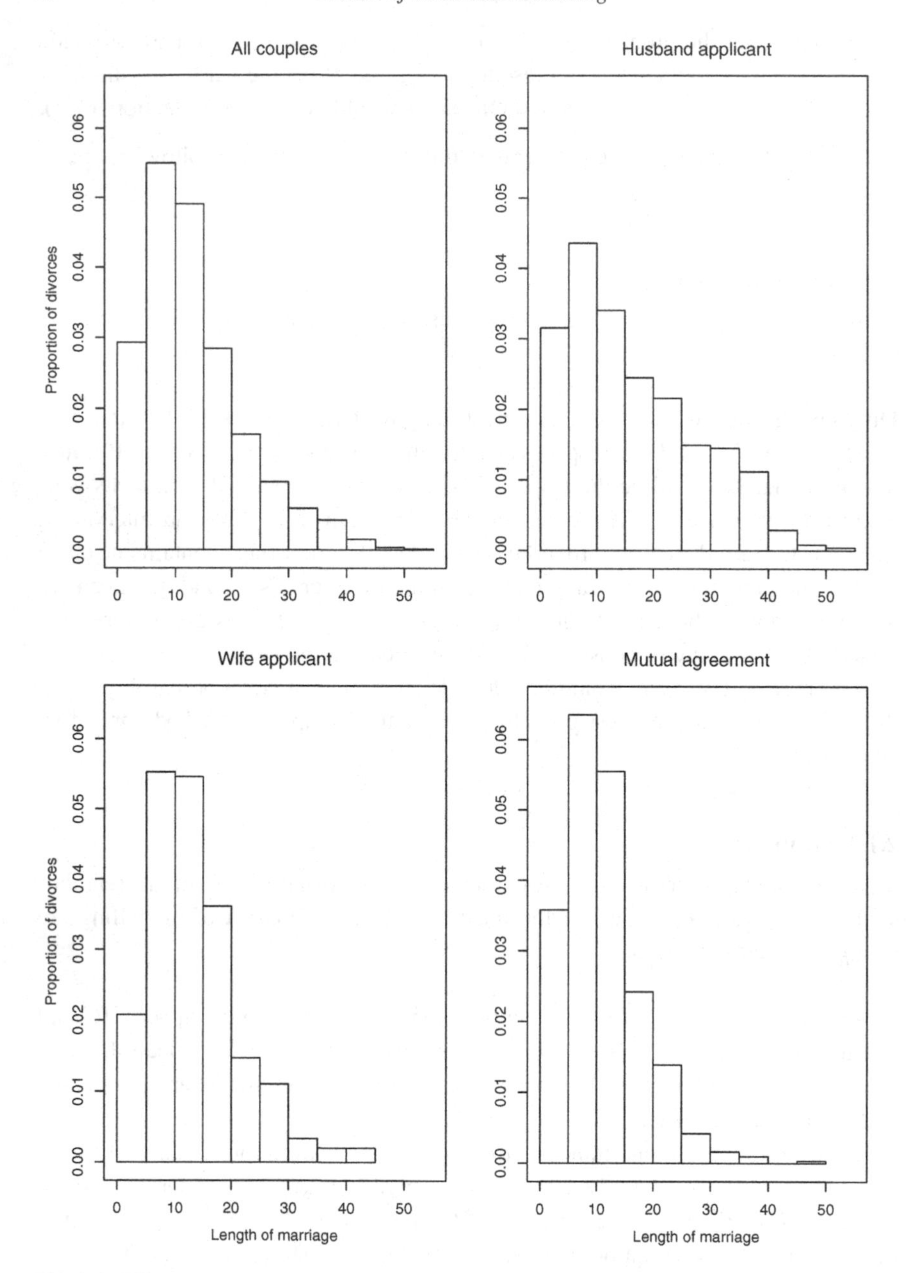

Fig. 2.1. Histograms showing the proportions of couples divorcing after various lengths of marriage in Liège in 1984, grouped into intervals of five years: all couples and separately by applicant.

the relationship between the response variable, duration of marriage, and husband's age. The scatterplot of length of marriage in relation to this age is given in Figure 2.2 (ignore, for the moment, the two diagonal lines). As might be expected, there is a rather strict upper relationship between the two variables. Length of marriage

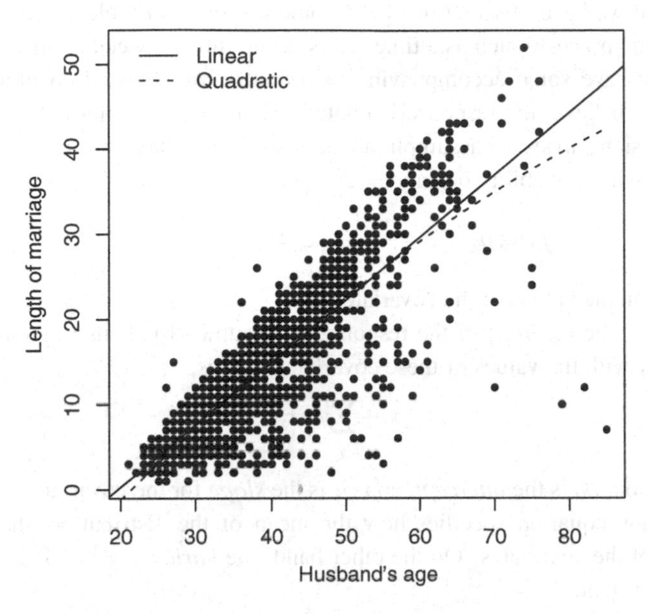

Fig. 2.2. A scatterplot showing the relationship between length of marriage before divorce in Liège in 1984 and the age of the husband, with the fitted normal distribution regression lines.

is, with few exceptions, constrained to be no greater than about age – 20 years. We can also notice the interesting fact that a few very old men married late in life and divorced rather quickly thereafter.

To relate this graph to histograms, consider the density of points along a vertical line for some fixed age of the husband. If we move this line to the left or right, we see that the mass of points along it shifts. This indicates that, if we produced histograms for different age groups, they would have different shapes, as did those for the three applicant groups. Thus, both of these graphical methods indicate that the conditional distribution of length of marriage is changing with the covariates, applicant and husband's age.

As we see from these graphs, the assumptions of normality and constant variance of the conditional distribution of length of marriage, given the applicant group or husband's age, do not appear to be fulfilled either. Nevertheless, I first shall attempt to fit such models to these data.

2.2 Linear regression

One of the most widely (mis)used tools in all of statistics is *linear regression*. This is often misnamed 'least squares' regression. However, least squares estimation refers to the study of a deterministic process, whereby the 'best' straight line is fitted through a series of points. In statistical modelling, the interpretation of linear regression is quite different, although the technical calculations remain the same.

2.2.1 Assumptions

Suppose that we have observations y_i of some response variable Y, say the state of a stochastic process, such as a time series or the time between recurrent events. As well, we have some accompanying explanatory variables or covariates, x_{ij}, to which, we believe, the response is related. Then, applying normal distribution linear regression carries the assumption that this response has a normal or Gaussian distribution with probability density

$$f(y_i;\boldsymbol{\mu},\sigma^2) = \frac{1}{\sqrt{2\pi\sigma^2}}\mathrm{e}^{-\frac{1}{2\sigma^2}(y_i-\mu_i)^2} \tag{2.1}$$

conditional on the values of the covariates.

In addition, the *mean* μ_i of the responses is assumed to change in some deterministic way with the values of these covariates, that is,

$$\mu_i = \beta_0 + \sum_j \beta_j x_{ij} \tag{2.2}$$

In this function, β_0 is the *intercept*, and β_j is the *slope* for the covariate, x_{ij}. Then, this regression equation specifies how the mean of the distribution changes for each value of the covariates. On the other hand, the *variance* σ^2 of Y is assumed to remain constant.

The model is not just the deterministic description of Equation (2.2). As an integral part of it, individual responses are dispersed randomly about the mean in the specific form of the normal distribution in Equation (2.1) with the given variance. This is illustrated in Figure 2.3. Such variability usually will be an integral part of the scientific phenomenon under study, not just measurement error.

This regression function is called *linear* by statisticians for the wrong reason: it is linear in the parameters β_j. This is irrelevant for scientific modelling. On the other hand, the shape of the curve may take certain restricted nonlinear forms, say if x^2 is included as a covariate, as we shall see below.

Once we understand that such a model is describing changes in a normal distribution, we easily can imagine various extensions:

- Other, more suitable, distributions can replace the normal distribution; these will most often be asymmetric.
- The dispersion (here, the variance) about the regression curve need not be held constant.
- The regression equation (2.2) may more appropriately be replaced by some wider class of nonlinear relationships.

These generally will permit more realistic analysis of the data at hand. I shall begin to examine them more closely in Section 2.4. However, first it will be useful to review in some more detail the standard models based on the normal distribution.

2.2.2 Fitting regression lines

Linear regression models are well known and can be fitted using any standard statistical software.

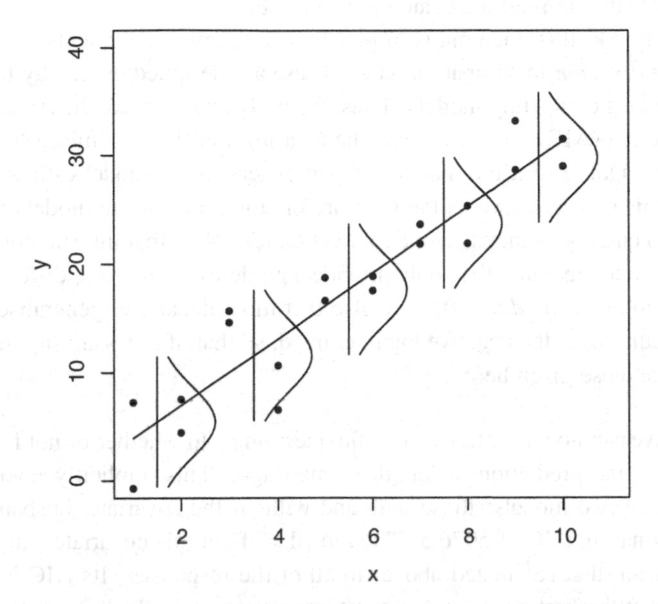

Fig. 2.3. A graphical representation of a simple linear normal regression showing the linear form of the regression function and the constant shape of the normal distribution about it.

Divorces Here, the response variable is a duration so that the normal distribution will, almost certainly, be inappropriate, as we already have seen from the graphical methods above. I shall, nevertheless, proceed first with models based upon it. Let us, then, attempt first to explore the relationship between the length of marriage and the age of the husband using linear regression. The estimated equation is

$$\mu_i = -14.4 + 0.73x_{i1}$$

where x_{i1} is the husband's age. The positive sign of the slope indicates, as might be expected, that mean length of marriage increases with the husband's age. Thus, according to this model, the length of marriage is estimated to increase, on average, by about three-quarters of a year for each additional year of husband's age. This relationship is plotted as the solid line on the scatterplot in Figure 2.2. It is not very convincing.

Likelihood

We can ask whether the inclusion of a given covariate actually does help to predict the response variable. As briefly outlined in Section 1.3, one intuitive way to do this is to look at how probable each model makes the observed data, called its *likelihood*. Often, this is easier to study if minus its logarithm is used. Then, smaller values indicate models 'closer' to the data. A problem with this procedure is that more complex models, those with more estimated parameters, even ones that are not really necessary, will generally make the data more probable. However, we usually prefer the simplest adequate model possible.

One solution to this dilemma is to penalise more complex models by using an *information criterion* to compare models. These are designed especially to help in selecting among competing models. Thus, the most widely used, the Akaike information criterion (AIC) involves minus the logarithm of the maximised likelihood, penalised by adding to it the number of parameters in the model estimated from the data. This penalty prevents the measure of suitability of the models from decreasing too quickly as they become more complex. Note that information criteria have no absolute meaning; they only provide a guide for *comparing* different models applied to the *same data*. Beware also that most linear and generalised linear software return *twice* the negative log likelihood, so that, if AICs are supplied, they will be twice those given here.

Divorces We can now turn to the specific question as to whether or not husband's age improves the prediction of length of marriage. This implicitly involves the comparison of two models: those with and without the covariate, husband's age. That with it has an AIC of 5076.5. The model without this covariate simply fits a common mean (that calculated above) to all of the responses. Its AIC is 6090.6, indicating that the first model was a great improvement on this. Of course, this is obvious from Figure 2.2; even if the regression model involving husband's age does not represent the responses very well, it does much better than simply assuming a common mean length of marriage for everyone.

Multiple regression

One possible way to proceed to more nonlinear forms of regression curves, remaining in the context of normal linear regression, is to add the square of a quantitative covariate to the regression equation. This is a simple case of *multiple regression*. Here, it will produce a nonlinear model even though statisticians call it linear!

Divorces If we add the square of husband's age, the estimated equation becomes

$$\mu_i = -20.6 + 1.04x_{i1} - 0.0035x_{i1}^2$$

This addition does not improve the fit nearly as much as did inclusion of the linear term in husband's age: the AIC is only reduced further to 5068.6. This relationship is plotted as the dashed line on the scatterplot in Figure 2.2; it can be seen to be only slightly curved.

The same analysis can be carried out for the age of the wife, say x_{i2}. In fact, these models fit better, with AICs of 5017.7 for the linear curve and 5009.7 for the quadratic. Perhaps surprisingly, if we combine the two models, with a quadratic relationship for both husband's and wife's age, we obtain a further substantial improvement. The AIC is 4917.1. However, the quadratic term for the wife's age is not necessary in this equation; eliminating it reduces the AIC to 4916.4.

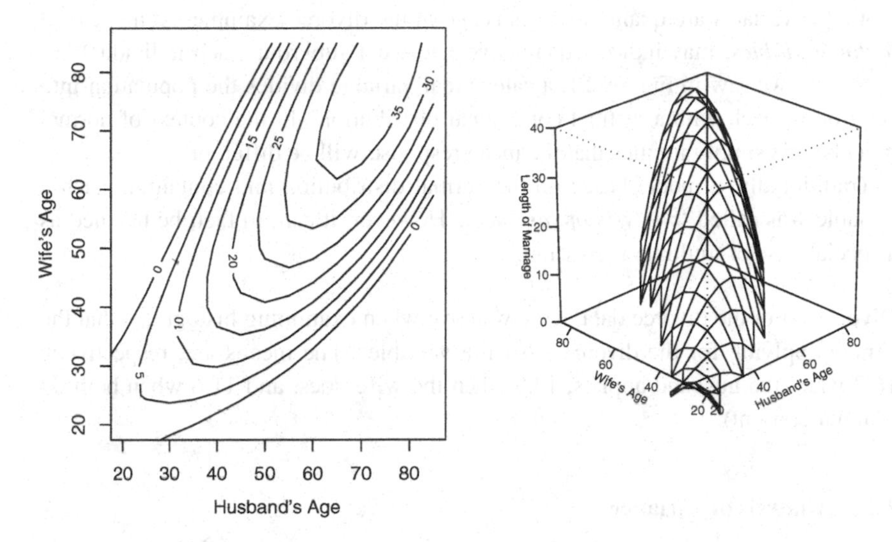

Fig. 2.4. Contour and three-dimensional plots of the model for mean length of marriage as it depends on the two ex-spouses' ages.

Interactions

We also can consider *interactions* between quantitative covariates, obtained by multiplying them together.

Divorces After some experimentation, we discover that the linear interaction is necessary, as well as the quadratic term for wife's age and the quadratic and cubic for husband's age. The final estimated equation, with a substantial reduction in AIC to 4710.3, is

$$\begin{aligned}\mu_i &= 13.42 - 1.64x_{i1} + 0.25x_{i2} + 0.018x_{i1}^2 - 0.034x_{i2}^2 \\ &\quad - 0.00039x_{i1}^3 + 0.0068x_{i1}x_{i2}\end{aligned}$$

It is more difficult to plot a regression function when there are two covariates, but it is still possible. Two ways, as contours and as a three-dimensional perspective plot, are shown in Figure 2.4 for this model of length of marriage as it depends on the ages of the husband and the wife.

Notice that this model is a completely arbitrary construction, obtained by empirical search. If, say, we were wanting to approximate some unknown nonlinear function by a Taylor's series expansion, we would want to use all terms up to a certain order, usually second. Here, I have included one third order term (x_{i1}^3) but not three others (including two further interactions). However, none of them improves the model significantly.

2.3 Categorical covariates

Not all covariates are quantitative, as is age in the divorce example. Some, called *factor variables*, may indicate qualitatively to what subgroup each individual belongs. As we saw in Figure 2.1, a categorical variable divides the population into subgroups, each with a distinct conditional distribution. In the context of normal models, this simply implies that the mean response will be different.

Traditionally, a model based on the normal distribution and containing such a variable was called '*analysis of variance*'. However, it can, in fact, be handled as a special case of multiple regression.

Divorces For the divorce data, we saw above when examining histograms that the person applying for the divorce is such a variable. The means are, respectively, 16.7 when the husband applies, 13.9 when the wife does, and 11.6 when both do (mutual consent).

2.3.1 Analysis of variance

With one categorical covariate, an analysis of variance model can be written

$$\mu_i = \mu + \alpha_k \tag{2.3}$$

where k indexes the categories of the covariate. However, as it stands, this model has one too many parameters. We must add a constraint. There is no unique way to do this. The choice generally will depend on interpretability. I shall consider two useful ways.

Another way to look at this problem is to realise that a categorical variable cannot, numerically, be summarised in one number. Instead, it requires a set of *indicator* or *dummy* variables, indicating to which category each observation belongs. Thus, as we shall see, Equation (2.3) can be written equivalently as the multiple regression of Equation (2.2) using such variables.

Baseline constraint

One way to add a constraint is to set $\alpha_k = 0$ for one value of k. This is called the *baseline constraint* because one category is chosen as a baseline of comparison for all of the others.

Now let us define indicator variables such that each can take the values 0 or 1, depending on whether or not the observation is in that particular category. However, this is slightly redundant: if we know that the value of the variable is not in any of the categories but one, then it must be in that remaining category. Thus, we only require one less indicator variable than the number of categories of the original variable. The category without an indicator variable is that with $\alpha_k = 0$. This yields one possible constraint on the parameters mentioned above.

These indicator variables can be used as covariates in the multiple regression of Equation (2.2). Fortunately, most software can handle factor variables automatically so that we do not need to set up the indicators ourselves. Care, however, must be taken in interpreting the results in terms of the constraint, or equivalently, the set of indicator variables, employed. Often, by default, the software chooses $\alpha_1 = 0$ or equivalently uses indicator variables for all but the first category.

Divorces For the divorce data, using the applicant as a factor variable with the baseline constraint yields $\hat{\mu} = 16.7$, the mean number of years of marriage for the first category, husband applying. Then, $\widehat{\alpha_1} = 0$, $\widehat{\alpha_2} = -2.8$, the difference in mean from the first category for wife applying, and $\widehat{\alpha_3} = -5.1$, the difference from the first category for mutual consent. Thus, the means given above are reproduced. The AIC is 6044.0, showing that this categorical variable does not help nearly as much in predicting the length of marriage as do the ages of the two ex-spouses.

Mean constraint

Instead of setting α_k to 0 for one of the categories (above it was the first, $k = 1$), another useful possibility for interpretation in many contexts is a constraint such that μ is the mean and the α_k are differences from it for each category. The appropriate constraint is $\sum \alpha_k = 0$. This are called the *mean constraint*, or sometimes the *conventional constraint*, because it was classically most often used in analysis of variance.

Here, the corresponding indicator variables are more complex. Let us start by specifying the values for all categories except the last. Then, an appropriate indicator variable will take the value 1 if the observation is in the given category, 0 if it is in another category except the last, and -1 if in the last. Again, there will be one fewer indicator variable than the number of categories. The value of α_k for the last category will be minus the sum of those for the other categories, using the fact that $\sum \alpha_k = 0$.

Again, these indicator variables can be used as covariates in the multiple regression of Equation (2.2), but many software packages can also do this automatically. However, generally they will only calculate the values for all but one of the categories, in the way just outlined.

Divorces For the divorce data, the values obtained using these constraints are $\hat{\mu} = 14.1$, the mean number of years of marriage, $\widehat{\alpha_1} = 2.6$, the difference from this mean for husband applying, $\widehat{\alpha_2} = -0.1$, the difference from the mean for wife applying, and $\widehat{\alpha_3} = -2.5$, the difference for mutual consent. Notice that this value, $\hat{\mu} = 14.1$, is not equal to the global mean calculated above. It is rather the unweighted mean of the means for each category.

With this parametrisation of the model, we see more easily that the length of marriage is about average when the wife applies, being considerably longer when the husband applies and about as much shorter when there is mutual consent. Notice that the differences in mean length of marriage are estimated to be the same between categories in the two parametrisations. Thus, in the first, the difference between husband and wife applying was -2.78; this is $-0.14 - 2.64 = -2.78$ in the second. Of course, the AIC is again 6044.0 because this is just a different parametrisation of the same model.

2.3.2 Analysis of covariance

More complex models will contain both quantitative and qualitative variables. Traditionally, this was called '*analysis of covariance*', but it is just still another case of multiple regression. A categorical covariate can be used to introduce a different curve for each of its category. Thus, for a straight line in a regression function like Equation (2.2), it will allow a different intercept for each category of the qualitative covariate:

$$\mu_i = \beta_{0k} + \sum_j \beta_j x_{ijk} \tag{2.4}$$

where, again, k indexes the categories of the qualitative covariate.

Divorces To continue the divorce example, we can model simultaneously the ages of the two spouses and the applicant for the divorce (husband, wife, or mutual). Thus, at this first stage, we are making the assumption that the mean length of marriage depends on the ex-spouses' ages in the same way for each type of application. This improves the model only slightly; the AIC is 4911.8 as compared to 4916.4 given above with the same intercept for all types of application (both models without interactions between the ages).

In order to be able easily to plot the regression curves, I shall use the simpler model with only the husband's age. (The AIC is 5064.2, a much worse model, as might be expected; again, it does not provide much improvement as compared to the model with the same intercept for all types of application, given above, which had 5068.6.) The three parallel curves, with different intercepts, are plotted in the left graph of Figure 2.5. There is not much separation between these lines.

Interactions

A still more complex model allows not only the intercepts but also the slopes to differ among the categories of the categorical variable. This model can be written

$$\mu_i = \beta_{0k} + \sum_j \beta_{jk} x_{ijk} \tag{2.5}$$

Here, the quantitative covariates are said to *interact* with the categorical covariate.

Divorces When both ages are included in the model for the divorce data, including necessary interactions, the AIC is reduced to 4695.5. Again, in order to be able to plot the regression curves easily, I shall use the model without the wife's age. (This has an AIC of 5049.0.) The curves are plotted in the right graph of Figure 2.5. They are quite different, with that for mutual consent levelling off more rapidly with age. (Here, the model could be simplified by eliminating the two parameters for interactions between type of application and the square of the husband's age.)

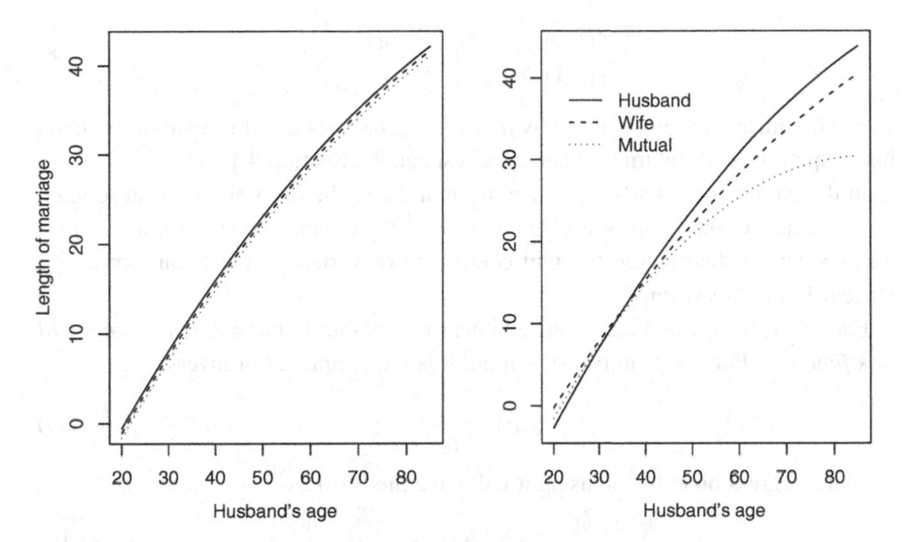

Fig. 2.5. The fitted normal distribution regression lines for the divorce data, separately for the three types of application. Left: parallel lines; right: different slopes.

2.4 Relaxing the assumptions

2.4.1 Generalised linear models

Two of the assumptions of normal models listed above (Section 2.2.1) easily can be relaxed by turning to *generalised linear models*. These provide a slightly wider choice of distribution and allow the mean to depend in a nonlinear way on the linear regression function. The standard distributional choices are normal, gamma, inverse Gauss, Poisson, and binomial. The modification to the regression equation involves some transformation of the mean, called the link function, say $g(\cdot)$:

$$g(\mu_i) = \beta_0 + \sum_j \beta_j x_{ij} \tag{2.6}$$

This must be monotone so that its inverse exists.

Gamma distribution

Survival data (Chapter 3), and other responses involving durations, can often usefully be modelled by the gamma distribution

$$f(t;\mu,\phi) = \frac{\phi^{\phi} t^{\phi-1} \mathrm{e}^{-\frac{\phi t}{\mu}}}{\mu^{\phi}\Gamma(\phi)} \tag{2.7}$$

Because we shall be using this distribution primarily to describe time, I have replaced y by t, as I also shall do in the following distributions.

An important special case, obtained by setting $\phi = 1$, is the exponential distribution, equivalently,

$$\begin{aligned} f(t;\mu) &= \frac{\mathrm{e}^{-\frac{t}{\mu}}}{\mu} \quad \text{or} \\ f(t;\lambda) &= \lambda\mathrm{e}^{-\lambda t} \end{aligned} \tag{2.8}$$

where the mean duration is given by $\mu = 1/\lambda$. I shall not use this distribution here, but frequently shall return to it later (see, especially, Section 4.1.2).

In the gamma distribution, μ is the mean and ϕ is the ratio of the mean squared to the variance, the reciprocal of the square of the *coefficient of variation.* Thus, an exponential distribution has unit coefficient of variation which can serve as a standard of comparison.

Each distribution in the generalised linear model family has a default, *canonical link function.* For the gamma distribution, it is the reciprocal or inverse,

$$g(\mu) = \frac{1}{\mu} \tag{2.9}$$

so that a regression equation using it will have the form

$$\frac{1}{\mu_i} = \beta_0 + \sum_j \beta_j x_{ij} \tag{2.10}$$

Most often, this is inappropriate and a log link

$$\log(\mu) = \beta_0 + \sum_j \beta_j x_{ij} \tag{2.11}$$

is more useful.

Thus, it is possible to change the link function for a given distribution. That used above, in Equation (2.2) with the normal distribution, is called the identity link:

$$g(\mu) = \mu \tag{2.12}$$

its canonical link.

Divorces Length of marriage can be thought of as the survival of the marriage. Then, for the divorce data, the gamma distribution may be appropriate. Here, I only shall consider the regression equation with a quadratic dependence on husband's age. The resulting curve is plotted as the solid line in Figure 2.6. The surprising form of this curve results, of course, from the inverse link function.

The AIC is 4890.1, very much better than any of the preceding models, even with more covariates and interactions among them. For comparison, the corresponding, previously obtained, curve from the normal distribution is plotted in the same graph, as the dotted line.

The first part of the curve from the gamma distribution with reciprocal link is similar to that from the normal distribution. But, in contrast to this latter curve, it reaches a peak at about 60 years and then goes back down for the older people. This may be as reasonable a representation of the data.

However, we do not yet know whether the improved fit results primarily from the more nonlinear form of the regression equation (the link function) or from the

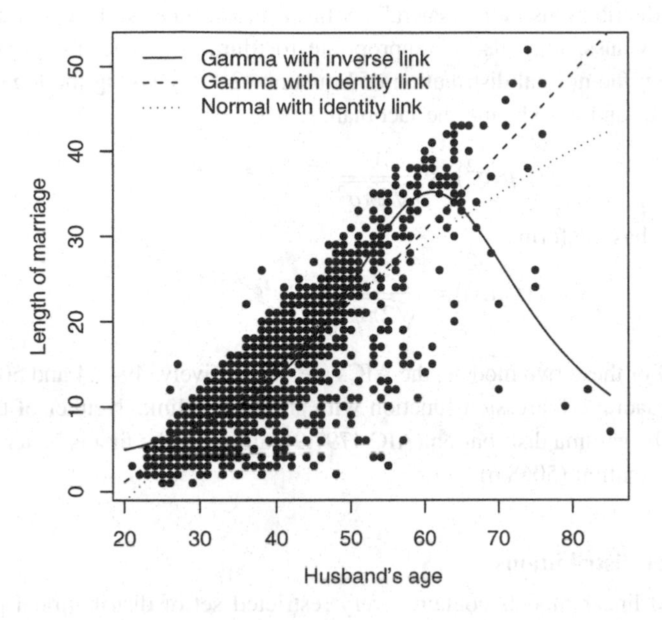

Fig. 2.6. The scatterplot of Figure 2.2 with the fitted gamma and normal regression lines.

changed distributional assumption (the gamma distribution). We best can examine this in steps.

When the identity link is used with the gamma distribution and the quadratic relationship in husband's age, the curve is shown as the dashed line in Figure 2.6; it bends in the opposite direction to that from the normal distribution. The equation from the normal distribution is

$$\mu_i = -20.6 + 1.04x_{i1} - 0.0035x_{i1}^2$$

whereas that from the gamma distribution is

$$\mu_i = -9.82 + 0.49x_{i1} + 0.0032x_{i1}^2$$

For the gamma distribution, the AIC is 4797.9, as compared to 5068.6 above for the normal distribution with an identity link. This demonstrates that most of the improvement comes from the changed distribution. This conclusion can be confirmed by using the reciprocal link with the normal distribution; this has an AIC of 5059.4.

Thus, we must conclude that the conditional distribution of this response variable is decidedly non-normal, at least when a quadratic dependence on husband's age is used. Appropriate choice of the distribution is essential when studying stochastic processes.

Log normal and inverse Gauss distributions

Two other distributions in the generalised linear model family, the log normal and the inverse Gauss, may also be appropriate for duration data. The first can be derived from the normal distribution of Equation (2.1) by taking the logarithm of the responses and introducing the Jacobian:

$$f(t;\mu,\sigma^2) = \frac{1}{t\sqrt{2\pi\sigma^2}} \mathrm{e}^{-\frac{1}{2\sigma^2}[\log(t)-\mu]^2} \tag{2.13}$$

The second has the form

$$f(t;\mu,\phi) = \frac{1}{\sqrt{2\pi\phi t^3}} \mathrm{e}^{-\frac{1}{2t\mu^2\phi}(t-\mu)^2} \tag{2.14}$$

Divorces For these two models, the AICs are, respectively, 4942.4 and 5074.6 for the same quadratic regression function with an identity link. Neither of these fits as well as the gamma distribution (AIC 4797.9), although the first is better than the normal distribution (5068.6).

2.4.2 Other distributions

Generalised linear models contain a very restricted set of distributional possibilities. For continuous responses, these are essentially the normal, log normal, gamma, and inverse Gauss distributions. For stochastic processes, many other distributions will also be important. Here, I shall look at a few of them.

Weibull distribution

For duration data, the Weibull distribution

$$f(t;\mu,\phi) = \frac{\phi t^{\phi-1}\mathrm{e}^{-(t/\mu)^\phi}}{\mu^\phi} \tag{2.15}$$

is especially important because of its simple properties (Section 3.2.1).

Divorces Here, this distribution, with the quadratic function of husband's age, has an AIC of 4643.0. This is a major improvement on the previous models. The regression equation is

$$\mu_i = -9.13 + 0.46x_i + 0.0041x_{i1}^2$$

similar to that for the gamma distribution. These two curves are plotted in Figure 2.7. We see that the Weibull curve is higher than the gamma.

This result is important because, for a model without covariates, the gamma distribution fits better than the Weibull. The AICs are, respectively, 5816.3 and 5850.5. Thus, the marginal distribution (gamma) is different than the conditional one (Weibull). The distributional assumptions can change as covariates are introduced into a model.

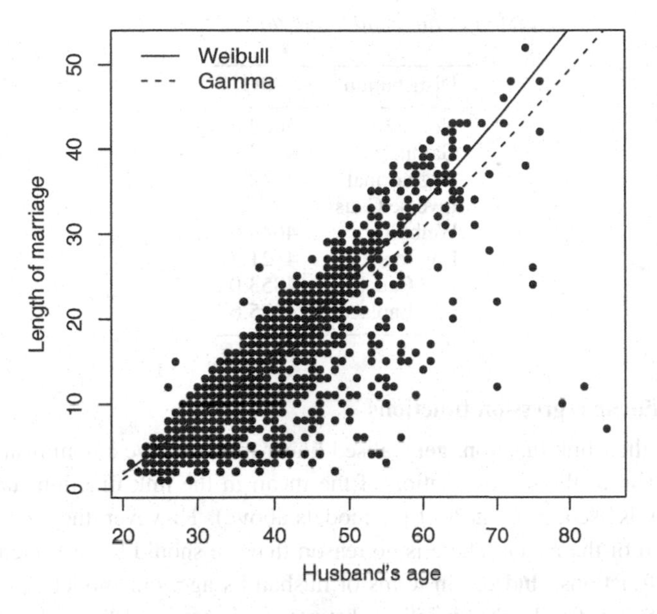

Fig. 2.7. The scatterplot of Figure 2.2 with the fitted gamma and Weibull regression lines. The dashed gamma line is the same as that in Figure 2.6.

Other distributions

Another distribution, the log logistic, has a similar shape to the log normal but with somewhat heavier tails:

$$f(t;\mu,\sigma) = \frac{\pi e^{\frac{\pi[\log(t)-\mu]}{\sqrt{3}\sigma}}}{t\sqrt{3}\sigma\left(1+e^{\frac{\pi[\log(t)-\mu]}{\sqrt{3}\sigma}}\right)^2} \tag{2.16}$$

Other possibilities include the log Cauchy

$$f(t;\mu,\phi) = \frac{\phi}{t\pi\{\phi^2+[\log(t)-\mu]^2\}} \tag{2.17}$$

and log Laplace

$$f(t;\mu,\phi) = \frac{1}{2t\phi}e^{-\frac{|\log(t)-\mu|}{\phi}} \tag{2.18}$$

with even heavier tails.

Divorces The log logistic distribution has an AIC of 4821.8, considerably better than the log normal, but not as good as the gamma and Weibull distributions. The log Cauchy has an AIC of 4853.0 and the log Laplace 4775.6, both also better than the log normal. However, none of these can compete with the Weibull distribution, although the second is better than the gamma. These results are summarised in Table 2.1.

Table 2.1. *Fits of various distributions to the divorce data with a quadratic regression in husband's age and identity link.*

Distribution	AIC
Normal	5068.6
Gamma	4797.9
Log normal	4942.4
Inverse Gauss	5074.6
Weibull	4643.0
Log logistic	4821.8
Log Cauchy	4853.0
Log Laplace	4775.6

2.4.3 Nonlinear regression functions

Because of their link function, generalised linear models have a nonlinear component arising from the transformation of the mean in the link function (unless the identity link is used, as in most of the models above). However, they are linear in that function of the mean. There is no reason that we should be restricted to such regression functions. Indeed, in terms of husband's age, our models above were nonlinear. Scientifically, the statistical distinction, in terms of the parameters, between linear and nonlinear models is generally that the former are approximations to the latter.

Logistic growth curve

Here, I shall look at one simple case of a regression function that is nonlinear both in the covariate and in the parameters. Suppose that response variable depends on a covariate in an S-shaped fashion, so that the mean follows the function

$$\mu_i = \frac{\alpha}{1 + \mathrm{e}^{-(\beta_0 + \beta_1 x_{i1})}} \tag{2.19}$$

instead of a quadratic one. This is called a *logistic growth curve* (Section 12.3.1).

Divorces There appears to be little theoretical reason for using this function with the divorce data; it implies that the mean length of marriage levels off to some constant value as the husband's age increases. With the Weibull distribution, the AIC is 4647.4, somewhat poorer than the quadratic regression function. For comparison, the two curves are plotted in Figure 2.8. We can see that they are similar for ages up to about 55, where most of the responses lie.

Because of the limitations in the design of this study outlined in Section 2.1, the results above are not meant to be definitive in any sense. However, they do show clearly how changing assumptions can alter the conclusions that may be drawn. As in all scientific modelling, care must always be taken in formulating models for stochastic processes.

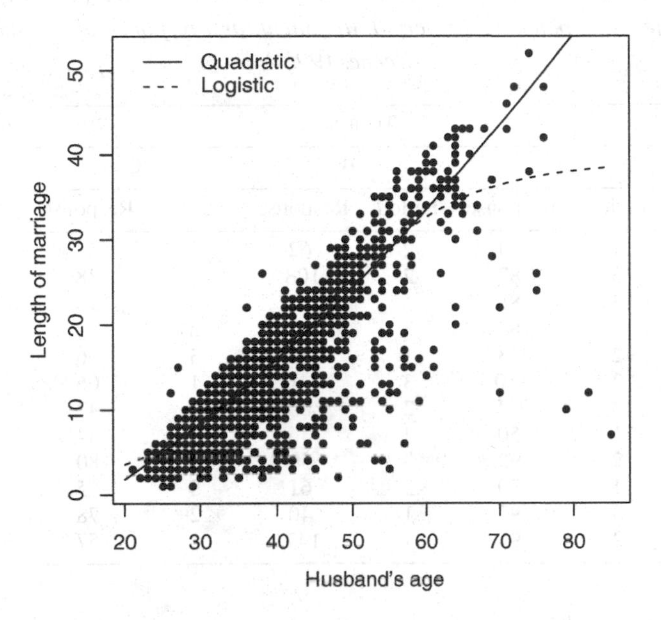

Fig. 2.8. The scatterplot of Figure 2.2 with the fitted linear and nonlinear Weibull regression lines. The solid line is the same as that in Figure 2.7.

Further reading

Good books on linear normal models are rare.

Several books on generalised linear models are available; these include Aitkin *et al.* (1989), Dobson (2002), Lindsey (1997), and McCullagh and Nelder (1989).

For nonlinear models, see Lindsey (2001).

Exercises

2.1 Perform a more complete analysis of the study of divorce in Liège.

(a) Develop an appropriate regression model using all necessary covariates, including the number of children, and any necessary interactions.

(b) The study also recorded the length of the court procedure. Analyse the dependence of this response variable on the covariates.

2.2 A medical three-period cross-over trial was performed to determine gastric half-emptying time in minutes. It involved 12 subjects, as shown in Table 2.2. This is an analysis of variance type of design for duration data.

(a) Find an appropriate model to determine whether or not there are treatment effects. Besides the covariates explicitly present in the table (treatment and period), you may want to consider the 'carry-

Table 2.2. *Gastric half-emptying time (min) for 12 subjects in a cross-over trial. Each line corresponds to the period, treatment, and response of one subject. (Keene, 1995)*

Treatment					
A		B		C	
Period	Response	Period	Response	Period	Response
1	84	2	62	3	58
3	87	2	108	1	38
3	85	1	85	2	96
1	82	3	46	2	61
2	83	1	70	3	46
2	110	3	110	1	66
3	215	2	86	1	42
1	50	3	46	2	34
2	92	3	50	1	80
1	70	2	61	3	55
3	97	1	40	2	78
2	95	1	147	3	57

over' effect of the treatment given in the previous period and also differences among the subjects participating in the trial.

(b) Try several members of the generalised linear model family with various link functions.

(c) Can you find a distribution outside the generalised linear model family that fits better?

Part II

Categorical state space

3

Survival processes

One of the simplest types of stochastic process involves the time until some terminating event of interest occurs, such as death of a biological organism or breakdown of a machine. Such an event signals the change into some usually irreversible state. Because of the importance of such models in medicine for the study of the survival of patients, their use is often known as survival analysis. In engineering applications, it is called reliability analysis.

3.1 Theory

As we have seen in Section 1.1.1, one possible centre of interest in studying a stochastic process may be the durations between events indicating changes of state of the process. We already have encountered, in Chapter 2, one special type of such data, in a more conventional context, the survival of marriages until divorce. There, the original state was 'married' and the event signalling a change was the divorce. For such durations, certain specific techniques usually will be necessary, many originally developed for the special case of survival data, but not covered in that chapter. Thus, here I shall look in more detail at some ways to model this type of process, turning to the more general case of series of several events in Chapters 4, 5, and 6.

3.1.1 Special characteristics of duration data

Durations between events differ from other types of responses in certain specific ways and the special case of survival data has particular characteristics of its own. It is important to look at them more closely before proceeding.

Interevent times

Duration times between events, whether involving irreversible or *absorbing* states (Section 3.1.1) or not, have several special characteristics:

(i) They must be non-negative and are usually strictly positive (unless several events can occur simultaneously).

(ii) Their distribution usually is positively skewed.

(iii) The recording may be *censored* (Section 3.1.2): some durations may be only partially recorded, although this still provides useful information.

As we shall see, response variables of this type can be studied in several closely related and equivalent ways, by looking at

(i) the durations between events,
(ii) the counts of events in given fixed time intervals, or
(iii) the cumulative number of events over time.

For survival data, such approaches clearly are only applicable to the combined results of a set of similar survival processes on different individuals.

Intensity of events

Let us first look a little more closely at the second approach listed above: the counts of events in fixed time intervals. If these intervals are allowed to become very small, we obtain, in the limit, the instantaneous rate of occurrence or the *intensity* of the event, to be described in detail in Sections 3.1.3 and 4.1.1. We shall see that the function chosen to define this intensity will be strictly equivalent to the corresponding distribution of the durations; both will contain the same information expressed in different ways.

Modelling the intensity directly, instead of the probability or density, as in most other areas of statistics, is often the most flexible approach. Although the intensity functions for certain common distributions are relatively complex, having non-linear forms, this is only a minor inconvenience with powerful modern statistical software and computers. Many different distributions, and their corresponding intensity functions, exist to model the durations of times until events.

Absorbing states

In many series, no special state stops the process; then, it is always possible for events to continue to occur, perhaps indicating passage among a number of different states (Chapters 4, 5, and 6). However, in this chapter, I shall look at a simpler, very special, situation; this arises when only two possible states exist. Then, the event of interest is the first and only passage from the first to the second state.

If, when a process enters a certain state, it can never leave it, such as death or irreparable breakdown, that state is said to be *absorbing*. In certain situations, including that which interests us here, the process necessarily ends when it enters this absorbing state and no further observations are made after this event occurs. (In any case, the process is no longer stochastic, but becomes deterministic at that time point.)

Thus, here we shall be interested in the time during which an individual 'survives' in some first state before passing into an absorbing state, from which there is no return, the *survival time*. One example is the divorce data in Chapter 2, whereby a marriage survives (at least legally) until a divorce occurs. (Other possibilities, not considered there, include death and remarriage.)

With such data, we shall only have, at most, one 'observation' per individual, because the process ends when the event of entering the absorbing state occurs. This contrasts with the situation that we shall study in subsequent chapters where a sequence of events in one or more series usually is observed. Nevertheless, even with survival data, individuals must be followed over time to determine when the event occurs.

When we observe a single stochastic process, without absorbing states, over time we must do so for long enough to record several events; otherwise, we shall not have enough information to model them. For a survival process, this is not possible because each event means entry to the absorbing state. Thus, in contrast to most other stochastic processes, we absolutely require observations on a number of individual series in order to obtain enough information to be able to proceed. We always must study a reasonably large collection of distinct survival processes and make strong uncheckable assumptions about the similarity among them with regard to occurrence of the event of interest (Section 1.1.5).

Time origin

The fact that a stochastic process leaves a given state after a certain length of time, of course, implies that a clear time origin must also be specified for the process. This is a critical point that must be considered carefully in any such study. For the divorces, the time origin was the date of marriage. In that example, there was implicitly one further state, unmarried, with the marriage event marking the change of state initiating the duration or survival time of interest. A similar situation occurs in many other contexts, such as in studies of the time to healing or to death after the start of a medical treatment.

However, this is not true in all situations. A special 'event', such as birth, may initiate a process without there being a preceding state. In either case, the initial event, important as it is, is not modelled explicitly, except as the indication of the start of the process of interest. (See Chapter 6 for models of a progressive series of states, where the previous state, in situations like the marriage and medical treatment examples, can be used.)

3.1.2 Incomplete data

By definition, responses that are measured as a duration occur over time. However, any research study must occupy a limited period. This has important implications for the collection of data, implying that it may not be possible to obtain all of the desired information.

Two major problems when observing any stochastic process are:

(i) The time and money of the observer are limited.
(ii) Individuals may, for various reasons, be susceptible to disappear from observation.

These obstacles are particularly acute when recording durations, so that special measures must be taken. In some special cases, there may be other possible reasons for not observing the event ending a duration:

(iii) The event may be impossible for certain individuals.
(iv) A competing event may occur before the event of interest.

Censoring

If the period in which the study is conducted is not reasonably long as compared to the lengths of the durations to be observed, then some of these durations will, almost invariably, be incomplete or *censored*. Thus, some individuals usually will survive longer than the period available for observation or may drop out for some reason before the decisive event occurs. Hence, complete survival information will be available for some individuals, but, for others, only the length of time under observation, and not the time of the event of interest, will be recorded.

Such censoring may happen in various planned ways; the specific criterion by which recording is to stop for all series will be (or should be!) under the control of the research worker. I shall examine various possibilities shortly.

The situation can be more difficult if individuals drop out for reasons supposedly not linked with the study or beyond the control of the research workers. If observation must stop because the subject drops out of the study, this may produce differential censoring in the population that can bias the results. Appropriate measures should be taken either to minimise or to model (Section 6.2) this type of censorship. Thus, an important question involves the reasons that observation of a given process had to stop, called the *stopping rule*.

An additional problem arises if certain individuals will never have the event for some reason. Thus, in many longitudinal studies, certain individuals tend to drop out (in similar ways to those described above for survival data). One may want to model the time until dropout as a survival process. However, most individuals will (hopefully) never drop out of a study. Thus, there will be a mixture of two subpopulations: those with a potential to drop out (who may or may not) and those who will always stay in the study until the end. In such cases, further special procedures will be required (Section 3.4).

Stopping rules

In the collection of duration data, two basic types of planned censoring have usually been distinguished:

(i) All recording may be stopped after a fixed interval of time, called Type I or time censoring, used most often in medical studies.
(ii) Recording may be continued until complete information is available on a fixed number of subjects, that is, until a prespecified number of events has occurred, called Type II or failure censoring, used more often in engineering studies and oncology trials.

As cases of censoring in stochastic processes, Aalen and Husebye (1991) suggest several other possibilities:

(iii) Censoring times are random variables, independent of the stochastic process.

(iv) Recording in a study stops either when a fixed number of complete durations is available or after a fixed time, whichever comes first.

(v) The study stops at the first event after a fixed period.

The second and the last two differ from the first and the third in that they depend on the stochastic processes themselves. However, because these three depend only on what already has been observed, and not on any as yet incompletely observed data, all five can be treated in the same way. They are all legitimate *stopping times* (see Section 4.1.1 for further details).

Suppose now that the last, incomplete, duration in Type I censoring is ignored, and not recorded. However, to know that the last duration recorded is the last complete one, one must look beyond it up to the fixed endpoint to ensure that there are no further events. This differs from the above cases, in that the cutting point depends on information after that point. It is not an appropriate stopping time. One also can see that this will bias downward the average duration time, because long intervals have a higher probability of being incomplete at the end.

The start of observation also must be at a legitimate stopping time. Interestingly, this implies that incomplete durations at the beginning can be ignored. This will involve some loss of information, but no bias. Ideally, the start is also the time origin of the process, but this is not always possible.

Time alignment

When studying a set of survival processes, such as the divorces in Chapter 2, the durations may not all start at the same time. Indeed, in that case, a retrospective study, they all ended at the same time. (In most cases, in a retrospective study, the terminating event may occur before or after the time of the study, as we shall see shortly.) The lengths of marriage for ten individual couples are represented in *chronological* time, in the top graph of Figure 3.1 (ignoring the crosses for the moment). In order to perform the analyses in Chapter 2, I implicitly aligned the durations as if they had all started at the same time. This is shown in the bottom graph (ignoring the triangles for the moment). In opposition to chronological time, this might be called *process* or *aligned* time.

However, these data contain no censored durations. Suppose that, instead of the length of marriage before divorce, we were interested in the time between marriage and birth of the first child, represented by the crosses in Figure 3.1; this is more typical of a retrospective study. (The short vertical lines now indicate the time of the study, not necessarily divorce.) Here, couples 5 and 7 did not have a child before the study occurred so that their durations are censored.

Still another situation may arise in a prospective study. There, it may not be possible to begin observation on all individual processes at the same time. For example, in a clinical trial, patients only enter when people with the condition of interest are found and agree to participate. This is called *staggered entry*; it is represented in the middle graph of Figure 3.1, again in chronological time. To perform an analysis, the same type of alignment as above must be performed so that here the result is the same, given in the bottom graph.

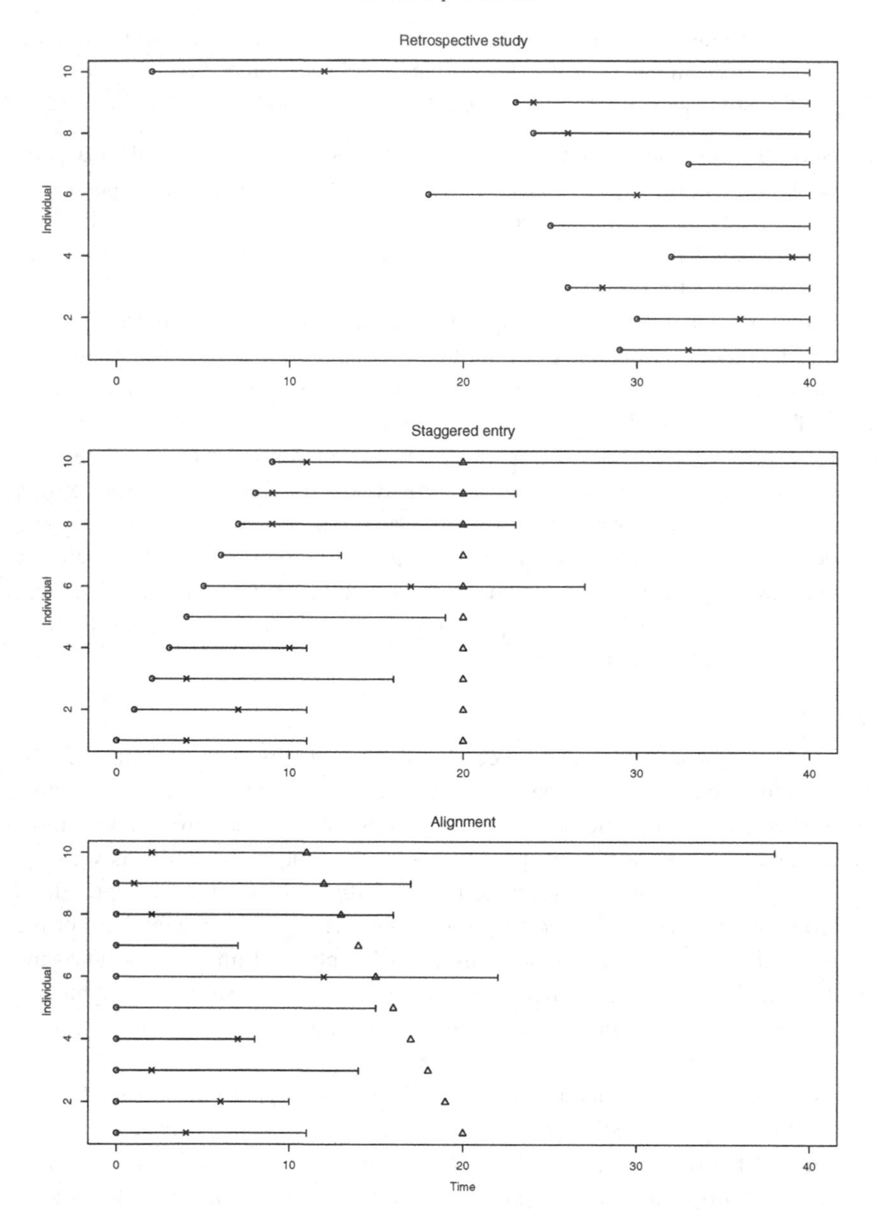

Fig. 3.1. Top: the durations of ten survival processes observed retrospectively. Middle: the durations of ten survival processes observed prospectively, with staggered entry. Bottom: the aligned processes for both situations. Circles indicate the start of the duration, short vertical lines the end (not necessarily observed), crosses some other event of interest, and triangles the end of the study.

As we have seen, in a prospective study, the stopping time is important. Suppose that Type I censoring was used in the clinical trial so that it occurs at time 20, as shown by the triangles in the middle graph of Figure 3.1. This results in individuals

6, 8, 9, and 10 being censored for the final event. In addition, with the fixed stopping time of Type I censoring in a study, individuals entering later will have a shorter potential observation period and a higher potential of being censored.

3.1.3 Survivor and intensity functions

As we have now seen, standard survival data have several specific characteristics. Two of importance here are that durations are non-negative and that some values may be censored. The first implies that only certain distributions will be suitable for the study of survivor curves. We already have encountered this problem in Section 2.4. The second, censoring, means that special methods may often be required in order to apply a model and to analyse the data.

In most areas of statistical modelling, the probability or density function is used directly, as in Chapter 2. Here, for analysis of durations, three other important functions will also be useful:

(i) The probability of a duration lasting at least a given time is called the *survivor function*

$$\begin{aligned} S(t) &= \Pr(T > t) \\ &= 1 - F(t) \\ &= \int_t^\infty f(u)\mathrm{d}u \end{aligned} \tag{3.1}$$

where $F(t)$ is the cumulative distribution function and $f(t)$ the corresponding density function (when it exists).

(ii) The instantaneous probability that the event will occur at a given time, so that the duration terminates, conditional on it having continued until then, or the *intensity* with which the event occurs,

$$\lambda(t)\mathrm{d}t = \Pr(t < T \leq t + \mathrm{d}t | T > t) \tag{3.2}$$

This is also called the *risk*, the *hazard*, the *failure rate*, or the *mortality rate*. This probability is conditional on the previous state, so that models based on the intensity will involve a form of state dependence (Section 1.2.3).

If the density function $f(t)$ exists,

$$\lambda(t) = \frac{f(t)}{S(t)} \tag{3.3}$$

For still another definition, see Equation (4.1).

(iii) The integrated intensity is

$$\begin{aligned} \Lambda(t) &= \int_0^t \lambda(u)\mathrm{d}u \\ &= -\log[S(t)] \end{aligned} \tag{3.4}$$

the latter relationship only holding if the density function exists (Daley and Vere-Jones, 1988, p. 106).

In most cases, any one of the four functions, $f(t)$, $S(t)$, $\lambda(t)$, and $\Lambda(t)$, defines the (univariate) distributional part of a model completely. However, in certain situations, the density function will not be defined so that Equation (3.3) cannot be used and the intensity function must be defined directly. Perhaps the most important case is when there are endogenous time-varying covariates, that is, covariates that depend on the previous history of the process.

From Equation (3.4), if the density exists,

$$S(t) = \mathrm{e}^{-\Lambda(t)} \tag{3.5}$$

which is a unit exponential distribution in terms of $\Lambda(t)$. (This is the survivor function corresponding to the density in Equation (2.8) with $\lambda = 1$ and t replaced by $\Lambda(t)$.) Thus, time in such survival processes can be transformed by means of the integrated intensity so that the (transformed) duration until the event has a unit exponential distribution.

3.1.4 Likelihood function

When it exists, the density function of a duration is the product of the intensity and the survivor function, as can be seen from rearranging Equation (3.3). This term will be included in the model for all complete durations: $t_i^e = t_i \leq t_i^c$, where t_i is the time that the ith process is under observation, t_i^e is the time to the event, and t_i^c is the censoring time. For a censored duration ($t_i^e > t_i^c = t_i$), we only know that no event occurred up until that point; this is given by the survivor function alone. Thus, censored responses will only contribute to the model a term that is the survivor function.

Then, for independent observations of durations t_i, the probability (density) of the sample, upon which the likelihood function can be based, will be

$$\begin{aligned} f(\mathbf{t};\boldsymbol{\theta}) &= \prod_i f(t_i;\boldsymbol{\theta})^{\mathrm{I}(t_i \leq t_i^c)} S(t_i;\boldsymbol{\theta})^{\mathrm{I}(t_i^e > t_i^c)} \\ &= \prod_i \lambda(t_i;\boldsymbol{\theta})^{\mathrm{I}(t_i \leq t_i^c)} S(t_i;\boldsymbol{\theta}) \end{aligned} \tag{3.6}$$

where $\mathrm{I}(\cdot)$ is the indicator function, taking the value 1 when its argument is true and 0 otherwise.

When the density does not exist, such as when endogenous time-varying covariates are present, this procedure will not work. Then, the model, upon which the likelihood function is based, must be constructed directly from the intensity function. I shall describe this procedure in Sections 3.5 and 4.1.

In Section 2.4, I gave an example of fairly standard analysis of survival data, that for divorces. Although those data contained no censored values, with modern software their presence introduces no new problems as long as the likelihood based on Equation (3.6) is used. Thus, I only shall look at one simple classical example of right-censored responses (Section 3.2), before proceeding to some more complex challenges of survival data in this chapter: interval censoring (Section 3.3), a subpopulation that does not have the event (Section 3.4), and time-varying covariates (Section 3.5).

3.1.5 Kaplan–Meier curves

In Section 3.1.3, we saw that one tool for describing the times between events is the survivor function. Just as a histogram can be used to obtain an idea of the form of a probability (density) function (Section 2.1.2 and Lindsey, 2004), it is often useful to have an empirical representation of the survivor function to compare with those given by parametric models, such as those used in Chapter 2. This is relatively simple if there is no censoring, but not so obvious if there is. The *Kaplan–Meier product limit estimate* (Kaplan and Meier, 1958) can provide us with this information whether or not there is censoring.

Let λ_j be the probability of having an event at time t_j, conditional on not having had it until then, that is, the hazard or intensity. Then, the probability of a sample will be a product of binomial distributions,

$$f(\mathbf{d},\mathbf{n};\boldsymbol{\lambda}) = \prod_{j=1}^{R} \binom{n_j}{d_j} \lambda_j^{d_j}(1-\lambda_j)^{n_j-d_j} \tag{3.7}$$

where n_j is the number not yet having had the event and still under observation, and hence still known to be at risk just prior to t_j, called the *risk set*, and d_j is the number having the event at time t_j. The likelihood function based on Equation (3.7) will yield the usual binomial maximum likelihood estimates, $\widehat{\lambda}_j = d_j/n_j$.

Then, the *product limit estimate* of the survivor function will be the product of the estimated probabilities of not having the event at all time points up to the one of interest:

$$\widehat{S(t)} = \prod_{j|t_j<t} \left(\frac{n_j - d_j}{n_j}\right) \tag{3.8}$$

The trick is that the risk set decreases in size, not only at an event, but also each time censoring occurs. This will provide a saturated model to which others can be compared.

The resulting values yield a step function, not a smooth curve. It may be plotted in various ways to explore what form of parametric model might fit the data. (I shall examine such possibilities in more detail in Section 4.2.3.)

As a preliminary step in any analysis of survival data, it always is useful to apply these procedures based on the Kaplan–Meier estimates. Care must, however, be taken: the distribution that appears suitable when no covariates are present will not necessarily remain so with the introduction of these variables into the model, making it conditional on them. We saw this occur in Section 2.4.2 with the gamma and Weibull distributions for the divorce data. Special care must be taken when the covariates are time-varying within the observed survival period.

3.2 Right censoring

Models for survival processes may be based on any distribution for positive-valued random variables. Symmetric distributions can be made skewed by applying a transformation to the durations, say by taking logarithms. In principle, the choice of distribution should be such that the model provides an appropriate mechanism to describe the process under study. Unfortunately, such theoretical information is not always available. This situation has been aggravated by the wide spread use of semi- and nonparametric methods; this has not helped to increase our knowledge about suitable forms of distribution in the contexts studied. Thus, often some convenient models must be selected. In such cases, it is especially important to be able to compare the fits of the models tried (Sections 1.3.2 and 2.2.2).

3.2.1 Families of models

Two special forms of intensity function have attracted particular interest because of their simplicity. They impose strong constraints on the stochastic process, implying that they are more often convenient than realistic.

Proportional hazards

If the intensity function can be written as a product of two functions such that the duration time only appears in the first factor and the covariates only in the second factor, then we have a *proportional hazards* or *multiplicative intensities* model:

$$\lambda(t_i;\boldsymbol{\theta},\boldsymbol{\beta},\mathbf{x}_i) = \lambda_0(t_i;\boldsymbol{\theta})h(\boldsymbol{\beta},\mathbf{x}_i) \tag{3.9}$$

Usually, $h(\cdot)$ is some non-negative regression function, often taking the value 1 when $\mathbf{x}_i = \mathbf{0}$, so that then $\lambda_0(\cdot)$ is a baseline intensity function. Generally, the two factors do not have any parameters in common and most often $h(\boldsymbol{\beta},\mathbf{x}_i) = \exp(\boldsymbol{\beta}^\top\mathbf{x}_i)$ to ensure the non-negativity ($\top$ indicates transpose).

Intensities for all individuals keep the same proportional relationship over all time, the proportionality constant depending on the explanatory variables. Thus, all individuals have the same shape of intensity function over time; only the magnitude differs.

The exponential and Weibull distributions, given, respectively, in Equations (2.8) and (2.15), are members of this family. On the other hand, if the base function, $\lambda_0(\cdot)$, is left unspecified, this formulation is called the semiparametric proportional hazards model (Cox, 1972).

In its semiparametric form, this model has gained great popularity in medical research, especially for its supposed lack of strong assumptions. This has had at least two unfortunate effects:

(i) Proportionality of intensities is, in fact, a strong assumption that is not often justified, especially in clinical research. In a clinical trial, the survival duration usually is calculated taking its time origin at randomisation: the time point at which the treatments are first administered. It is difficult to believe that most treatments will have immediate and constant effects such that the intensities will remain in the same proportion to each other over the duration of a study. On the other hand, in the study of natural populations in epidemiology, it may be quite plausible, in certain situations, that risks are proportional in various natural subgroups.

(ii) Vast amounts of data on survival for a wide variety of illnesses under various treatments have been collected and analysed using such models. If the models provided a reasonable approximation to reality, which is far from certain given the first point, then useful information on the effectiveness of some treatment may have been provided. However, very little or nothing has been learned about functional forms of the variations in the risk of death over time for these diseases.

A more scientific approach would be to select possible models based on the shape that the intensity (risk) function for the given illness might be expected to have over time and to check which are empirically most appropriate in each given situation.

Accelerated failure times

Now suppose that the duration for an individual with covariates $\mathbf{x}_0 = \mathbf{0}$ can be represented by the random variable T_0 and that, for those with nonzero covariate values, it is modified in the following way:

$$T_i = \frac{T_0}{h(\boldsymbol{\beta}, \mathbf{x}_i)} \tag{3.10}$$

The effect of explanatory variables is to accelerate or decelerate the time to completion of the duration. This is called an *accelerated life* or *failure time* model.

This relationship implies that the intensity function can be written

$$\lambda(t_i; \boldsymbol{\theta}, \boldsymbol{\beta}, \mathbf{x}_i) = \lambda_0[h(\boldsymbol{\beta}, \mathbf{x}_i)t_i; \boldsymbol{\theta}]h(\boldsymbol{\beta}, \mathbf{x}_i) \tag{3.11}$$

Again, usually $h(\boldsymbol{\beta}, \mathbf{x}_i) = \exp(\boldsymbol{\beta}^\top \mathbf{x}_i)$, in which case $\lambda_0(\cdot)$ is the baseline intensity function when $\mathbf{x}_i = \mathbf{0}$. Then, from Equation (3.10),

$$\log(T_i) = \log(T_0) - \boldsymbol{\beta}^\top \mathbf{x}_i \tag{3.12}$$

so that this family often can be modelled by a linear regression function for the mean log duration. Because the distribution of $\log(T_i)$ for different covariate values varies only by a translation, its variance (if it exists) is constant (Cox and Oakes, 1984, p. 65). Of course, the variance of the actual durations T_i will not be constant.

These models have been used most widely in engineering, for example in the study of failure rates of machinery. The most common distributions in this context are the Weibull, log normal, and log logistic. Notice that the first of these is also a member of the proportional hazards family. Many of the commonly used intensity functions in this family have a mode of high risk. Then, the covariates determine whether this mode occurs early or late. Thus, the shape of the intensity function is modified by covariates.

This family has at least one drawback: in many situations, an intensity function with a mode is not a reasonable assumption. Most phenomena do not have low risk early and late in life with a higher risk in between. The Weibull distribution avoids this criticism, and also has proportional hazards, helping to explain its popularity.

3.2.2 Intensity and survivor functions

The intensity functions for the common parametric distributions can be derived from Equation (3.3). Several important distributions have survivor and intensity functions that can be obtained explicitly: the exponential distribution, with constant intensity,

$$S(t;\mu,\phi) = \mathrm{e}^{-t/\mu} \tag{3.13}$$

$$\lambda(t;\mu) = \frac{1}{\mu} \tag{3.14}$$

the log logistic distribution

$$S(t;\mu,\sigma) = \frac{1}{1+\mathrm{e}^{\frac{\pi[\log(t)-\mu]}{\sqrt{3}\sigma}}} \tag{3.15}$$

$$\lambda(t;\mu,\sigma) = \frac{\pi \mathrm{e}^{\frac{\pi[\log(t)-\mu]}{\sqrt{3}\sigma}}}{t\sqrt{3}\sigma\left\{1+\mathrm{e}^{\frac{\pi[\log(t)-\mu]}{\sqrt{3}\sigma}}\right\}} \tag{3.16}$$

showing that it is an accelerated failure time model, and the Weibull distribution

$$S(t;\mu,\phi) = \mathrm{e}^{-(t/\mu)^{\phi}} \tag{3.17}$$

$$\lambda(t;\mu,\phi) = \frac{\phi t^{\phi-1}}{\mu^{\phi}} \tag{3.18}$$

Taking the latter as the baseline intensity function, we can write

$$\begin{aligned} \lambda(t_i;\mu,\boldsymbol{\beta},\phi) &= \frac{\phi[h(\boldsymbol{\beta},\mathbf{x}_i)t_i]^{\phi-1}}{\mu^{\phi}}h(\boldsymbol{\beta},\mathbf{x}_i) \\ &= \left[\frac{h(\boldsymbol{\beta},\mathbf{x}_i)}{\mu}\right]^{\phi}\phi t_i^{\phi-1} \end{aligned} \tag{3.19}$$

showing that the Weibull distribution is a member of both the accelerated failure times and the proportional hazards families.

In contrast to these three cases, usually, the survivor function in the denominator of Equation (3.3) involves an integral that only can be evaluated numerically. However, most statistical software supply functions for calculating cumulative distribution functions from which these integrals can be obtained.

As we have just seen, the exponential intensity function is constant over time. The gamma and Weibull intensity functions both either increase or decrease monotonically. The gamma intensity, with the parametrisation of Equation (2.7), tends from below or above to ϕ/μ depending on whether $\phi > 1$ or $\phi < 1$, whereas the Weibull intensity function increases without limit or decreases toward zero, depending on the value of ϕ (Cox and Lewis, 1966, p. 136). In contrast, intensity functions with a mode include the log normal, log logistic, log Cauchy, log Laplace, and inverse Gauss.

One useful generalisation of several of these distributions is the generalised

Table 3.1. *Remission times (weeks; asterisks indicate censoring) from acute leukæmia under two treatments. (Gehan, 1965, from Freireich* et al.*)*

6-mercaptopurine										
6	6	6	6*	7	9*	10	10*	11*	13	16
17*	19*	20*	22	23	25*	32*	32*	34*	35*	
Placebo										
1	1	2	2	3	4	4	5	5	8	8
8	8	11	11	12	12	15	17	22	23	

gamma distribution,

$$f(t;\mu,\phi,\kappa) = \frac{\kappa\phi^{\phi}t^{\phi\kappa-1}\mathrm{e}^{-\phi(t/\mu)^{\kappa}}}{\mu^{\phi\kappa}\Gamma(\phi)} \tag{3.20}$$

When the parameter $\kappa = 1$, this yields a gamma distribution, when $\phi = 1$, a Weibull distribution, when both are 1 an exponential distribution, and when $\kappa \to \infty$, a log normal distribution.

Where possible, a model should be chosen with an intensity function of an appropriate shape for the phenomenon under study. Thus, in some situations, the intensity may be high for early and late times and lower in between, called a 'bathtub' function. Such forms easily can be obtained by directly modelling the intensity function.

In the following example, I again shall compare the distributions already used in Chapter 2, but now classified according to their family.

Leukæmia clinical trial Some classical data, involving the times that cases of acute leukæmia were maintained in remission under two treatments, are given in Table 3.1. In this clinical trial, conducted sequentially so that patients were entering the study staggered over time, treatment with 6-mercaptopurine was compared to a placebo. The results in the table are from one year after the start of the study, with an upper limit of the observation time of about 35 weeks. Right censoring only occurred in the treatment group. The Kaplan–Meier estimates of the survivor functions for these two groups are plotted in the left graph of Figure 3.2. We can see that there is a considerable difference between the two treatment groups.

As in Section 2.4.2, I have fitted various distributions to these data. Here, there is little difference in fit among them, as can be seen in Table 3.2. The Weibull, log normal, and gamma fit best, representatives from all types of models! Clearly, there is not enough information in these data to ascertain precisely the form of the intensity function.

Reassuringly, all distributions show the presence of a treatment effect. With a log link for the gamma distribution, the treatment effect is estimated to be 1.28, almost identical to that with the Weibull distribution. The generalised gamma distribution

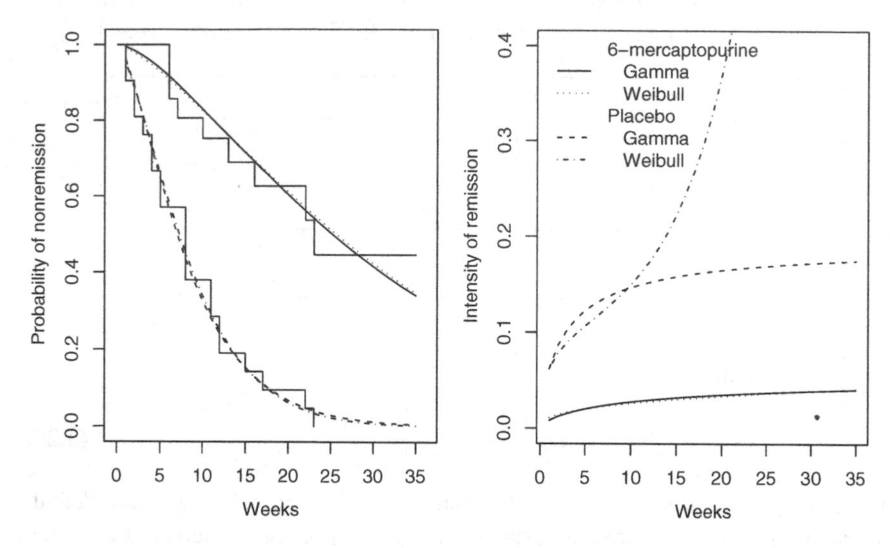

Fig. 3.2. The Kaplan–Meier curves with the gamma and Weibull survivor functions (left) and gamma and Weibull intensity functions (right) for the data of Table 3.1.

Table 3.2. *AICs for various models classified by family and fitted to the leukæmia data of Table 3.1.*

	Null	Treatment effect
Proportional hazards		
Exponential	117.8	110.5
Weibull	118.4	109.6
Accelerated failure times		
Weibull	118.4	109.6
Log normal	117.4	109.7
Log logistic	117.4	110.7
Log Cauchy	120.1	116.6
Log Laplace	117.7	111.8
Other		
Inverse Gauss	118.6	112.0
Gamma	118.2	109.4
Generalised gamma	118.4	110.4

has $\phi = 3.29$ and $\kappa = 0.66$, not indicating any simpler distribution, but it does not fit as well as the simpler models.

For the gamma distribution, $\hat{\phi} = 1.65$ and, for the Weibull distribution, $\hat{\phi} = 1.37$, both some distance from the constant intensity of an exponential distribution ($\phi = 1$). These values both indicate an increasing risk of remission over time. As can be seen in the right graph of Figure 3.2 for the gamma and Weibull distributions, the intensity curves are very similar for the treatment but differ greatly after about 12 weeks for the placebo. This is because there are only three events in this group after this time, providing little information about the form of the intensity curve. On the other hand, the left graph shows that the fitted survivor curves all follow closely the nonparametric ones.

3.3 Interval censoring

In many situations, the times of events are not observed very precisely; in any case, they can never be observed 'exactly' in continuous time. This is called *interval censoring*. Often, ignoring such censoring does not make much difference for conclusions, if the size of the censoring intervals is small compared to the rate of events (Lindsey, 1998). However, it is relatively simple to handle the possibility of interval censoring when necessary.

3.3.1 Probability models

In order to allow for interval censoring, the probability of a sample given in Equation (3.6) must be modified. We must integrate over the time intervals within which the events occur instead of using the density, say for the centre point of the interval. This easily can be done by replacing the density by the difference of cumulative distribution functions, yielding

$$f(\mathbf{t};\boldsymbol{\theta}) = \prod_i [F(t_i^r;\boldsymbol{\theta}) - F(t_i^l;\boldsymbol{\theta})]^{\mathrm{I}(t_i^r \le t_i^c)} S(t_i^r;\boldsymbol{\theta})^{\mathrm{I}(t_i^r > t_i^c)} \tag{3.21}$$

where t_i^l and t_i^r are respectively the left and right limits of the ith interval. I shall use this formula with a number of the distributions that we already have encountered in Chapter 2 and above.

Breast cancer A set of heavily interval-censored data, shown in Table 3.3, arose from a study of the replacement of mastectomy by excisional biopsy followed by irradiation in the treatment of breast cancer, the primary reason being an enhanced cosmetic outcome. Adjuvant chemotherapy can improve relapse-free and overall survival, so that the study was designed to determine whether it affects the rate of deterioration of cosmetic state. Thus, the goal of the study was to determine what differences exist in cosmetic deterioration between the groups without and with chemotherapy in addition to radiotherapy.

Here, with the same probability distributions as in Chapter 2, it is necessary to allow both the location and the dispersion parameters to vary between the two treatments. The results are displayed in Table 3.4. They clearly indicate that both the location and the dispersion are different in the two treatment groups. This implies that a regression model with the intensity functions in the two treatment groups proportional to each other, the proportional hazards model, is not supported by these data. For all models (except the exponential), the total change in log likelihood, when difference in treatment is taken into account, is about nine, with two parameters added. The Weibull, gamma, log logistic, and log Laplace distributions all give similar results, slightly better than the others.

Table 3.3. *Intervals, in months, within which early breast cancer patients under two treatments were subject to cosmetic deterioration. (Finkelstein and Wolfe, 1985)*

Radiotherapy						Radiotherapy–chemotherapy					
t_i^l	t_i^r	t_i^l	t_i^r	t_i^l	t_i^r	t_i^l	t_i^r	t_i^l	t_i^r	t_i^l	t_i^r
45	∞	25	37	37	∞	8	12	0	5	30	34
6	10	46	∞	0	5	0	22	5	8	13	∞
0	7	26	40	18	∞	24	31	12	20	10	17
46	∞	46	∞	24	∞	17	27	11	∞	8	21
46	∞	27	34	36	∞	17	23	33	40	4	9
7	16	36	44	5	11	24	30	31	∞	11	∞
17	∞	46	∞	19	35	16	24	13	39	14	19
7	14	36	48	17	25	13	∞	19	32	4	8
37	44	37	∞	24	∞	11	13	34	∞	34	∞
0	8	40	∞	32	∞	16	20	13	∞	30	36
4	11	17	25	33	∞	18	25	16	24	18	24
15	∞	46	∞	19	26	17	26	35	∞	16	60
11	15	11	18	37	∞	32	∞	15	22	35	39
22	∞	38	∞	34	∞	23	∞	11	17	21	∞
46	∞	5	12	36	∞	44	48	22	32	11	20
46	∞					14	17	10	35	48	∞

Table 3.4. *AICs for various models fitted to the cosmetic deterioration data of Table 3.3.*

	No difference	Location	Dispersion	Location and dispersion
Exponential	155.0	152.3	—	—
Weibull	151.2	146.7	150.5	144.9
Gamma	151.0	147.5	151.0	144.8
Inverse Gauss	152.1	148.4	152.9	146.8
Log normal	151.1	150.0	148.5	145.8
Log logistic	150.8	148.8	147.8	145.0
Log Cauchy	154.4	148.9	151.1	147.8
Log Laplace	151.8	147.3	148.9	144.8

The Kaplan–Meier estimates of the survivor curves for the two treatment groups are plotted in Figure 3.3, along with those for the log Laplace distribution which is typical of the four (and, in fact, of all seven). The parametric curves, with a total of four parameters, follow the nonparametric ones of Equation (3.7), with the 56 estimated parameters, very closely.

An important conclusion from this graph is that the intensities are not proportional: the survivor curves cross each other. Those for the log Laplace and gamma distributions are plotted in Figure 3.4. The survivor curves cross at about one year whereas, more importantly, the intensities cross at about six months. At the beginning, the risk of deterioration is less under the double therapy, but later increases

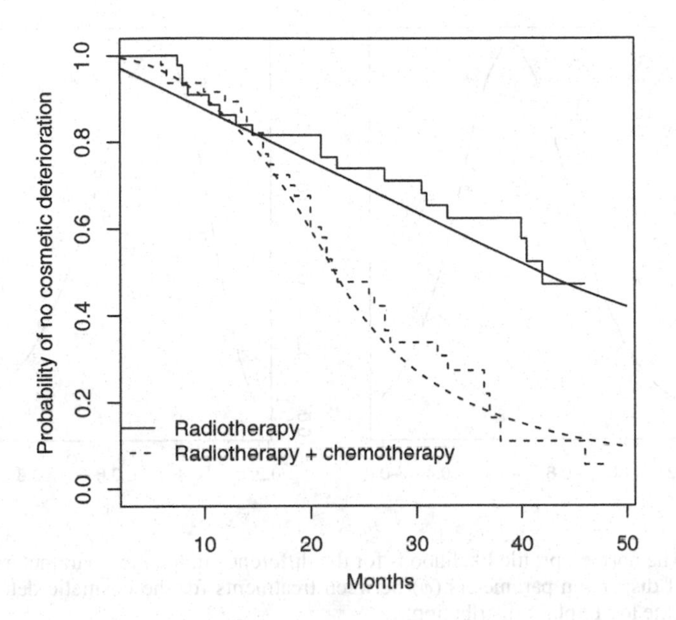

Fig. 3.3. The Kaplan–Meier and log Laplace survivor curves for the cosmetic deterioration data of Table 3.3.

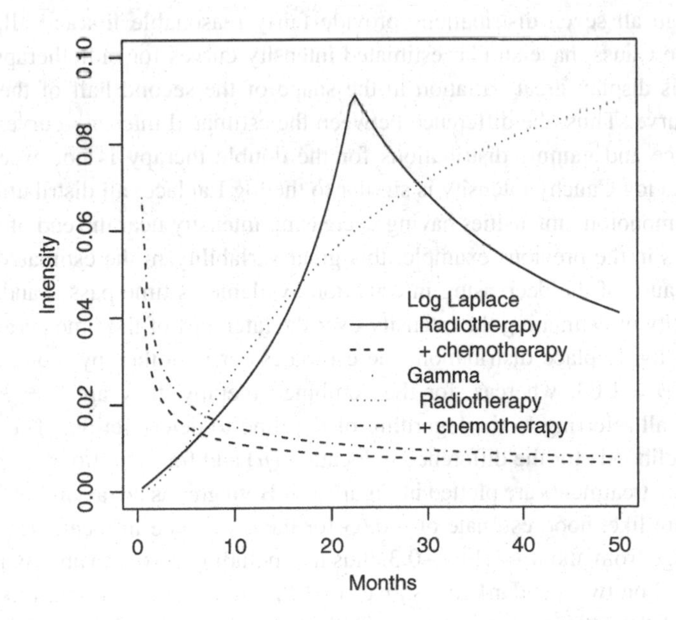

Fig. 3.4. The log Laplace and gamma intensity curves for the cosmetic deterioration data of Table 3.3.

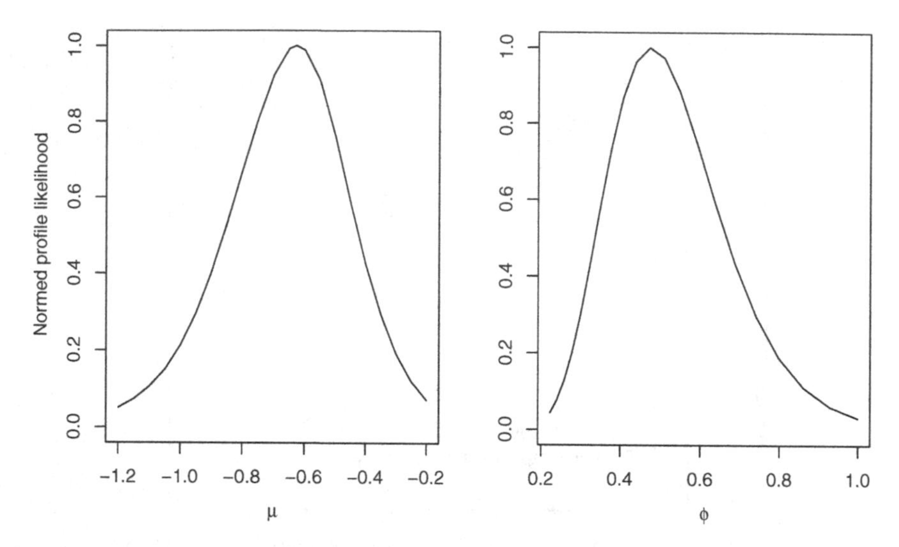

Fig. 3.5. The normed profile likelihoods for the difference in location parameters (μ) and the ratio of dispersion parameters (ϕ) between treatments for the cosmetic deterioration data using the log Laplace distribution.

to much higher levels than under radiotherapy alone, although this is not estimated too well because of the paucity of data.

Although all seven distributions provide fairly reasonable fits and all, except the inverse Gauss, have similar estimated intensity curves for radiotherapy alone, the models display great variation in the shape of the second half of the double therapy curve. Thus, the difference between the estimated intensity curves for the log Laplace and gamma distributions for the double therapy is not exceptional. Indeed, the log Cauchy intensity is similar to the log Laplace; all distributions that allow nonmonotone intensities having decreasing intensity near the end of the time period. As in the previous example, this great variability in the estimated curves arises because of the decreasing information available as time passes and, hence, the difficulty of estimating the intensity over the later part of the time period.

For the log Laplace distribution, the estimates for radiotherapy alone are $\hat{\mu} = 3.73$ and $\hat{\phi} = 1.03$, whereas, for the combined therapy, they are $\hat{\mu} = 3.10$ and $\hat{\phi} = 0.49$, all referring to the logarithm of the time to deterioration. The normed profile likelihoods for the difference in location (μ) and for the ratio of dispersions (ϕ) between treatments are plotted in Figure 3.5. Both graphs are asymmetric. With a maximum likelihood estimate of -0.63 for the difference in location, plausible values range from about -1.1 to -0.3, thus not including zero. An approximate interval based on two standard errors gives $(-1.0, -0.3)$. The maximum likelihood estimate of the ratio of dispersions is 0.48, with a reasonable likelihood interval from about 0.25 to 0.9, hence not including one. The (transformed) interval based on two standard errors gives (0.3, 0.85). Neither approximate interval is very accurate.

The location parameter of the log Laplace distribution models the median (log) survival time. Thus, the estimated median times to deterioration are, respectively, 41.7 and 22.2 months (or a ratio of 0.53) for radiation therapy alone versus the combination. From the above results on precision, plausible values of the ratio of medians vary from about 0.33 to 0.74. On the other hand, the variability of times among patients is also much larger under the first treatment. Although it provides a longer median time, it is less reliable. Finkelstein and Wolfe (1985) also find a significant difference in the rate of deterioration but their logistic model does not allow the survivor curves to cross, as seen in their Figure 4. Thus, they were not able to detect the difference in variability between the two groups.

3.4 Finite mixtures

In Section 3.1.1, I suggested that sometimes not all individuals under study may be susceptible to having the event of interest. In such a situation, the group that is censored ($t_i^e > t_i^c$) is composed of a mixture from two populations: those who did not yet have the event when observation stopped and those who would never have it.

3.4.1 Probability models

Again, we need to modify Equation (3.6). Suppose that the probability of being able to have the event is π. Then, under these conditions, the sample probability becomes

$$f(\mathbf{t};\boldsymbol{\theta},\pi) = \prod_i \pi f(t_i;\boldsymbol{\theta})^{\mathrm{I}(t_i \leq t_i^c)}[\pi S(t_i;\boldsymbol{\theta}) + 1 - \pi]^{\mathrm{I}(t_i > t_i^c)} \tag{3.22}$$

The mixture probability π may possibly depend on covariates.

Recidivism One situation in which not all individuals under study may have the event of interest involves people released from prison. If the event is recidivism (return to prison), this model is (hopefully) especially true. Schmidt and Witte (1988, pp. 26–28) provide data on the months before recidivism of all people released from prison in North Carolina, USA, from 1 July 1977 to 30 June 1978 (to be called the 1978 group) and from 1 July 1979 to 30 June 1980 (to be called the 1980 group). The published times were rounded to the nearest month.

Censoring occurs in these data because the study stopped after 81 months, counting from 1 July 1977. People released in the 1978 group were censored after between 70 and 81 months, depending on their month of release, if recidivism had not already occurred. In a similar way, those released in the 1980 group were censored after between 46 and 57 months.

Many covariates were recorded but these were not published. There were 9327 releases in the 1978 group and 9540 in 1980. However, many had information missing, especially on alcohol or drug abuse. These were removed in the published

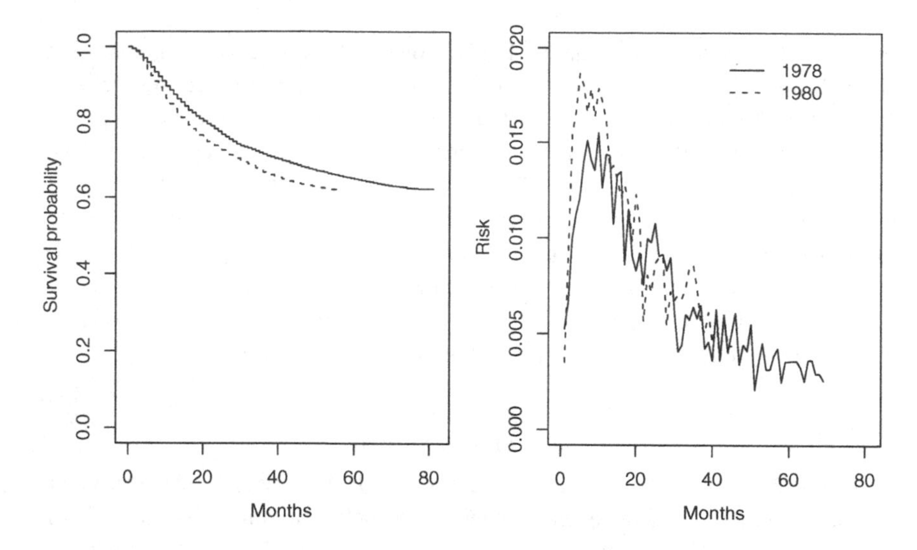

Fig. 3.6. The Kaplan–Meier and empirical intensity functions for the recidivism data.

data, leaving respectively 4618 and 5739 people. One person in 1978 and seven in 1980 who returned almost immediately to prison will also be ignored in the following analyses.

The Kaplan–Meier curves for the two years are plotted in the left graph of Figure 3.6. We see that the probability levels off above 0.6. Thus, there may be about this proportion of people who will not recidive. There is also some difference between the two years with events occurring more rapidly at the beginning in the 1980 group.

These are grouped (interval censored) data with a fair number of events in each time category and right censoring occurs only near the end of the observation period, so that calculation of an empirical estimate of the intensity function is straightforward. For each month, it is the proportion of those who have not yet had the event (the risk set) who have it that month. This is plotted in the right graph of Figure 3.6. Although the curves are rather irregular, the generally tendency is clear: the risk of recidivism is low shortly after release but rises rapidly in the first ten months, higher for 1980 than 1978, dropping off afterwards.

Notice, however, that this is an intensity function for the whole population, not just those potentially at risk, that is, those in the subpopulation with recidivist tendencies. If we could exclude the latter from the risk set, the intensities would be higher.

Thus, we shall require a model with an intensity function that is not monotone (increasing or decreasing), but that has a mode. Many of the common distributions, such as the exponential, given in Equation (2.8), gamma, given in Equation (2.7), and Weibull, given in Equation (2.15), do not have this characteristic. As mentioned in Section 3.2.1, some distributions that do have this form are the in-

Table 3.5. *Fits of some models for the recidivism data.*

Distribution	Standard	Mixture	
		Without years	With years
Exponential	22482.3	22052.2	22038.6
Gamma	22386.7	21948.2	21918.8
Weibull	22351.7	21977.1	21945.7
Inverse Gauss	22792.6	21955.9	21948.5
Log normal	22086.1	21876.2	21862.4
Log logistic	22239.1	21907.9	21890.2
Log Cauchy	23249.6	22370.4	22346.8
Log Laplace	22482.8	22050.2	22026.9

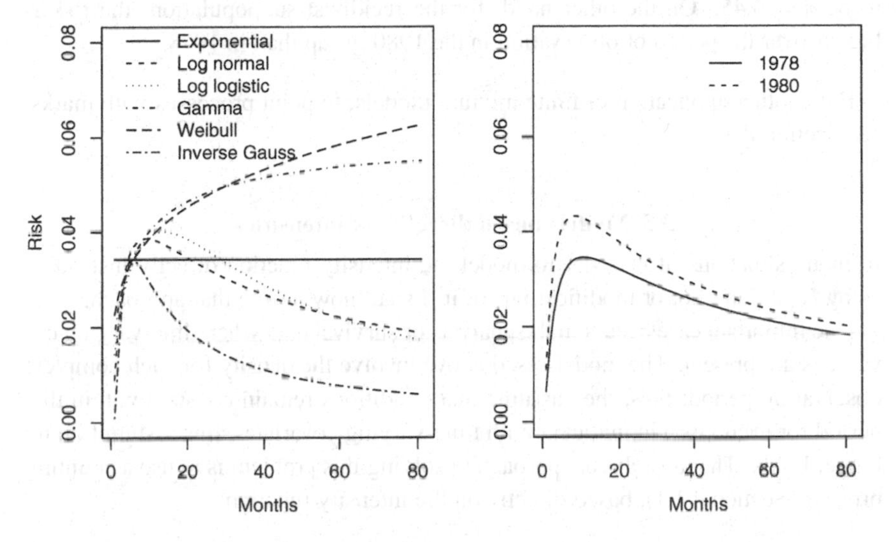

Fig. 3.7. Some parametric intensity functions for the recidivism data without year (left) and differences for the two years using the log normal distribution (right).

verse Gauss given in Equation (2.14), the log normal in Equation (2.13), the log logistic in Equation (2.16), the log Cauchy in Equation (2.17), and the log Laplace in Equation (2.18).

Let us first fit a set of standard survival distributions to these data, ignoring the difference between years and the possibility of a subpopulation without recidivism. The AICs for these models are given in the left column of Table 3.5. Only two of the five distributions having an intensity function with a mode fit better than the others, with the log normal best.

When we add the possibility of a subpopulation of nonrecidivists by using the finite mixture model, the fit improves greatly, as seen in the middle column of Table 3.5. The log normal distribution is still the best. The intensity functions from a selection of these models are plotted in the left graph of Figure 3.7. These are quite different, at least after the first ten months. Those for the log normal and log logistic, the best fitting ones, most resemble the empirical intensity function plotted in the right graph of Figure 3.6. This shows the importance of choosing an appropriate model.

Finally, we can examine differences between the two years. The risk might be different, shown by a difference in μ, or the proportion of possible recidivists, through π, or both. Here, only the risk shows a difference, with AICs given in the right column of Table 3.5. The two intensity functions for the log normal distribution are plotted in the right graph of Figure 3.7. Notice, however, that, compared to the empirical intensity functions in Figure 3.6, these are much higher because they refer only to the risk in the recidivist subpopulation rather than to that in the whole population.

Thus, the probability of being a recidivist is the same for the two years, estimated to be $\hat{\pi} = 0.45$. On the other hand, for the recidivist subpopulation, the risk is higher over the period of observation in the 1980 group than in 1978.

For another application of finite mixture models, to point processes with marks, see Section 8.3.

3.5 Models based directly on intensities

In many situations, it is easier to model the intensity function directly instead of using Equation (3.6) or modifications of it. I shall now turn to that approach.

One important case occurs in the analysis of survival data when time-varying covariates are present. The models used above involve the density for each complete observation period; thus, they assume that conditions remain constant within that period for each given individual. With time-varying covariates, this assumption no longer holds. The most direct approach to solving this problem is to use a counting process (Section 4.1.1), based directly on the intensity function.

3.5.1 Durations and counts of events

From Equations (3.3) and (3.4), the density for a duration can be expressed in terms of the intensity function:

$$\begin{aligned} f(t) &= \lambda(t)S(t) \\ &= \lambda(t)\mathrm{e}^{-\Lambda(t)} \end{aligned} \tag{3.23}$$

Empirically, the time t_i during which an individual i is observed is always divided up, by interval censoring, into R_i small intervals, say Δt_{ij}. Let Δn_{ij}, taking values 1 or 0, indicate whether or not an event was observed in each of these small periods. For survival data, these will all be 0 except perhaps the last, which will only be 1 if the event of entering the absorbing state occurs and 0 if the duration is censored.

For such small intervals, the integrated intensity $\Lambda(t)$ over an interval Δt is approximately equal to the intensity times the interval width:

$$\begin{aligned} \int_{t-\Delta t/2}^{t+\Delta t/2} \lambda(u)\mathrm{d}u &= \Lambda(t+\Delta t/2) - \Lambda(t-\Delta t/2) \\ &\doteq \lambda(t)\Delta t \end{aligned} \tag{3.24}$$

where t is the centre of the interval Δt. Substituting this equation into Equation (3.23) gives

$$\begin{aligned} f(t_i)\mathrm{d}t_i &= \lambda(t_i)\mathrm{e}^{-\sum[\Lambda(t_{ij}+\Delta t_{ij}/2)-\Lambda(t_{ij}-\Delta t_{ij}/2)]}\mathrm{d}t_i \\ &\doteq \lambda(t_i)\mathrm{e}^{-\sum\lambda(t_{ij})\Delta t_{ij}}\Delta t_{iR_i} \\ &= \prod_{j=1}^{R_i}[\lambda(t_{ij})\Delta t_{ij}]^{\Delta n_{ij}}\mathrm{e}^{-\lambda(t_{ij})\Delta t_{ij}} \end{aligned} \tag{3.25}$$

recalling that Δn_{ij} has, at most, one nonzero (unit) value.

The result has exactly the form of a Poisson distribution

$$f(n;\lambda) = \frac{(\lambda\Delta t)^n \mathrm{e}^{-\lambda\Delta t}}{n!} \tag{3.26}$$

so that Δn_{ij} has, to close approximation, this distribution with mean given by Equation (3.24). (Because Δn_{ij} can only be 0 or 1, the factorial is unnecessary.) Notice that the last line in Equation (3.25) holds even for censored data; in that case, all Δn_{ij} are zero and the first factor disappears completely, leaving the survivor function, as required.

Equation (3.26) carries the assumption that the intensity is constant within the interval Δt, but in general the mean number of events in an interval equals the integrated intensity over that interval (Section 4.1). Because the intensity is conditional on not yet having changed state, these Poisson probabilities are independent and can be multiplied together in Equation (3.25) to obtain the probability of the duration. As mentioned above, this is a form of state dependence model (Section 1.2.3).

The logarithm of the intensity $\lambda(\cdot)$ then can be modelled as some regression function, perhaps using Equation (3.3) to define the distribution. This will be described in more detail in Section 4.1; see particularly Equations (4.13) and (4.20).

Myeloid leukæmia Consider a study of chronic myeloid leukæmia. This disease accounts for 20–30 per cent of all leukæmia patients in Western countries, with an incidence of about one in 100 000 persons per year. It has a chronic or stable phase followed by transition to an accelerated or blast phase that resembles a refractory acute leukæmia. The second phase usually is quickly followed by death. Thus, the patients may be in one of three ordered phases: stable, blast, or dead.

Interest in this study centred on finding a marker to predict a future transition so that aggressive therapy, such as a bone marrow transplant, could be initiated beforehand in order to prolong the chronic phase. One possible marker is adenosine deaminase (ADA), an enzyme important in purine catabolism, that has been found in high concentrations in leukæmia cells. This study tried to determine whether patients had high levels of ADA just before entering the accelerated phase. Klein *et al.* (1984) provide the data.

Here instead, I shall look at the total survival time, using ADA and the transition

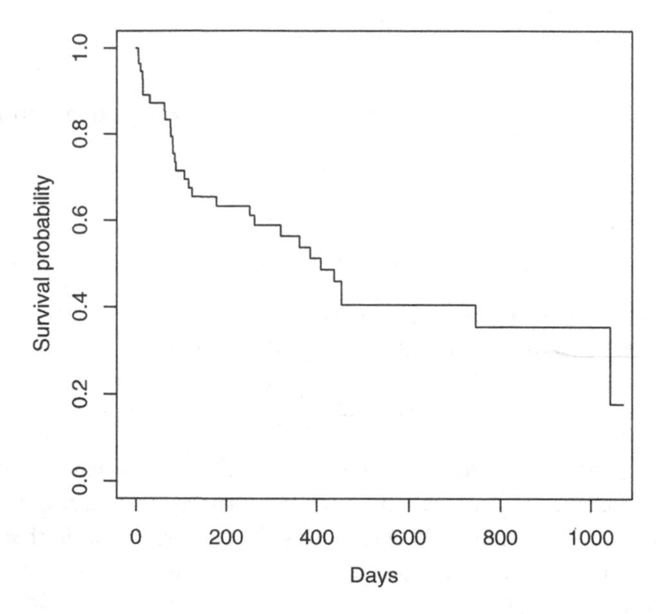

Fig. 3.8. The Kaplan–Meier curve for the time to death, ignoring ADA level and the transition from the chronic to accelerated phase for the study of chronic myeloid leukæmia.

from the chronic to the accelerated phase as time-varying covariates. The level of ADA (10^{-8} mol of inosine/h per million blood cells), in leukæmia cells was measured periodically. I arbitrarily shall assume that the value at the beginning of a period holds over the whole period until the next measurement. There is a total of 18 259 days of observation, but only 30 events. The Kaplan–Meier curves for the time to death is plotted in Figure 3.8.

As previously, I shall base the models on several distributions: the exponential, Weibull, gamma, log normal, log logistic, and log Cauchy models, but here I shall model their (log) intensity functions directly. In the exponential model, only the constant intensity can change with the time-varying covariates. However, in the other models, the shape parameter may also depend on them. As mentioned above, the log intensity can be calculated from the density and cumulative distribution functions using Equation (3.3).

The AICs for various models, with and without the effects of ADA and phase, are given in Table 3.6. The log logistic and log normal distributions fit best. There is an effect of adenosine deaminase, with no difference in effect between the two phases, that is, no interaction between ADA and phase. From both the log logistic and log normal models, the location coefficient (for duration, not intensity) for ADA is estimated to be -0.23: the risk of dying increases with higher levels of adenosine deaminase. The corresponding coefficients for ADA for the shape parameter, on the log scale, are 0.0054 and 0.0046, both referring to the standard deviation.

The estimated intensity curves are plotted in Figure 3.9 at mean levels of ADA.

Table 3.6. *AICs for various models fitted to the myeloid leukæmia data.*

	Exponential	Weibull	Gamma	Log normal	Log logistic	Log Cauchy
Null	223.3	220.9	220.9	220.4	220.6	221.6
Location depends on						
Phase	221.1	216.6	217.0	216.5	219.3	222.5
ADA	216.1	213.4	216.9	205.7	204.7	212.7
Phase+ADA	215.3	209.4	215.1	203.6	205.1	213.6
Phase*ADA	216.2	210.4	216.0	202.9	206.5	214.1
Location and shape depend on						
ADA	—	214.4	213.4	206.0	205.2	213.0
Phase+ADA	—	214.8	207.3	201.0	201.2	207.3

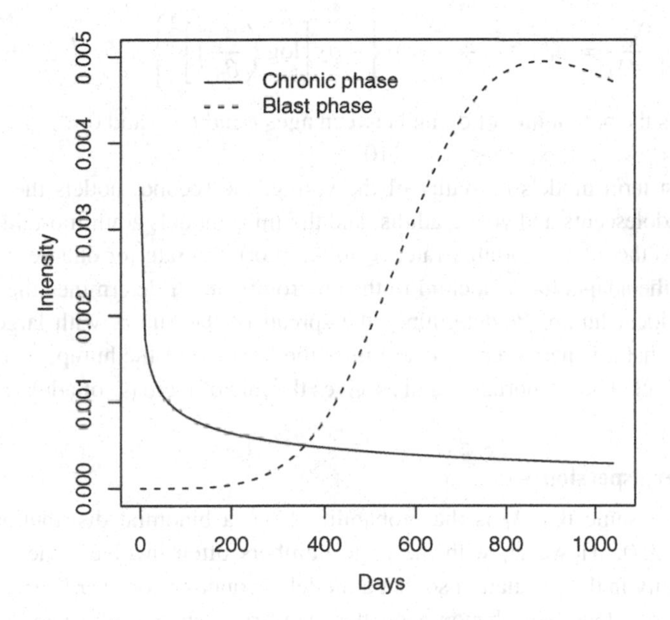

Fig. 3.9. The log logistic intensity function for the chronic and accelerated phases of chronic myeloid leukæmia, plotted at the mean level of ADA (11.1×10^{-8} mol of inosine/h per million blood cells).

Recall, however, that these levels are changing over time for each subject; the Kaplan–Meier curve in Figure 3.8 ignores the ADA level, whereas the plotted intensity curves from the model are for the mean level. We see that the risk of death decreases over time as long as one stays in the chronic phase, but increases quickly in the blast phase.

In Section 6.1.4, I shall approach the modelling of these data from another point of view, more appropriate to answering the original question. There, I shall use three states and two types of events: the transitions from the chronic to the accelerated phase and to death, rather than just one absorbing event, death.

3.6 Changing factors over a lifetime

Over the complete life history of an organism, various factors will enter into play to influence the risk of mortality. It usually is preferable to model relatively short periods so that such changes need not be included in the models. However, attempts have been made to model the mortality rate over a complete human life span. This can be complex because of the various factors that enter into play at different ages.

3.6.1 Complex regression function

One model that has been suggested (Dellaportas *et al.*, 2001) to model a complete human lifetime is

$$\frac{\lambda_t}{1-\lambda_t} = \beta_1^{(t+\beta_2)^{\beta_3}} + \beta_4 \exp\left\{-\beta_5 \left[\log\left(\frac{t}{\beta_6}\right)\right]^2\right\} + \beta_7 \beta_8^t \tag{3.27}$$

where λ_t is the probability of dying between ages t and $t+1$ and $0 < \beta_1, \beta_2, \beta_3, \beta_4, \beta_7 < 1$, $0 < \beta_5, \beta_8 < \infty$, $15 < \beta_6 < 110$.

The first term models mortality of the young, the second models the accident hump of adolescents and young adults, and the third models adult mortality. Here, β_1 refers to the infant mortality rate, β_2 to the mortality rate for one-year olds, β_3 measures the adaptation of a child to the environment, β_4 determines the severity of the accident hump, β_5 determines the spread of the hump, with large values indicating that it is narrower, β_6 determines the location of the hump, β_7 gives the baseline level of adult mortality, and β_8 gives the rate of increase of adult mortality.

3.6.2 Overdispersion

We might assume that λ_t is the probability from a binomial distribution, as in Equation (3.7). However, with the large numbers often involved, there may be heterogeneity in the population so that a model accounting for *overdispersion* may be necessary. This is a phenomenon that can arise when inadequate covariates are available in a model based on the binomial or Poisson distribution, so that the responses are hence more variable than might be expected otherwise.

One possible model for overdispersion uses the beta-binomial distribution:

$$f(d|n;\lambda,\phi) = \binom{n}{d} \frac{B[d+\phi\lambda, n-d+\phi(1-\lambda)]}{B[\phi\lambda, \phi(1-\lambda)]} \tag{3.28}$$

where $B(\cdot)$ is the beta function. For further details on overdispersion, see Section 4.3.2.

Table 3.7. *English and Welsh female mortality, 1988–1992 (population in 100s). (Dellaportas* et al., *2001)*

Age	Population	Deaths	Age	Population	Deaths
0	16820	11543	38	16846	1516
1	16664	940	39	17076	1693
2	16447	538	40	17559	1905
3	16344	420	41	18445	2207
4	16100	332	42	18375	2517
5	15818	250	43	18122	2565
6	15645	254	44	17771	2918
7	15547	228	45	16998	3077
8	15498	208	46	15632	3119
9	15446	215	47	14992	3369
10	15143	182	48	14532	3677
11	14825	200	49	14084	3740
12	14539	215	50	13755	4130
13	14367	204	51	13654	4564
14	14430	294	52	13735	5017
15	14964	339	53	13613	5417
16	15768	412	54	13359	5786
17	16705	535	55	13139	6567
18	17445	561	56	13062	7173
19	18228	592	57	13062	8068
20	18332	591	58	13146	8809
21	19304	640	59	13254	10148
22	19644	623	60	13306	11390
23	20156	653	61	13321	12789
24	20517	668	62	13282	13999
25	20784	689	63	13223	15528
26	20843	698	64	13230	17368
27	20670	712	65	13290	19277
28	20214	799	66	13442	20991
29	19630	795	67	13701	23665
30	19038	787	68	14082	26365
31	18446	935	69	13374	27664
32	17880	978	70	12495	28397
33	17455	977	71	11742	29178
34	17148	1131	72	10988	30437
35	16903	1219	73	10295	32146
36	16714	1270	74	10524	35728
37	16680	1435			

Lifetime human mortality Appropriate data for such a model will be the number of deaths at a given age and the population of that age that is at risk. Thus, the mortality for English and Welsh females from 1988 to 1992 is given in Table 3.7. The population is calculated as the average of that resident on 30 June of each of the five years. The number of deaths is the total for the five years. The mortality rate is plotted in Figure 3.10.

The saturated binomial model, with a different rate for each age group, has an AIC of 429.7. That with the mortality following Equation (3.27) has 485.6, indicat-

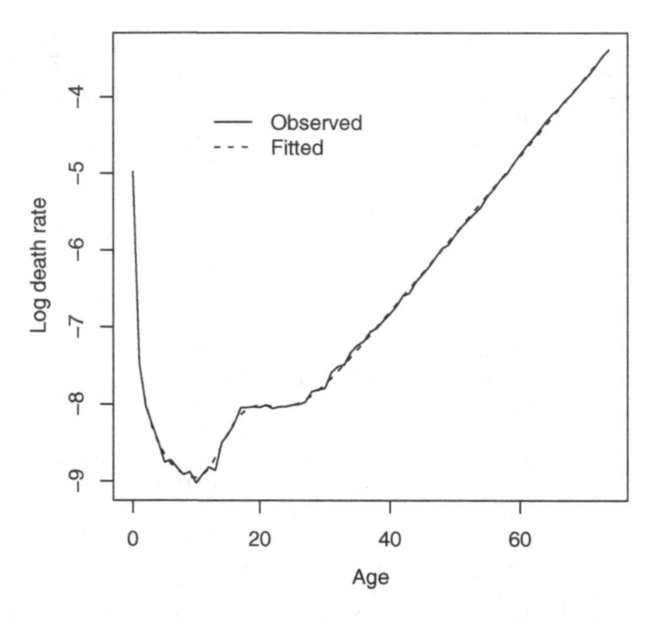

Fig. 3.10. The observed and fitted mortality rates for the females in England and Wales, from Table 3.7.

ing considerable lack of fit, although this is not surprising given the large numbers involved. The beta-binomial model using Equation (3.27) has an AIC of 448.8. The fitted curves for these two models appear very similar. The curve is plotted in Figure 3.10. It follows the observed mortality rate rather closely, although clearly it does not fit as well as the highly parametrised (nonparametric) saturated model.

Further reading

Many good texts are available on survival data. The reader may like to consult the classical works by Kalbfleisch and Prentice (1980) and Lawless (1982), as well as Collett (1994), Cox and Oakes (1984), Klein and Moeschberger (1997), or Marubini and Valsecchi (1994).

Exercises

3.1 Survival times (weeks) were recorded for patients with acute myelogenous leukæmia, along with white blood cell counts (wbc) in 1000s and AG factors, as shown in Table 3.8. Patients were classified into the two groups according to morphological characteristics of their white blood cells: AG positive had Auer rods and/or granulature of the leukaemia cells in the bone marrow at diagnosis, whereas AG negative did not. No patients had palpable enlargement of the liver or spleen at diagnosis. The largest white

Table 3.8. *Survival times (weeks) and white blood cell counts (wbc, in 1000s) for patients with acute myelogenous leukæmia. (Feigl and Zelen, 1965)*

Time	wbc	Time	wbc	Time	wbc	Time	wbc
Positive AG-factor							
65	2.3	156	0.8	100	4.3	134	2.6
108	10.5	121	10.0	4	17.0	39	5.4
56	9.4	26	32.0	22	35.0	1	100.0
5	52.0	65	100.0	56	4.4	16	6.0
143	7.0	1	100.0				
Negative AG-factor							
65	3.0	17	4.0	7	1.5	16	9.0
3	10.0	4	19.0	2	27.0	3	28.0
4	26.0	3	21.0	30	79.0	4	100.0
22	5.3	8	31.0	43	100.0		

Table 3.9. *Survival times (days; asterisks indicate censoring) of 90 patients suffering from gastric cancer under two treatments. (Gamerman, 1991, from Stablein* et al.*)*

Chemotherapy + radiation				Chemotherapy			
17	167	315	1174*	1	356	524	977
42	170	401	1214	63	358	535	1245
44	183	445	1232*	105	380	562	1271
48	185	464	1366	125	383	569	1420
60	193	484	1455*	182	383	675	1460*
72	195	528	1585*	216	388	676	1516*
74	197	542	1622*	250	394	748	1551
95	208	567	1626*	262	408	778	1690*
103	234	577	1736*	301	460	786	1694
108	235	580		301	489	797	
122	254	795		342	499	955	
144	307	855		354	523	968	

blood cell counts were actually greater than 100 000. Notice that none of the survival times are censored.

(a) Do either of these variables help to predict survival time?
(b) Try transformations of the white blood cell counts.

3.2 A total of 90 patients suffering from gastric cancer was assigned randomly to two groups. One group was treated with chemotherapy and radiation, whereas the other only received chemotherapy. The resulting survival times in days are shown in Table 3.9.

(a) Is there any evidence that radiation lengthens survival times?
(b) Can the intensity functions be assumed to be proportional?

Table 3.10. *Survival from nonHodgkin's lymphoma (weeks; asterisks indicate censoring) for patients with two types of symptoms. (Dinse, 1982)*

Asymptomatic			Symptomatic
50	257	349*	49
58	262	354*	58
96	292	359	75
139	294	360*	110
152	300*	365*	112
159	301	378*	132
189	306*	381*	151
225	329*	388*	276
239	342*		281
242	346*		362*

3.3 The Eastern Cooperative Oncology Group in the USA conducted a study of lymphocytic nonHodgkin's lymphoma. Patients were judged either asymptomatic or symptomatic at the start of treatment, where symptoms included weight loss, fever, and night sweats. Survival times (weeks) of patients were classified by these symptoms, as shown in Table 3.10. Patients with missing symptoms are not included.

(a) Do the symptoms provide us with a means of predicting differences in the survival time?

(b) How robust are your conclusions to change of model?

3.4 Fifty female black ducks *Anas rubripes* from two locations in New Jersey, USA, were captured by the Fish and Wildlife Service of the USA over a four-week period from 8 November to 14 December 1983. The ducks were then fitted with radio emitters and released at the end of the year. Of these, 31 were born in the year (age zero) and 19 the previous year (age one). Body weight (g) and wing length (mm) were recorded. Usually, these are used to calculate a condition index: the ratio of weight to wing length. The status of each bird was recorded every day until 15 February 1984 by means of roof-mounted antennæ on trucks, strut-mounted antennæ on fixed-wing aeroplanes, and hand-held antennæ on foot and by boat. The recorded survival times are given in Table 3.11.

(a) What variables influence survival?

(b) Try constructing appropriate new variables from those provided in order to improve your model.

3.5 Danish homosexuals were diagnosed for HIV antibody positivity at six widely-spaced time points between December 1981 and May 1989. Many of the 297 people were not present for all visits. An additional complicating problem is that the time origin, when all individuals were uninfected, is unknown; the data are doubly censored. Following Carstensen (1996), assume that the time origin is the same for all individuals and pro-

Table 3.11. *Survival times (days; asterisks indicate censoring) of 50 female black ducks,* Anas rubripes. *(Pollock* et al., *1989)*

Age	Weight	Wing	Time	Age	Weight	Wing	Time
1	1160	277	2	0	1040	255	44
0	1140	266	6*	0	1130	268	49*
1	1260	280	6*	1	1320	285	54*
0	1160	264	7	0	1180	259	56*
1	1080	267	13	0	1070	267	56*
0	1120	262	14*	1	1260	269	57*
1	1140	277	16*	0	1270	276	57*
1	1200	283	16	0	1080	260	58*
1	1100	264	17*	1	1110	270	63*
1	1420	270	17	0	1150	271	63*
1	1120	272	20*	0	1030	265	63*
1	1110	271	21	0	1160	275	63*
0	1070	268	22	0	1180	263	63*
0	940	252	26	0	1050	271	63*
0	1240	271	26	1	1280	281	63*
0	1120	265	27	0	1050	275	63*
1	1340	275	28*	0	1160	266	63*
0	1010	272	29	0	1150	263	63*
0	1040	270	32	1	1270	270	63*
1	1250	276	32*	1	1370	275	63*
0	1200	276	34	1	1220	265	63*
0	1280	270	34	0	1220	268	63*
0	1250	272	37	0	1140	262	63*
0	1090	275	40	0	1140	270	63*
1	1050	275	41	0	1120	274	63*

visionally take it to be December 1979. The data are presented in this form in Table 3.12.

(a) Develop a suitable model to describe these data.

(b) Is the assumption that the time origin is December 1979 reasonable?

3.6 Consider data from an experiment on the toxicity of zinc for fish (species not given). Six groups of 50 fish each were subjected to three levels of zinc at concentration ratios, 67:100:128, with either one or two weeks acclimation in the test aquaria before introduction of the zinc. Observations were made over a period of 10 days, beginning when the zinc was introduced, yielding a $3 \times 2 \times 11$ contingency table, as shown in Table 3.13. Pierce *et al.* (1979) used log zinc concentration -4, with values, 0.2047, 0.6052, and 0.8520.

(a) Develop a suitable model to describe these data.

(b) Compare the relative importance of acclimation time and zinc concentration.

(c) Is there any evidence that some fish are resistant to zinc poisoning?

Table 3.12. *Left and right bounds of intervals (mon) between visits within which subjects changed from HIV-negative to positive and the corresponding frequencies n_i, with time measured starting in December, 1979. (Carstensen, 1996)*

t_i^l	t_i^r	n_i	t_i^l	t_i^r	n_i
0	24	24	28	∞	8
0	39	2	39	57	3
24	28	4	39	113	2
24	39	1	39	∞	15
24	57	10	57	88	5
24	88	3	57	113	1
24	113	4	57	∞	22
24	∞	61	88	113	1
28	39	4	88	∞	34
28	88	1	113	∞	92

Table 3.13. *Survival times of fish subjected to three levels of zinc and two acclimation times. (Pierce* et al., *1979)*

	Acclimation time					
	One week			Two weeks		
	Zinc concentration					
Day	Low	Med	High	Low	Med	High
1	0	0	0	0	0	0
2	3	3	2	0	1	3
3	12	17	22	13	21	24
4	11	16	15	8	8	10
5	3	5	7	0	5	4
6	0	1	1	0	0	1
7	0	0	2	0	0	0
8	0	1	0	0	0	0
9	0	0	0	0	0	0
10	0	0	0	0	0	0
≥ 11	21	7	1	29	15	8

3.7 Find a good series of mortality data for your country and try fitting Equation (3.27) to it. Are the parameter estimates similar to those found above for England and Wales?

4

Recurrent events

We now are ready to examine processes in which more than one event can occur. For the moment, only a small number of possible outcomes or states will be allowed. Indeed, in this chapter, I shall concentrate on the simplest case, with only two states. One of the states will dominate, with the other occurring only occasionally. The latter will be a *point event*, occurring from time to time and yielding a *recurrent event* process (Section 1.1.2). After survival processes, this is one of the simplest types of stochastic process. I shall leave the more complex case of several types of events to the following chapters.

In contrast to survival processes, here the event will not involve entry to an absorbing state. Indeed, the second state will be so transitory (at a point in time) that the event does not really signal a change of state at all. Thus, a binary indicator can be used, such that, at any time point, it will be 1 if the event occurs and 0 otherwise. Hence, this can also be called a *binary point process*. Generally, only the times of events, called the *arrival times*, or the times between events, called the *interarrival times*, are recorded, the latter being similar to survival data.

In the simplest cases, the varying interarrival times between recurrent events can be assumed to be independent and identically distributed. In such a situation, we shall have a *renewal process* (Section 4.1.4). This term comes from industry where certain machines or parts must be replaced or renewed at varying intervals of time. However, models for renewal processes have much wider application. Nevertheless, generally there will be some sort of dependencies in the process, either systematically over time or among successive periods, so that renewal models will often be too simple.

As for survival data, with recurrent events it is usually especially important to establish a time origin. If this is birth, then the time to the first event generally will be quite distinct from subsequent repetitions of the event. Often, it is convenient to start the process from the time of the first occurrence of the event. If this is unknown, the possible models that can be fitted may be limited; thus, a birth process (Section 1.2.5), conditioning on the number of previous events, is usually unreasonable because the actual number from the start of the process is unknown.

4.1 Theory

In Section 3.1.1, we saw how durations between events can be studied in several equivalent ways, by considering

(i) the durations themselves,
(ii) the count of events in given fixed time intervals, or
(iii) the cumulative number of events over time.

In the second approach, we group the data into fixed time periods of sufficient length and look at the counts of occurrence of the events. However, by applying such a procedure, we lose information if we also have available the actual times between successive events.

I developed the basic relationship between the first two approaches in Section 3.5.1. Here, I shall begin by looking at the third approach in some detail.

4.1.1 Counting processes

Let us look first at a general theory for events occurring over time, more widely applicable than just to recurrent events. Less mathematically inclined readers who feel familiar with the results in Section 3.5.1, can skip to Section 4.1.2.

Basic concepts

Suppose that various events can recur randomly over time. The cumulated number N_t of events from some specified time origin up to any time t is known as a *counting process*. Let $\lambda_{jk}(t|\mathcal{F}_{t-})$ be the intensity of transition from state j to state k (for recurrent events, the state might perhaps be specified simply by the total number of events so far) defined by

$$\begin{aligned} \lambda_{jk}(t|\mathcal{F}_{t-})\mathrm{d}t &= \Pr(\text{the event in } (t, t+\mathrm{d}t)|\mathcal{F}_{t-}) \\ &= \Pr(\mathrm{d}N_t = 1|\mathcal{F}_{t-}) \end{aligned} \tag{4.1}$$

where $\mathcal{F}_{t-}$ is the complete history up to, but not including, t (technically, it is called the *filtration*). Notice that, as usual, this is a *conditional* intensity.

The above formulation should not be confused with the unconditional 'rate' function, occasionally defined by some authors to be the first derivative of $\mathrm{E}(N_t)$. This will only be identical to the intensity function of Equation (4.1) in very special cases. The conditional intensity specifies the process uniquely, whereas such a rate function, in general, does not. (In renewal processes, $\mathrm{E}(N_t)$ is called the renewal function and its first derivative, the renewal intensity; see Section 4.1.4.)

In the formulation of Equation (4.1), I am implicitly assuming that more than one event cannot occur in the time interval $\mathrm{d}t$, at least if it is very small. In other words, the process is *orderly*: $\Pr(\mathrm{d}N_t > 1|\mathcal{F}_{t-})$ becomes extremely small as $\mathrm{d}t$ approaches zero. Technically, we say that

$$\Pr(\mathrm{d}N_t > 1|\mathcal{F}_{t-}) = o(\mathrm{d}t) \tag{4.2}$$

as $dt \to 0$ for all t. (One simple and often reasonable approach to modelling processes where simultaneous events are possible is to use a marked point process; see Section 8.3).

If T_n is the arrival time of the nth event, then these times can be related directly to the counting process:

$$\Pr(N_t \geq n|\mathcal{F}_{t-}) = \Pr(T_n \leq t|\mathcal{F}_{t-}) \tag{4.3}$$

Thus, both the processes N_t and T_n contain the same information. Of course, the interarrival times also can be obtained directly from T_n and the counts of events in various fixed time intervals can be obtained from N_t.

A *stopping time* T_s is a time that depends only on $\mathcal{F}_{t-}$, and not on events after t: if $T_s \leq t$, then $T_s \in \mathcal{F}_{t-}$ so that the occurrence of T_s can be determined by observing the process only up until time t.

If the numbers of events that occur in disjoint time intervals are independent, the process has *independent increments*. If the distribution of the number of events in any interval of time only depends on the length of that interval, the process has *stationary increments*.

Let us now consider a special stochastic process M_t such that

$$\begin{aligned} \mathrm{E}(M_t) &< \infty \\ \mathrm{E}(M_{t+k} - M_t|\mathcal{F}_{t-}) &= 0 \end{aligned} \tag{4.4}$$

If these conditions hold, M_t is called a *martingale*. For a counting process,

$$\begin{aligned} M_t &= N_t - \mathrm{E}(N_t|\mathcal{F}_{t-}) \\ &= N_t - \textstyle\int_0^t \lambda_{jk}(u|\mathcal{F}_{u-})\mathrm{d}u \end{aligned} \tag{4.5}$$

fulfils these conditions. The *integrated intensity* $\Lambda(t|\mathcal{F}_{t-}) = \int_0^t \lambda_{jk}(u|\mathcal{F}_{u-})du$ in the second line is called the *compensator* of the counting process; it is the local conditional mean of N_t. M_t is the *innovation*, containing all information about N_t not available in $\mathcal{F}_{t-}$; it is the analogue of a residual for classical independent responses.

Some simple special cases

Important simplifications occur when the intensity depends only on the complete history through time or N_t: $\lambda_{jk}(t|N_t = n_t)$. Several special cases follow.

- The ordinary homogeneous *Poisson process* (Section 4.1.2) has constant intensity

$$\lambda(t|N_t = n_t) = \lambda \tag{4.6}$$

  already given in Equation (3.14). The intensity does not depend on N_t, but is always the same. This is the only counting process with stationary, independent increments.

- A *nonhomogeneous Poisson process* has an intensity which is some function only of time, the arrival time,

$$\lambda(t|N_t = n_t) = \lambda(t) \tag{4.7}$$

For a given distribution, such a model can be constructed using Equation (3.3). Cases given above include the Weibull distribution in Equation (3.18) and the log logistic in Equation (3.16). Such processes can always be made homogeneous by a suitable transformation of time, sometimes called operational time; see Equation (3.5) above and Equation (4.19) below. However, this definition can be broadened to allow changes over time due to covariates.

- The *pure birth process* (Section 1.2.5) or *Yule process* has an intensity proportional to the number of previously observed events n_t

$$\lambda(t|N_t = n_t) = n_t\lambda \tag{4.8}$$

  This process obviously cannot start with initial condition $N_0 = 0$.
- A *nonhomogeneous birth process* has an intensity proportional to the number of previous events and also some function of time

$$\lambda(t|N_t = n_t) = n_t\lambda(t) \tag{4.9}$$

- In contrast to a nonhomogeneous Poisson process, with intensity depending on the arrival time, a *renewal process* has an intensity depending only on the interarrival time, some function of the time since the last recurrent event,

$$\lambda(t|N_t = n_t) = \lambda(t - t_{n_t}) \tag{4.10}$$

  It starts afresh after each event (Section 4.1.4).
- A *semi-Markov* or *Markov renewal process* has the form of the intensity function depending on some function of the time since the last event

$$\lambda_{jk}(t|N_t = n_t) = \lambda_{jk}(t - t_{n_t}) \tag{4.11}$$

  However, in contrast to a renewal process, this function differs according to the states (here denoted as from j to k) between which it moves at the event (Section 6.1.4).

Those processes with an intensity depending on time are nonstationary. As we see, this dependence may be on the elapsed time, either total (the arrival time) or since the previous event (the interarrival time), or on the number of previous events, or both. In more complex cases, any of the above processes easily may be extended to depend also on time-varying covariates external to the process. The extension to endogenous covariates, influenced by the history of the process, is more difficult, usually requiring the covariate(s) and the process of interest to be modelled simultaneously as a multivariate process.

Modelling intensities

Construction of a model directly by means of the duration density and survivor functions, as in Equation (3.6), is only straightforward when the conditions for a series remain constant over each interarrival time between events. As we saw in Section 3.5, if there are time-varying covariates that change within the period, then these cannot easily be incorporated into the probability distribution. If the covariates are endogenous, then the density is not even defined. Instead, the intensity of the process should be modelled; of course, this can also be done in other simpler contexts as well.

Models for intensities easily can be constructed in terms of a counting process (Lindsey, 1995b). As we have seen, N_t is a random variable over time that counts the number of events that have occurred up to time t. From Equation (4.1), the corresponding intensity of the process is $\lambda_{jk}(t|\mathcal{F}_{t-})$. Then, the kernel of the log likelihood function for observation over the interval $(0,T]$ is given by the stochastic integrals

$$\log[\mathrm{L}(\boldsymbol{\beta})] = \int_0^T \log[\lambda_{jk}(t|\mathcal{F}_{t-};\boldsymbol{\beta})]\mathrm{d}N_t - \int_0^T \lambda_{jk}(t|\mathcal{F}_{t-};\boldsymbol{\beta})\mathrm{I}(t)\mathrm{d}t \tag{4.12}$$

where $\mathrm{I}(t)$ is an indicator function, with value 1 if the process is under observation at time t and 0 otherwise. Because the second term on the right is the integrated intensity and because $\mathrm{d}N_t$ is zero except when there is an event, this is essentially the logarithm of the likelihood function in Equation (3.6).

Now, in any empirical situation, even a continuous-time process will be observed only at discrete time intervals: once an hour, once a day, once a week. Suppose that these are sufficiently small in relation to the rate at which events are occurring so that, at most, one event is observed to occur in any interval. Of course, there will be a finite nonzero theoretical probability of more than one, unless the event is absorbing or a transition to another state. With R intervals of observation, not all necessarily the same length, Equation (4.12) becomes, by numerical approximation of the integrals,

$$\log[\mathrm{L}(\boldsymbol{\beta})] \doteq \sum_{t=1}^{R} \log[\lambda_{jk}(t|\mathcal{F}_{t-};\boldsymbol{\beta})]\Delta n_t - \sum_{t=1}^{R} \lambda_{jk}(t|\mathcal{F}_{t-};\boldsymbol{\beta})\mathrm{I}(t)\Delta t \tag{4.13}$$

where Δt is the width of the tth observation interval and Δn_t is the observed change in the count during that interval, with possible values 0 and 1 if Δt is sufficiently small. We already have met another form of this in Equation (3.25).

In this approach to modelling durations, the stochastic process is viewed as a form of discrete-valued time series with the probability of the present state being conditional on the previous sequence of states. Thus, it is a form of state dependence model (Section 1.2.3). These conditional probabilities are assumed to be independent and hence can be multiplied together to yield the distribution for each duration.

As we saw in Section 3.5.1, Equation (4.13) is the kernel of the log likelihood for the Poisson distribution of ΔN_t, with mean $\lambda_{jk}(t|\mathcal{F}_{t-};\boldsymbol{\beta})\Delta t$. Conditional on the previous state, it is the likelihood for a Poisson process (Section 4.1.2). Any observable counting process is locally a Poisson process, conditional on its previous history, if events cannot occur simultaneously.

The structure that one places on the function $\lambda_{jk}(t|\mathcal{F}_{t-};\boldsymbol{\beta})$ will determine what stochastic process is being modelled. One particularly simple case occurs if it is multiplicative. Then, we have a proportional hazards model (Section 3.2.1) and the regression function will be log linear. Thus, if the logarithm of the elapsed time since the previous event is included in a Poisson log linear regression,

$$\log[\lambda_{jk}(t|\mathcal{F}_{t-};\boldsymbol{\beta})] = \beta_0 + \beta_1 \log(t - t_{n_t}) \tag{4.14}$$

we obtain a Weibull (renewal) process where $\beta_0 = \log(\phi) - \phi\log(\mu)$ and $\beta_1 = \phi - 1$ in Equation (3.18). If the intensity is allowed to jump at each event time, we have the semiparametric proportional hazards model (Section 3.2.1). However, there is no reason why the function should be log *linear*, so that the intensity function for any distribution can be so modelled, say based on Equation (3.3).

4.1.2 Poisson process

After this general introduction to models for series of events, let us now return specifically to recurrent events and look at the simplest model for them.

Poisson distribution

In a *Poisson process*, point events occur completely randomly over time. Thus, the numbers of events Δn_t observed in disjoint intervals of time of fixed length Δt have independent Poişson distributions as given in Equation (3.26). Because Δt may not necessarily be very small, Δn_t may be greater than 1. Here, λ is the constant rate or intensity of occurrence of the events over all time and, hence, $\lambda\Delta t$ is the average number of events in any time period of length Δt.

This model carries the assumptions:

(i) The conditional probability of an event in $(t, t+\Delta t)$, given the previous history, is $\lambda\Delta t + o(\Delta t)$.

(ii) Thus, the occurrence of events in $(t, t+\Delta t)$ is independent of what happened before t.

(iii) The process is orderly: the conditional probability of more than one event in $(t, t+\Delta t)$ is $o(\Delta t)$, so that the corresponding probability of no event is $1 - \lambda\Delta t + o(\Delta t)$.

See Equation (4.2) for the meaning of $o(\Delta t)$.

Thus, in terms of a counting process, $N_0 = 0$, the process has stationary, independent increments, and the number of events in any interval of length Δt has a Poisson distribution with mean $\lambda\Delta t$. When λ is fixed for all time points and the above relationships are true for any size of Δt, the Poisson process is homogeneous as outlined above in Section 4.1.1.

Exponential distribution

In a homogeneous Poisson process, the durations between recurrent events, the interarrival times, are independent, each having the same exponential distribution, as in Equation (2.8),

$$f(t;\lambda) = \lambda e^{-\lambda(t-t_{n_t})} \tag{4.15}$$

where λ is the same constant-intensity parameter as in the Poisson distribution of Equation (3.26). Thus, the mean duration between events is given by $\mu = 1/\lambda$, the reciprocal of the intensity.

This relationship between the Poisson and exponential distributions can be derived by noting that $T > t - t_{n_t} = \Delta t$ only occurs if there is no event in Δt, that is, between t_{n_t} and t. Thus, $\Delta N_t = 0$ and the observed number of events is still n_t. In a Poisson process, this is given by

$$\begin{aligned} \Pr(T > t - t_{n_t}) &= \Pr(\Delta N_t = 0) \\ &= \mathrm{e}^{-\lambda(t - t_{n_t})} \end{aligned} \tag{4.16}$$

the survivor function of the exponential distribution, given already in another form in Equation (3.13).

Such a process is *memoryless*:

$$\begin{aligned} \Pr(T > s + t | T > t) &= \frac{\mathrm{e}^{-\lambda(t+s)}}{\mathrm{e}^{-\lambda t}} \\ &= \Pr(T > s) \quad \text{for all } s, t \geq 0 \end{aligned} \tag{4.17}$$

the only one possessing this property. The risk of an event is constant, so that the distribution of the remaining duration does not change over time. Thus, every instant of such a process is a regeneration point (Section 1.1.4).

We also may note that the waiting time to the nth event in a Poisson process has a gamma distribution,

$$f(t; \lambda, n) = \lambda \mathrm{e}^{-\lambda t} \frac{(\lambda t)^{n-1}}{(n-1)!} \tag{4.18}$$

This is a reparametrisation of Equation (2.7) with $\phi = n$ and $\lambda = \phi/\mu$, written to show its relationship to the exponential distribution.

Modifications of Poisson processes

The superposition of several Poisson processes is still a Poisson process. Indeed, the sum of independent random variables can be Poisson distributed only if every component is Poisson distributed. If a Poisson process is *thinned* by removing points randomly or if the events of the process are randomly shifted in time, the process is still Poisson. (See Daley and Vere-Jones, 1988, pp. 76, 241.)

On the other hand, the conditions for a given process to be Poisson should be examined carefully. Processes exist having marginal distributions for the interarrival times between recurrent events that are exponential, but also having dependencies among these times. Similarly, a process may have numbers of recurrent events in fixed time intervals that are Poisson distributed, but where the counts in disjoint intervals are not independent. These are not Poisson processes. (See Cox and Isham, 1980, p. 45.)

Care also must be taken in simultaneous modelling of several replications of a Poisson process. If one assumes that the intensities of the different series are identical when they are not, one will find a trend of decreasing intensity over time (Proschan, 1963). If one is not prepared to introduce a distinct parameter for each series, one possibility is to use a frailty model (Section 7.3).

Nonhomogeneous Poisson processes

In a nonhomogeneous Poisson process, the intensity function changes over time, as we saw in Equation (4.7). In the simplest cases, it is a function of time alone

$\lambda(t)$, but it may also change due to outside influences (measured by covariates). Although the homogeneous Poisson process is a special case of a renewal process (Section 4.1.4), in general a nonhomogeneous Poisson process will not be a renewal process. The intensity usually depends on the arrival time, not just on the interarrival time.

As we saw in Section 3.1.3, the distribution of times between events can be transformed to a unit exponential distribution. The time transformation to construct a homogeneous Poisson process with unit intensity from a nonhomogeneous one (without covariates) with intensity function $\lambda(t)$ is the integrated intensity of Equation (3.5)

$$t^*(\Delta t) = \int_{t_{n_t}}^{t_{n_t}+\Delta t} \lambda(u)\mathrm{d}u \tag{4.19}$$

where $t^*(\Delta t)$ is the *operational time*. For a homogeneous Poisson process, this is simply $t^*(\Delta t) = \lambda \Delta t$.

When the intensity is a function of time, the distribution of the number of recurrent events Δn_t in a given time interval Δt is

$$f(\Delta n_t|t, t+\Delta t) = \frac{\mathrm{e}^{-\Lambda(t,t+\Delta t)}\Lambda(t,t+\Delta t)^{\Delta n_t}}{\Delta n_t!} \tag{4.20}$$

where the integrated intensity is

$$\Lambda(t, t+\Delta t) = \int_{t}^{t+\Delta t} \lambda(u)\mathrm{d}u \tag{4.21}$$

We saw one special case of this in Section 3.5.1.

If the intensity is constant (λ), then $\Lambda(t, t+\Delta t) = \lambda \Delta t$. However, if, say, the intensity is a log linear function of time, such that

$$\lambda(t) = \mathrm{e}^{\lambda t} \tag{4.22}$$

then

$$f(\Delta n_t|t, t+\Delta t) = \frac{\mathrm{e}^{-\frac{1}{\lambda}[\mathrm{e}^{\lambda(t+\Delta t)}-\mathrm{e}^{\lambda t}]}\left\{\frac{1}{\lambda}\left[\mathrm{e}^{\lambda(t+\Delta t)}-\mathrm{e}^{\lambda t}\right]\right\}^{\Delta n_t}}{\Delta n_t!} \tag{4.23}$$

If the intensity is oscillating, perhaps seasonally, such as

$$\lambda(t) = \lambda_0[1+\lambda_1 \sin(\omega t)] \tag{4.24}$$

then

$$\Lambda(t, t+\Delta t) = \lambda_0\left[\Delta t + \frac{2\lambda_1}{\omega}\sin\left(\omega\frac{2t+\Delta t}{2}\right)\sin\left(\omega\frac{\Delta t}{2}\right)\right] \tag{4.25}$$

and so on. As might be expected, the number of events will depend on the time point when the interval starts as well as on the width of the interval.

The situation is more complex when the intensity depends on the previous history of the process. Consider, say, a linear birth or contagion process (Section 8.1) given by

$$\lambda(t) = \lambda_0 + n_t\lambda_1 \tag{4.26}$$

a generalisation of Equation (4.8) where λ_0 could be the rate of immigration. This results in a negative binomial distribution for the counts of events in fixed intervals of time (Bartlett, 1955, pp. 54–56). Other ways in which a negative binomial distribution can arise will be described in Sections 4.3.2, 7.4, and 8.1.3.

4.1.3 Departures from randomness

A homogeneous Poisson process describes a completely random series of recurrent events. All of the other processes presented so far in this chapter describe some ways in which nonrandomness may occur. However, departures from this assumption can occur in quite a number of different ways. It may be useful here to summarise some of the most important ones (adapted from Cox, 1955).

(i) The intensity function may show a trend over time (Section 4.2.1), a nonhomogeneous Poisson process.

(ii) The intensity function may be a function of some exogenous time-varying covariates (Sections 3.5 and 6.3).

(iii) The intensity may depend on the time elapsed since the previous event.

  (a) This dependence may or may not be the same in all interarrival intervals.

  (b) There may or may not be independence among intervals.

  If both of these conditions hold, we have a renewal process (Sections 4.1.4 and 4.4).

(iv) The current intensity may depend on previous events in some way, such as on

  (a) the number of previous events, a birth process (Sections 1.2.5 and 8.1) or

  (b) the immediately preceding interarrival time(s) (Section 8.2).

(v) An event may signal a change of state, with the intensity function depending on the state, such as in a Markov chain (Chapter 5) or a Markov renewal process (Chapter 6).

(vi) Events may tend to occur together.

  (a) There may be clustering of events. Thus, a primary event may stimulate a series of secondary events to occur. In simpler cases, there may be positive dependence among successive events, as in a Markov chain (Section 5.1.2) or changes in the intensity of events as in a hidden Markov model (Section 7.2).

  (b) The process may not be orderly, so that it may be possible for more than one event to occúr at each point in time. One approach is through marked point processes (Section 8.3).

(vii) The intensity may be varying randomly over time in some way, a doubly stochastic process (Chapter 7 and Section 8.4).

(viii) The series may be a superposition of several empirically indistinguishable processes.

(ix) Only some of the events may be recorded, so that the series may be *thinned* in some systematic or random way.

(x) The determination of the times of occurrence of events may be subject to random measurement error.

Naturally, several of these cases may need to be combined in constructing models in certain situations.

4.1.4 Renewal processes

A Poisson process can be generalised in many ways. Several possibilities were listed in Section 4.1.1. In terms of intervals between recurrent events, an immediate generalisation is to allow the durations between events to have some distribution other than the exponential. If the distributions are identical and independent for all intervals between recurrent events, this is a *renewal process*. Each event is a regeneration point (Section 1.1.4). These assumptions imply that the distribution of numbers of events in fixed intervals of time will no longer be constant over time and, indeed, generally will have a complex form.

Asymptotics

Much of renewal theory concerns asymptotics (Smith, 1958) and is of little use to us here. Thus, the *elementary renewal theorem* states that, in the limit for infinite observation time, the intensity of events equals the reciprocal of the mean duration between events. As we have seen, for a Poisson process this holds for all times and not just in the limit.

The expected value of the number of events until time t,

$$\Lambda(t) = \mathrm{E}(N_t) \tag{4.27}$$

is called the *renewal function* and its first derivative is the *renewal intensity*. For a renewal process, these define uniquely the distribution of interarrival times by the integral equation of renewal theory (Smith, 1958),

$$\Lambda(t) = F(t) + \int_0^t \Lambda(t-u)\mathrm{d}F(u) \tag{4.28}$$

Notice that, in contrast to Section 4.1.1, this is not a conditional integrated intensity.

Stationarity

For renewal processes, we encounter an interesting problem with the concept of stationarity. The random intervals between recurrent events are independent and identically distributed, hence ‘stationary’. However, here the word is not being used in the same sense as the definition of strict stationarity in Section 1.1.4. There, it was defined in terms of displacing a fixed interval width in time along the series. Here, it is in terms of shifting across fixed numbers of events, not fixed time intervals. For these processes, a sequence of intervals of fixed width, starting from an arbitrary time origin, generally is not stationary. (See Cox and Isham, 1980, p. 25.)

Notice that the renewal intensity function derived from Equation (4.27) will only be stationary over time if the renewal function is linear in time: a Poisson process. The mean number of counts of events in a small time interval of fixed width will not be constant (except for the Poisson process), depending on its position in the series, so that a renewal process is not stationary in terms of counts of events. (See Cox and Lewis, 1966, p. 61.)

Recurrence times

If observation does not start at an event, the distribution of the time to the first event will be different from that between subsequent events (Section 1.1.1), except for the Poisson process. If observation starts at a random time point, the time to the first observed event is called the *forward recurrence time* or the *residual* or *excess lifetime*. In such situations, the mean waiting time to the first event will depend both on the mean time between events and on the variance of the time between events, increasing with the latter. This is because, as the variance increases, the random observation point becomes more likely to fall within a very long interval.

If the process is in equilibrium, the density for this forward recurrence time will be

$$f_1(t-t_{n_t}) = \frac{S(t-t_{n_t})}{\mu} \tag{4.29}$$

where $S(t-t_{n_t})$ is the survivor function for the time between events and μ is the mean time between events. As might be expected, for the Poisson process, this is just

$$f_1(t-t_{n_t}) = \frac{\mathrm{e}^{-\lambda(t-t_{n_t})}}{1/\lambda} \tag{4.30}$$

the exponential distribution, the same distribution as for all interarrival times. If the equilibrium assumption can be made, Equation (4.29) often can be useful for constructing models when observation does not start at an event.

The *backward recurrence time* is defined in a similar way as the time since the last event before observation at a random time point. The intensity function for a renewal process only depends on this time.

Now consider the sum of the forward and backward recurrence times when a random observation time point is chosen, that is, the length of the duration in which that point falls. The expected value of this sum is given by

$$\mathrm{E}(T_F+T_B) = \mathrm{E}(T) + \frac{\mathrm{var}(T)}{\mathrm{E}(T)} \tag{4.31}$$

where T_F and T_B are the forward and backward recurrence times, and T the time between any random pair of successive events. Thus, this sum has a larger mean than $\mu = \mathrm{E}(T)$, the mean of all interarrival times, because there is more chance of the start of observation falling in a longer period between events. This is called the *inspection paradox*. (See Thompson, 1988, p. 27, and Wolff, 1989, p. 66.)

Variability

The Poisson process can serve as a valuable point of comparison for other processes. Thus, its exponential distribution of interarrival times has a *coefficient of variation* (ratio of the standard deviation to the mean duration) equal to one. Equivalently, counts of recurrent events in fixed time intervals in a Poisson process have a *coefficient of dispersion* (ratio of the variance to the mean count) equal to one. Then, for this process, because the variance of the durations equals the square of their mean,

$$\mathrm{E}(T_F + T_B) = 2\mathrm{E}(T) \tag{4.32}$$

from Equation (4.31).

More generally, for a renewal process, if these coefficients are greater than one, it is overdispersed and events are more irregular and clustered than the Poisson process. If they are less than one, the process is underdispersed, being more regular and equally spaced than a Poisson process.

Types of failure

In contrast to a Poisson process, the intensity function of other renewal processes will be changing over time between recurrent events. If it is increasing, this can be interpreted as a process of wearing out. However, the process starts again, as if new, at each event, so that wearing out only applies within the independent time intervals, not to the process as a whole.

Several types of failures can be distinguished in a wearing out process with replacement of units (Marshall and Proschan, 1972).

- The failure rate is increasing if $S(t - t_{n_t} + c)/S(t - t_{n_t})$ is decreasing in t for all $c > 0$.
- The failure rate is increasing on average if the average integrated intensity, given by $-\log[S(t - t_{n_t})]/(t - t_{n_t})$, is increasing in t.
- New is better than used if

$$S(t_2 - t_{n_t}) \geq \frac{S(t_1 + t_2 - t_{n_t})}{S(t_1 - t_{n_t})} \tag{4.33}$$

  The chance that a new unit will survive to age t_2 is greater than that an old unit, having survived to t_1, will continue an additional time t_2.
- New is better than used on average if

$$\mathrm{E}(t - t_{n_t}) \geq \frac{1}{S(t - t_{n_t})} \int_0^\infty S(t - t_{n_t} + u)\mathrm{d}u \tag{4.34}$$

  The expected life of a new unit is greater than the expected remaining life of an old unfailed one.

The earlier conditions in this list imply the later ones. If units improve after a wearing in period, the inequalities can be reversed.

As for survival analysis (Chapters 2 and 3), a wide variety of distributions is available to model the durations between recurrent events. Indeed, all of the techniques used there are suitable here. However, the situation sometimes may be simpler because there will be few or no censored times. On the other hand, it will often be necessary to model dependencies among the recurrent events.

In addition, weaker assumptions may be required. Repeated events are available from the same process so that we do not need to assume that a collection of different processes (individual subjects) are identical in order to obtain sufficient information to proceed. Of course, we must assume that the process is not fundamentally changing over time, at least in ways for which no information is available to incorporate into a model.

4.2 Descriptive graphical techniques

We have now met some of the theory of recurrent events forming a point process. As we have seen, at least when a process is orderly so that events do not occur simultaneously, it can be defined equivalently in terms of

(i) a counting process of the cumulative number of recurrent events occurring over time,
(ii) the intensity function giving the rate of occurrence of recurrent events, or
(iii) the probability density of intervals between recurrent events, the interarrival times, replaced by the survivor function for censored intervals.

The simplest assumption is that a process is Poisson. In such a process, the risk of an event remains constant both between two events and over the longer period covering several events. Observed processes can diverge from this in many ways. Some of the simplest to examine first are:

- Although the intensity function may remain constant between any two events, this constant intensity may differ among pairs of consecutive events: one kind of nonhomogeneous Poisson process. We may want to model such changes as some appropriate *trend* over time.
- There may be no trend over time in the long term, but the intensity function may evolve between pairs of consecutive events, always in the same way: a renewal process.
- The way in which the intensity function evolves between two consecutive events may change over the long term: a nonhomogeneous renewal process.

Thus, we need to know how to detect such deviations from a Poisson process and how to model them.

4.2.1 Detecting trends

Let us first see how we might detect whether or not a process is nonhomogeneous. If the time intervals between successive recurrent events are independent and identically distributed, the counts of occurrence of events in relatively long time periods of constant length should not be changing systematically over time. If they were

Table 4.1. *Times (seconds) between vehicles passing a point on the road. (Bartlett, 1963; read across rows)*

2.8	3.4	1.4	14.5	1.9	2.8	2.3	15.3	1.8	9.5
2.5	9.4	1.1	88.6	1.6	1.9	1.5	33.7	2.6	12.9
16.2	1.9	20.3	36.8	40.1	70.5	2.0	8.0	2.1	3.2
1.7	56.5	23.7	2.4	21.4	5.1	7.9	20.1	14.9	5.6
51.7	87.1	1.2	2.7	1.0	1.5	1.3	24.7	72.6	119.8
1.2	6.9	3.9	1.6	3.0	1.8	44.8	5.0	3.9	125.3
22.8	1.9	15.9	6.0	20.6	12.9	3.9	13.0	6.9	2.5
12.3	5.7	11.3	2.5	1.6	7.6	2.3	6.1	2.1	34.7
15.4	4.6	55.7	2.2	6.0	1.8	1.9	1.8	42.0	9.3
91.7	2.4	30.6	1.2	8.8	6.6	49.8	58.1	1.9	2.9
0.5	1.2	31.0	11.9	0.8	1.2	0.8	4.7	8.3	7.3
8.8	1.8	3.1	0.8	34.1	3.0	2.6	3.7	41.3	29.7
17.6	1.9	13.8	40.2	10.1	11.9	11.0	0.2		

changing, the distribution of time intervals would depend on the (total) time since the process began. In the simplest case, the mean time interval would be changing over time as a nonhomogeneous Poisson process. Thus, one of the first things that can be verified is that no time trend exists in the data.

Cumulative events and counts of events

One convenient way to look for changes in trend is to plot the counting process (Section 4.1.1) of the cumulated number of recurrent events against time. Another way is to group the data into larger time intervals and to study the counts of events in each interval.

Road traffic In a study of road traffic in Sweden in the 1960s, the times of vehicles passing a specific point on a two-lane rural road were recorded. The intervals between these times are given in Table 4.1. In the 2023.5 s of observation, 129 vehicles passed by. This means that vehicles were passing at a rate of 0.063/s or 3.80/min. This is an estimate of the average intensity (Section 3.1) of recurrent events over that period.

In such a relatively short observation period, we might expect that there will be no 'long term' time trends. On the other hand, it is quite possible that the intensity function varies over time between two consecutive events because of the minimum distance possible between cars and because of clumping due to occasional slower moving vehicles.

The plot of the cumulative number of events is shown in Figure 4.1 for these data. Notice that, like a Kaplan–Meier curve (Section 3.1.5), this is a step function, here going up by one each time an event occurs. For these data, the plotted line is reasonably straight, showing little indication of a change in rate of events over the period of observation.

To examine the counts of events for these data, measured in seconds, we can use intervals of, say, one minute. One simple way to do this is to plot a histogram for the numbers of events using the successive time intervals as breaks. This is shown

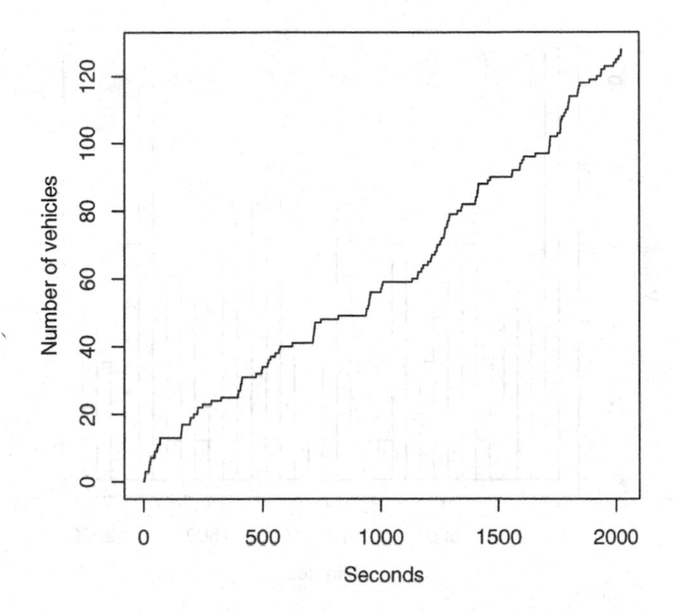

Fig. 4.1. The observed counting process of vehicles on a Swedish road, from Table 4.1.

in Figure 4.2. The number of vehicles passing in one minute varies from 0 to 11. Thus, there is considerable variability in the counts of events.

4.2.2 Detecting time dependence

The probability of an event occurring may depend on the previous history of the process in many different ways. One simple situation occurs when it depends on the time since the previous event. Thus, a second simple departure from a homogeneous Poisson process arises if some (nonPoisson) renewal process is required. It also will be useful to be able to examine this graphically.

The *autointensity function* is defined to be the conditional probability density of an event occurring at time $t+h$ given that one occurred at time t, for various values of h (Guttorp, 1995, pp. 229, 238–240). This can be estimated by grouping the times into small intervals, the *window* or *bandwidth*, say Δ, and counting the number of pairs of events separated by $h \pm \Delta/2$. One way in which this estimate can be calculated is by applying a histogram routine to the times between all pairs of events (not just consecutive ones).

Road traffic For the vehicle data, the autointensity function is plotted in Figure 4.3 for a window of eight seconds. We see that there is a high probability of two vehicles passing closely together, that is, at intervals of less than 20 s. If a second vehicle has not passed in that time, the intensity remains fairly constant for the

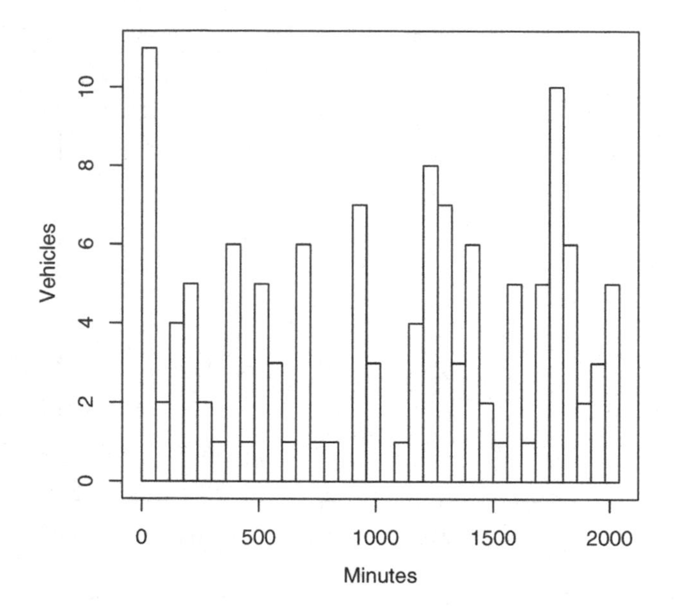

Fig. 4.2. The numbers of vehicles in successive minutes, from Table 4.1.

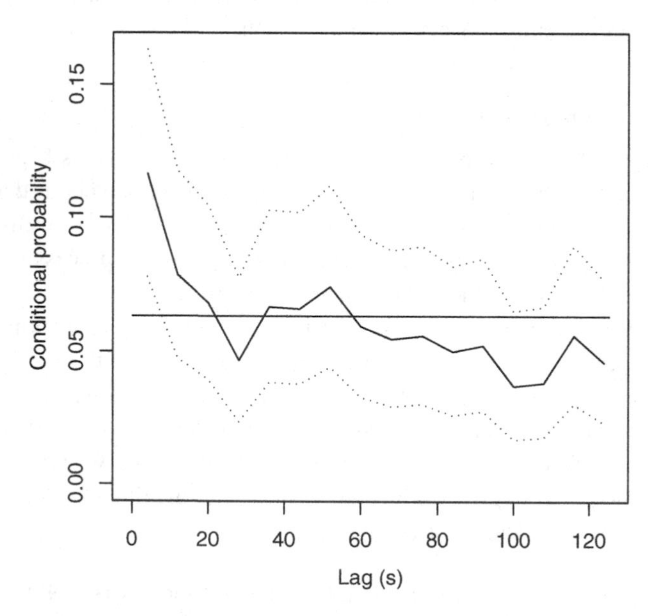

Fig. 4.3. The autointensity function, with an interval of two standard errors, for the vehicle data of Table 4.1, with the constant average rate of 0.063/s (horizontal line).

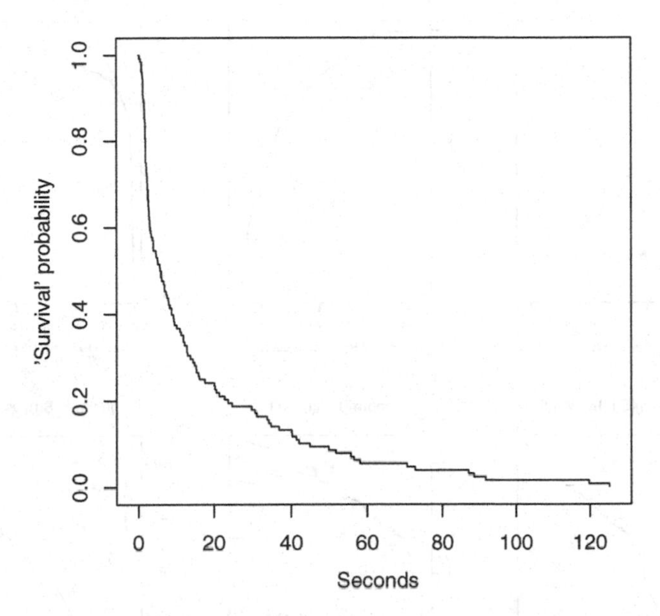

Fig. 4.4. The Kaplan–Meier survivor curve for times between vehicles passing, from Table 4.1.

longer times, following reasonably closely the average rate of 0.063/s calculated above. Thus, there appears to be time dependence in these data.

4.2.3 Kaplan–Meier curves

If we believe that a renewal process might be an appropriate model for the durations between events, one visual way of determining which parametric distribution might be suitable is to plot transformations of the empirical survivor curve, obtained from the Kaplan–Meier estimates (Section 3.1.5), against transformations of time.

Road traffic The Kaplan–Meier curve for the road traffic data is plotted in Figure 4.4. Notice that, in order to calculate this curve, I have aligned all interarrival times to start at zero when an event occurs, as described in Section 3.1.2 for survival data. Thus, I am implicitly assuming a renewal process whereby all interarrival times have independent and identical distributions. The resulting curve drops very quickly, a consequence of there being a large number of short intervals between vehicles during congestion. However, there are also some very long times, resulting in the relatively flat part of the curve for longer intervals.

A collection of transformations of the Kaplan–Meier survivor curve is presented in Figure 4.5. From these, we see that the exponential, log normal, and log logistic distributions appear to be most suitable.

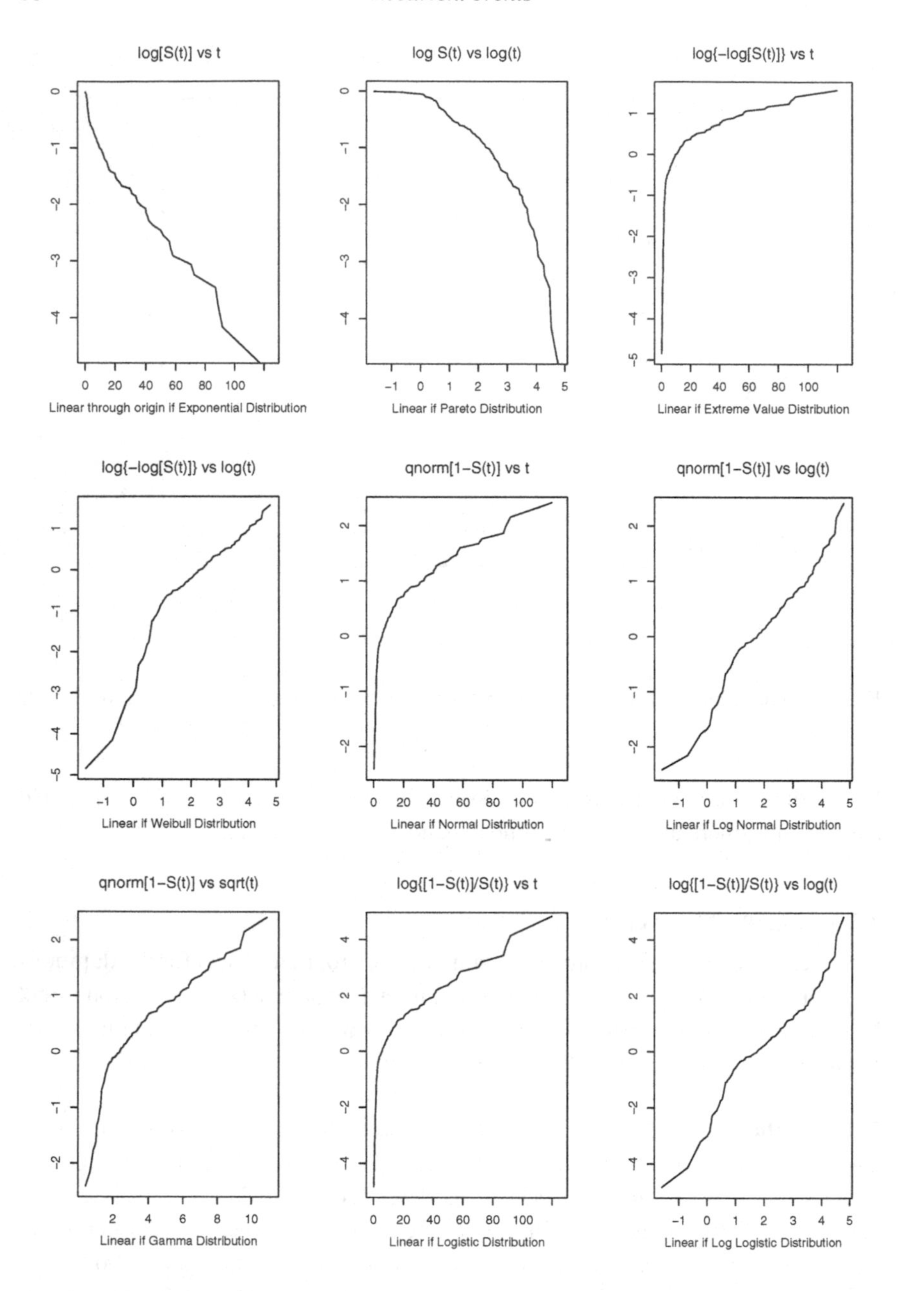

Fig. 4.5. Plots of transformations of the Kaplan–Meier survivor curve against transformations of time for the times between vehicles passing, from Table 4.1.

4.3 Counts of recurrent events

As we saw above, a Poisson process is characterised both by the counts of events in fixed time intervals and by the times between events. In an analogous way, for any series of recurrent events, even though we do not assume that it is a Poisson process, we can study either the counts of events in fixed intervals or the times between events. The former approach was not possible with survival data, at least not for a single individual. Let us consider it first.

Road traffic For the vehicle data in Section 4.2.1 above, we estimated the intensity of passing to be $\hat{\lambda} = 3.80$ vehicles/min; however, we saw in Figure 4.3 that the intensity function may not be constant.

4.3.1 Poisson regression

If we assume a Poisson process, we can use Poisson regression (log linear models). This provides one way to look at changes in trend over time (nonhomogeneity) through a model rather than simply descriptively, as in Section 4.2.1 above.

Road traffic We have seen in Section 4.2.2 that the time at which a vehicle passes appears to depend on the timing of the previous vehicle. Let us nevertheless fit a Poisson process. The Poisson regression with constant mean applied to the counts of the histogram for the vehicle data in Figure 4.2 has an AIC of 86.1 compared to 87.0 when the mean is allowed to change (log) linearly with time. (A model with a quadratic term in time included fits slightly better, but this is probably because the two minutes with the most vehicles happen to be near the two ends of this short series.) Thus, this form of trend does not appear to be present. Notice that this is a global type of trend, as in a nonhomogeneous Poisson process, not one involving interarrival times, as in a renewal process.

Now let us see whether or not the counts in Figure 4.2 could have a Poisson distribution, as they should if they arise from a Poisson process. To do this, we can create a histogram from the histogram: we need to look at the numbers of one minute intervals in which each number of vehicles (from 0 to 11) passed. These are plotted as relative frequencies in Figure 4.6. We see that there is a large number of minutes with only one vehicle and more than might be expected with a large number (10 or 11) vehicles. The first is a characteristic of uniformly spaced traffic whereas the second indicates bunching, perhaps due to congestion. The latter was what we detected in Figure 4.3. These are all indications that the distribution may not be Poisson.

We can check the Poisson assumption by fitting a Poisson distribution to the frequencies in this latter histogram. The mean number of vehicles per minute is estimated to be 3.76. This is slightly different than the estimate of 3.80/min obtained above from the raw data; the difference is due to the information lost in grouping into one minute intervals.

The fitted curve for the Poisson distribution is also shown in Figure 4.6. It has an AIC of 86.1 (as above) as compared to 87.4 for the multinomial (saturated) model, showing that it fits reasonably well. However, this does not necessarily imply that the fit is acceptable. Another fairly simple model may fit even better.

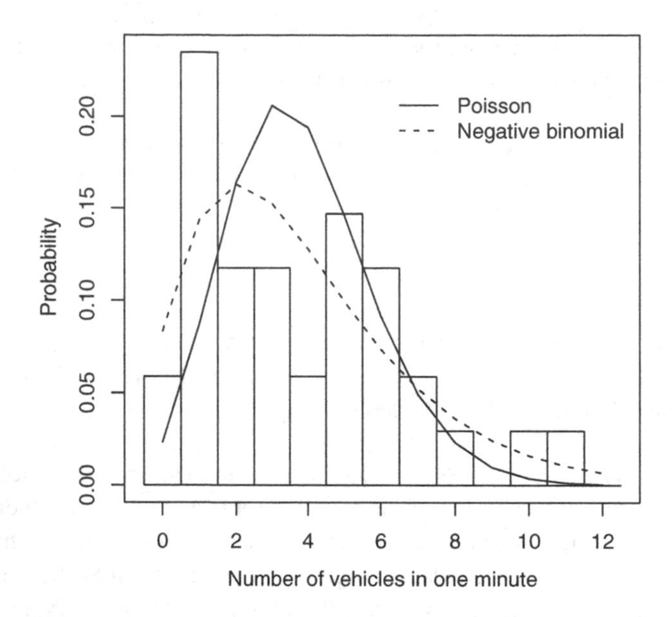

Fig. 4.6. The numbers of minutes with given counts of vehicles, from Table 4.1 and Figure 4.2, with the fitted Poisson and negative binomial distributions.

4.3.2 Over- and underdispersion

One way to attempt to improve on the Poisson assumption is to look at the variability of the counts. As we saw in Section 4.1.4, the Poisson distribution carries the strong assumption that the mean equals the variance, or that the coefficient of dispersion equals one. If there is too much variability in the numbers of recurrent events per unit time, the distribution is said to be *overdispersed* as compared to the Poisson distribution. We already have seen an example of overdispersion, but with respect to the binomial distribution, in Section 3.6.2.

One way in which the responses may be overdispersed is if the mean number of counts per fixed time interval is not constant over the observation period, but we do not know when it is changing. Then, we might assume that this intensity parameter, which I shall now call ψ instead of λ, has a random distribution. For convenience (see Section 14.1.2), this often is chosen to be a gamma distribution, as in Equation (2.7), but with a slightly different parametrisation:

$$p(\psi; \upsilon, \alpha) = \frac{\psi^{\alpha-1} e^{-\frac{\psi}{\upsilon}}}{\upsilon^{\alpha} \Gamma(\alpha)} \tag{4.35}$$

The marginal distribution of the counts can then be obtained from the Poisson distribution $f(\Delta n;\psi)$ by integrating out the varying unobservable quantity ψ:

$$\begin{aligned} f(\Delta n;\upsilon,\alpha) &= \int_0^\infty f(\Delta n;\psi)p(\psi;\upsilon,\alpha)\mathrm{d}\psi \\ &= \frac{1}{\upsilon^\alpha\Gamma(\alpha)\Delta n!}\int_0^\infty \mathrm{e}^{-\psi}\psi^{\Delta n}\psi^{\alpha-1}\mathrm{e}^{-\frac{\psi}{\upsilon}}\mathrm{d}\psi \\ &= \frac{\Gamma(n+\alpha)}{\Gamma(\alpha)\Delta n!}\left(\frac{1}{1+\upsilon}\right)^\alpha\left(\frac{\upsilon}{1+\upsilon}\right)^{\Delta n} \end{aligned} \tag{4.36}$$

The result is called the *negative binomial* distribution.

Road traffic For the data on vehicles passing on the road, grouped into counts as above, the mean count is 2.6 and the variance is 5.3, giving a coefficient of dispersion of 2.0. In a similar way, the mean time between vehicles is 15.8 s, with a standard deviation of 23.7, so that the coefficient of variation is 1.5. Both indicate some overdispersion.

In Figure 4.6, overdispersion is indicated by the occasional larger counts due to the clumping of congestion. On the other hand, if the vehicles are too uniformly distributed over time, the distribution will be *underdispersed.* This corresponds to the large number of minutes with one vehicle. Thus, these responses seem to show both characteristics!

When the negative binomial distribution is fitted to these data, the AIC is 81.3, a considerable improvement over 86.1 for the Poisson process. The curve is plotted in Figure 4.6; this model does not satisfactorily account for the large number of minutes with one vehicle passing, because it cannot handle underdispersion. (However, models allowing underdispersion do not fit better.) Thus, we can conclude that this is a more satisfactory fit and that the process is not homogeneous.

4.4 Times between recurrent events

As we saw above, when the intervals of times between recurrent events are independent and identically distributed, we have a renewal process. In this section, I shall look at how to check the assumption of a Poisson process from the point of view of the durations between events and at how to fit other renewal processes when the Poisson process is not an acceptable model.

4.4.1 Renewal processes

We have already seen in previous chapters that a wide choice of parametric distributions is available for duration data. Here, I shall examine a few of them. I shall first look at the exponential (the Poisson process) and gamma distributions, given respectively in Equations (2.8) and (2.7), and then at a number of other possible distributions.

Road traffic In Section 4.2.3 above, I used Kaplan–Meier curves to obtain some indication as to what models might be suitable for a renewal process applied to

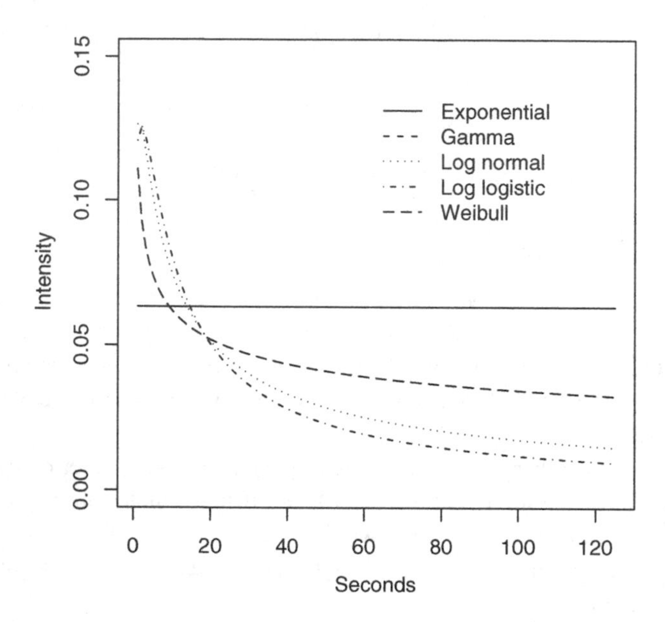

Fig. 4.7. Several intensity functions for the vehicles passing on the road, from Table 4.1.

the vehicle data. Let us see whether or not fitting them provides support for the graphical conclusions.

The exponential distribution, corresponding to a Poisson process, has an AIC of 482.4, with $\hat{\mu} = 15.81$ or $\hat{\lambda} = 0.063$, as above. The gamma distribution has 475.6, much better. (This was not obvious from the graphs in Figure 4.5.) The estimated mean duration has the same value and the dispersion parameter is $\hat{\phi} = 0.67$ so that the intensity is decreasing over time. Recall that, if $\phi = 1$ in the gamma distribution, we have an exponential distribution, so that it is some distance from that distribution.

The intensity functions for these two distributions are plotted in Figure 4.7. As we know, the Poisson process, represented here by the exponential distribution, has a constant intensity function. In contrast, the gamma intensity is estimated to be decreasing sharply over time ($\hat{\phi} < 1$).

As we have seen in Section 2.4.1, the log normal distribution of Equation (2.13) can be derived from the normal distribution by taking the logarithm of the responses. For these data, the AIC is 460.9, much better than the previous two. The intensity curve also is plotted in Figure 4.7. Although it has a mode, this occurs very early and is not really visible in the plot.

The log logistic distribution of Equation (2.16) has a similar shape to the log normal but with somewhat heavier tails. Here, this model does not fit as well as the log normal: 465.0. The intensity is plotted in Figure 4.7.

Finally, I shall look at another distribution commonly used for durations: the

Weibull, given in Equation (2.15). Like the gamma distribution, this reduces to an exponential when $\phi = 1$. For these data, the AIC is 471.7, slightly better than the gamma, but considerably worse than the log normal. Here, $\hat{\phi} = 0.75$, again showing deviation from the exponential distribution, with a decreasing intensity function.

It is interesting to notice that two distributions with an intensity function having a mode fit best among those tried. This mode occurs early in the interarrival period, so that the low very early intensity, before the mode, may allow for the physical difficulty of vehicles being very close together.

We clearly can conclude that the intensity of recurrent events is not constant over the time between events, as it would be in a Poisson process, but is diminishing. Thus, we have different forms of evidence from the count and the duration analyses that these responses do not follow a Poisson process.

4.4.2 Nonhomogeneous renewal processes

If a renewal process does not fit well to a series, one reason may be that there is a trend over time, similar to that which may occur in a nonhomogeneous Poisson process. Thus, a parameter of the distribution, usually the location parameter, might depend, say, on the (arrival) time at which the interval starts or on the number of previous events (a form of birth process). If time-varying covariates are available, these also may be useful.

Road traffic For the vehicle data, the interarrival times are plotted against the event numbers and the arrival times in Figure 4.8. There does not appear to be any trend over this short time period. On the other hand, clustering does appear to be present.

This is confirmed by fitting models as above, but with the parameters of the distributions allowed to vary with either of these covariates (event number or arrival time). In all cases, the AIC increases.

Further reading

Feller (1950, Ch. 13) provides a good introduction to recurrent events; see also Cox and Lewis (1966), although they concentrate on a variety of *ad hoc* frequentist procedures. A fundamental book on counting processes, at an advanced level, is Andersen *et al.* (1992); see also Karr (1991). For Poisson processes, see Kingman (1992). Cox and Isham (1980) and Karr (1991) give rather theoretical presentations of point processes. A standard text on renewal theory is still Cox (1962); see also Grimmett and Stirzaker (1992, Ch. 10) and Karlin and Taylor (1975, Ch. 5).

Snyder and Miller (1991) provide a practical discussion of many types of point processes and Guttorp (1995) of a wide variety of stochastic processes. Berg (1993, pp. 75–94) shows how the motility of the bacterium *Escherichia coli* can be modelled by a Poisson process.

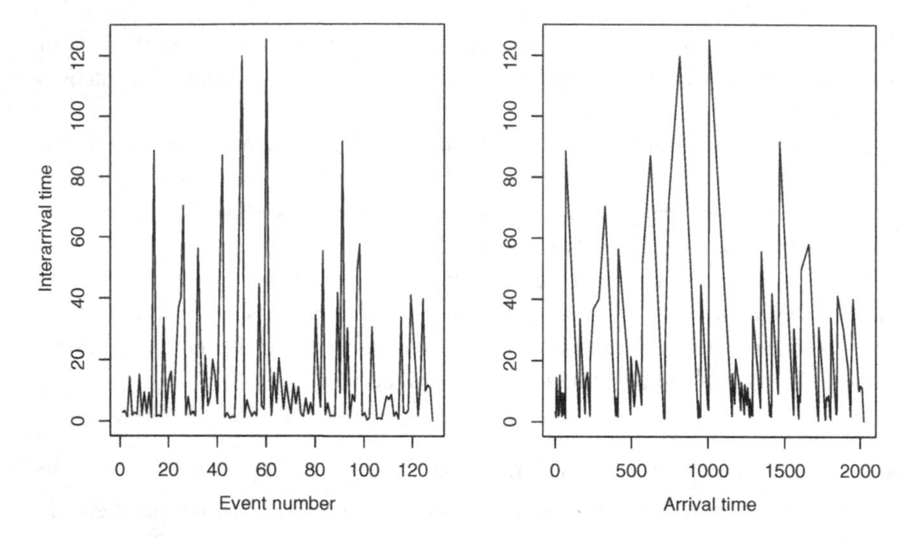

Fig. 4.8. The relationships of interarrival times to event number (left) and to arrival time (right) for the vehicles passing on the road, from Table 4.1.

Table 4.2. *Deaths by horse kicks in the Prussian army, 1875–1894, for 14 corps. (Andrews and Herzberg, 1985, p. 18)*

Corps	Year
G	0 2 2 1 0 0 1 1 0 3 0 2 1 0 0 1 0 1 0 1
I	0 0 0 2 0 3 0 2 0 0 0 1 1 1 0 2 0 3 1 0
II	0 0 0 2 0 2 0 0 1 1 0 0 2 1 1 0 0 2 0 0
III	0 0 0 1 1 1 2 0 2 0 0 0 1 0 1 2 1 0 0 0
IV	0 1 0 1 1 1 1 0 0 0 0 1 0 0 0 0 1 1 0 0
V	0 0 0 0 2 1 0 0 1 0 0 1 0 1 1 1 1 1 1 0
VI	0 0 1 0 2 0 0 1 2 0 1 1 3 1 1 1 0 3 0 0
VII	1 0 1 0 0 0 1 0 1 1 0 0 2 0 0 2 1 0 2 0
VIII	1 0 0 0 1 0 0 1 0 0 0 0 1 0 0 0 1 1 0 1
IX	0 0 0 0 0 2 1 1 1 0 2 1 1 0 1 2 0 1 0 0
X	0 0 1 1 0 1 0 2 0 2 0 0 0 0 2 1 3 0 1 1
XI	0 0 0 0 2 4 0 1 3 0 1 1 1 1 2 1 3 1 3 1
XIV	1 1 2 1 1 3 0 4 0 1 0 3 2 1 0 2 1 1 0 0
XV	0 1 0 0 0 0 0 1 0 1 1 0 0 0 2 2 0 0 0 0

Exercises

4.1 The numbers of deaths by horse kicks in the Prussian army from 1875 to 1894 for 14 corps are given in Table 4.2. G indicates the guard corps. This and corps I, VI, and XI have different organisations than the others.

(a) Can you detect any trends with time?
(b) Are there systematic differences among the corps?

Table 4.3. *Numbers of compensable accidents per day for one shift in a section of a mine in the UK over 223 days. (Maguire* et al., *1952)*

1 0 0 1 0 0 0 0 0 0 0 0 0 0 0 0 0 0 0 0 0 0 0 0 0 0 3 0 1 0 0 2 0 1 2 5 2 1 0
0 1 0 1 1 0 3 0 1 0 0 6 0 0 1 0 0 0 1 0 1 0 0 1 0 0 0 0 3 0 1 0 0 0 1 0 0 0 0
2 0 0 0 3 0 1 1 1 1 1 0 0 0 0 0 0 0 3 0 2 0 1 0 2 0 3 0 1 0 0 0 0 0 0 0 1 0 0
0 0 0 0 0 1 0 2 1 1 0 0 0 0 0 0 0 4 1 0 0 0 0 0 0 0 1 0 0 2 0 2 1 1 0 0 1 0 2
0 0 0 1 0 0 1 0 0 0 0 0 0 0 0 0 0 0 1 0 0 0 0 0 0 0 0 0 1 0 0 0 2 0 0 1 0 1 0
3 0 0 0 0 0 0 0 0 0 0 0 0 0 1 1 0 0 0 1 0 0 0 0 0 0 0 1

Table 4.4. *Monthly numbers of cases of poliomyelitis in the USA, 1970–1983. (Zeger, 1988)*

	Month											
Year	J	F	M	A	M	J	J	A	S	O	N	D
1970	0	1	0	0	1	3	0	2	3	5	3	5
1971	2	2	0	1	0	1	3	3	2	1	1	5
1972	0	3	1	0	1	4	0	0	1	6	14	1
1973	1	0	0	1	1	1	1	0	1	0	1	0
1974	1	0	1	0	1	0	1	0	1	0	0	2
1975	0	1	0	1	0	0	1	2	0	0	1	2
1976	0	3	1	1	0	2	0	4	0	2	1	1
1977	1	1	0	1	1	0	2	1	3	1	2	4
1978	0	0	0	1	0	1	0	2	2	4	2	3
1979	3	0	0	2	7	8	2	4	1	1	2	4
1980	0	1	1	1	3	0	0	0	0	1	0	1
1981	1	0	0	0	0	0	1	2	0	2	0	0
1982	0	1	0	1	0	1	0	2	0	0	1	2
1983	0	1	0	0	0	1	2	1	0	1	3	6

(c) These data have often been aggregated over corps and years and used as a good example of a Poisson distribution. Is this reasonable?

4.2 The data in Table 4.3 provide the numbers of accidents in a section of a mine in the UK, recorded over a number of days.

(a) Fit a Poisson process to these data.

(b) Check the assumptions of this model to see whether or not they are reasonable.

4.3 Table 4.4 shows the monthly numbers of cases of poliomyelitis in the USA over a 14 year period.

(a) What is the simplest model that can describe these data?

(b) Does the number of cases vary with month?

(c) Has the number of cases per month changed over the years?

4.4 In a study to determine whether or not air traffic density varied over time, aircraft entrance times (s) were recorded for a low altitude transitional control sector (called 454LT) northwest of the Newark, USA, airport. Those

Table 4.5. *Aircraft entrance times (s) at Newark airport on 30 April 1969 from 12:00 to 20:00. (Hsu, 1979; read across rows)*

467	761	792	812	926	1 100	1 147
1 163	1 398	1 462	1 487	1 749	1 865	2 004
2 177	2 208	2 279	2 609	2 682	2 733	2 818
2 837	2 855	2 868	3 089	3 209	3 223	3 233
3 272	3 399	3 595	3 634	3 650	3 851	4 176
4 304	4 391	4 453	4 539	4 748	4 839	5 049
5 202	5 355	5 551	5 598	5 640	5 702	5 935
6 000	6 192	6 435	6 474	6 600	6 810	6 824
7 168	7 181	7 202	7 218	7 408	7 428	7 720
7 755	7 835	7 958	8 307	8 427	8 754	8 819
8 904	8 938	8 980	9 048	9 237	9 268	9 513
9 635	9 750	9 910	9 929	10 167	10 254	10 340
10 624	10 639	10 669	10 889	11 386	11 515	11 651
11 727	11 737	11 844	11 928	12 168	12 657	12 675
12 696	12 732	13 092	13 281	13 536	13 556	13 681
13 710	14 008	14 151	14 601	14 877	14 927	15 032
15 134	15 213	15 491	15 589	15 600	15 631	15 674
15 797	15 953	16 089	16 118	16 215	16 394	16 503
16 515	16 537	16 570	16 597	16 619	16 693	17 314
17 516	17 646	17 770	17 897	17 913	17 922	18 174
18 189	18 328	18 345	18 499	18 521	18 588	19 117
19 150	19 432	19 662	19 758	19 789	19 831	19 978
20 119	20 312	20 346	20 449	20 455	20 604	20 675
20 817	20 898	21 245	21 386	21 562	22 022	22 056
22 095	22 182	22 554	22 764	22 955	22 993	23 025
23 117	23 321	23 341	23 650	23 766	23 879	23 888
24 458	24 889	24 930	24 967	25 224	25 312	25 477
25 498	25 712	25 721	25 884	25 919	25 985	26 196
26 459	26 468	26 494	26 505	26 554	26 906	27 003
27 437	27 661	27 675	27 697	27 721	27 734	27 802
27 971	28 116	28 746				

for one afternoon are shown in Table 4.5. There were airways passing through the sector in many directions and at different altitudes, which might contribute to apparent randomness. However, at the same time, locations of aircraft were controlled by the air traffic controllers, which might produce systematic separation.

(a) Do these data follow a Poisson process? (Assume that there was an aircraft at 12:00.)

(b) Is there any evidence of a change in air traffic density over this period?

(c) Is there clustering of aircraft at any time?

4.5 Table 4.6 gives the results of an experiment recording the successive times (s) of initiation of mating between flies for two different species.

(a) Do these durations follow a Poisson process?

Table 4.6. *Times (seconds) of initiations of mating between flies of two different species. (Aalen, 1978)*

Ebony flies							
143	180	184	303	380	431	455	475
500	514	521	552	558	606	650	667
683	782	799	849	901	995	1131	1216
1591	1702	2212					
Oregon flies							
555	742	746	795	934	967	982	1043
1055	1067	1081	1296	1353	1361	1462	1731
1985	2051	2292	2335	2514	2570	2970	

Table 4.7. *Times (days of service) at which valve seats were replaced on 41 diesel engines in a service fleet. (Nelson, 1995; Lawless and Nadeau 1995)*

761	593
759	573 589
98 667	165 408 604 606
326 653 653 667	249 594
665	344 497 613
84 667	265 586 595
87 663	166 206 348 389
646 653	601
92 653	410 581 601
651	611
258 328 377 621 650	608
61 539 648	587
254 276 298 640 644	367 603
76 538 642	202 563 570 585
635 641	587
349 404 561 649	578
631	578
596	586
120 479 614	585
323 449 582	582
139 139 589	

(b) Is there a difference over time in the intensity of mating between the two species?

4.6 The times at which valve seats were replaced on 41 diesel engines in a service fleet are shown in Table 4.7. The engines had 16 valves, but we do not know which ones were replaced. The last time for each engine is censored.

(a) Is a renewal process an appropriate model for these data?
(b) How does the average number of valves replaced change with age?
(c) Does the replacement rate increase with time?

Table 4.8. *Lengths of migrating motor complex periods in 19 fasting individuals. The last interval is always censored. (Aalen and Husebye, 1991; read across rows)*

112	145	39	52	21	34	33	51	54	
206	147	30							
284	59	186	4						
94	98	84	87						
67	131								
124	34	87	75	43	38	58	142	75	23
116	71	83	68	125	111				
111	59	47	95	110					
98	161	154	55	44					
166	56	122							
63	90	63	103	51	85				
47	86	68	144	72					
120	106	176	6						
112	25	57	166	85					
132	267	89	86						
120	47	165	64	113	12				
162	141	107	69	39					
106	56	158	41	41	168	13			
147	134	78	66	100	4				

4.7 The mixture and transport of the intraluminal content of the small bowel is performed by muscular activity called motility. During fasting, a cyclical pattern occurs. Bowel pressure can be measured by manometry. Modern digital ambulatory techniques allow such measurements easily to be made. These involve two transducers mounted on a thin catheter in the proximal small bowel.

In one study, 19 healthy adults (15 men and four women) were given a standardised mixed meal of 405 kcal at 18:00. Following a fed state of from four to seven hours, they reach a fasting period during which motility (muscular activity) of the small bowel could occur. Periodically, a fasting cycle occurs, with a migrating motor complex (MMC) phase. The data in Table 4.8 record the lengths of times between occurrences of this phase, called MMC periods, starting at the first occurrence of such a phase. Measurement started at 17:45 and stopped at 7:25 the next day so that all final periods are censored.

(a) Is there any evidence of differences among individuals?
(b) Does a given interval depend in some way on the previous history of the individual?

4.8 Times between successive failures of air-conditioning equipment in 13 Boeing 720 aircraft are presented in Table 4.9. The planes were in a fleet for which records were kept. After roughly 2000 h of service, four of the planes received major overhauls. Because this may have affected the length of the corresponding failure intervals, these particular durations

Table 4.9. *Numbers of operating hours between successive failures of air-conditioning equipment in 13 aircraft. Dashes indicate major overhauls. (Proschan, 1963)*

Aircraft number												
1	2	3	4	5	6	7	8	9	10	11	12	13
194	413	90	74	55	23	97	50	359	50	130	487	102
15	14	10	57	320	261	51	44	9	254	493	18	209
41	58	60	48	65	87	11	102	12	5		100	14
29	37	186	29	104	7	4	72	270	283		7	57
33	100	61	502	220	120	141	22	603	35		98	54
181	65	49	12	239	14	18	39	3	12		5	32
	9	14	70	47	62	142	3	104			85	67
	169	24	21	246	47	68	15	2			91	59
	447	56	29	176	225	77	197	438			43	134
	184	20	386	182	71	80	188				230	152
	36	79	59	33	246	1	79				3	27
	201	84	27	—	21	16	88				130	14
	118	44	—	15	42	106	46					230
	—	59	153	104	20	206	5					66
	34	29	26	35	5	82	5					61
	31	118	326		12	54	36					34
	18	25			120	31	22					
	18	156			11	216	139					
	67	310			3	46	210					
	57	76			14	111	97					
	62	26			71	39	30					
	7	44			11	63	23					
	22	23			14	18	13					
	34	62			11	191	14					
		—			16	18						
		130			90	163						
		208			1	24						
		70			16							
		101			52							
		208			95							

were omitted from each series. Such data provide information on the distribution of failure intervals for air conditioning systems that may be used for predicting reliability, scheduling maintenance, and providing spare parts.

(a) Are there differences in failure rate among the aircraft?

(b) Does the failure rate evolve over time?

4.9 Kidney patients using a portable dialysis machine may have recurrent infections. These occur at the point of insertion of the catheter. When this happens, the catheter must be removed and the infection cured, after which the catheter can be reinserted. There can be right censoring if the catheter must be removed for other reasons. Patients were followed for two recurrence times, one or both of which might be censored. Age, sex, and type of disease (GN, AN, PKD) are available as risk factors, as shown in Table 4.10.

Table 4.10. *Times (units not specified; asterisks indicate censoring) to recurrence of infections in kidney patients. (McGilchrist and Aisbett, 1991)*

Times		Sex	Disease	Age
8	16	M	Other	28
23	13*	F	GN	48
22	28	M	Other	32
447	318	F	Other	31–32
30	12	M	Other	10
24	245	F	Other	16–17
7	9	M	GN	51
511	30	F	GN	55–56
53	196	F	AN	69
15	154	M	GN	51–52
7	333	F	AN	44
141	8*	F	Other	34
96	38	F	AN	35
149*	70*	F	AN	42
536	25*	F	Other	17
17	4*	M	AN	60
185	177	F	Other	60
292	114	F	Other	43–44
22*	159*	F	GN	53
15	108*	F	Other	44
152	562	M	PKD	46–47
402	24*	F	Other	30
13	66	F	AN	62–63
39	46*	F	AN	42–43
12	40	M	AN	43
113*	201	F	AN	57–58
132	156	F	GN	10
34	30	F	AN	52
2	25	M	GN	53
130	26	F	GN	54
27	58	F	AN	56
5	43*	F	AN	50–51
152	30	F	PKD	57
190	5*	F	GN	44–45
119	8	F	Other	22
54*	16*	F	Other	42
6*	78	F	PKD	52
63	8*	M	PKD	60

(a) Which factors are related most closely to risk of infection?
(b) Is the risk of a second infection higher than that of the first one?

5

Discrete-time Markov chains

Up until now, I only have considered processes in which one type of event can occur. In Chapter 3, the event signalled an irreversible change of state; in Chapter 4, it could recur. Except in the general theory of Section 4.1.1, I have made no assumptions about any dependencies among these events over time; the only condition was in the definition of the intensity as the probability of the event happening, given that it had not yet occurred.

In this chapter, I shall begin the study of stochastic processes involving several kinds of events. I shall consider one way in which to introduce dependence assumptions, by conditioning on what happened before, here simply on the previous state(s). Thus, the probabilities of the various possible outcomes that can be observed at a given time will depend on the previous history of the process in one specific way. I shall model explicitly the *probability of transition* between states. In the simplest case, the present state will depend only on that immediately preceding it (Section 1.2.2), a *first-order Markov chain*. This procedure will yield *state dependence* models (Section 1.2.3).

I shall assume that each event occurs in a given constant interval of time and that these intervals are small enough so that only one event, however defined, can occur in each. In addition, in contrast to previous chapters, I shall assume that the small intervals are all of equal size, so that I am working in *discrete time*. (I shall leave the presentation of the equivalent Markov chains in *continuous* time until Section 6.1.3.)

Because of the Markov and discrete time assumptions, suitable data can easily be modelled as Markov chains after summary in multidimensional contingency tables. If there are only two states, logistic regression can be used. Otherwise, log linear models will be required.

For the special case of only two states, so that the outcome is binary, other approaches include studying either the numbers of events occurring in fixed time intervals or the times between successive events. I already have looked at some of these possibilities in the previous chapter, when recurrent events are independent, and shall return to them in the following chapters to examine dependencies.

Markov chains have many applications. Below, I shall examine examples in industry (accidents), political science (voting behaviour), sociology (social mobility), medicine (changing blood pressure), and demography (geographical migration).

5.1 Theory

A great deal is known about the theoretical properties of Markov chains. They are relatively easy to study because generally only linear algebra is required. Here, I shall outline some of the more important results.

Throughout this chapter, I shall be assuming that the stochastic process has a finite number of possible discrete states and that observations are made at equally-spaced, discrete, time intervals. Thus, a finite number of different types of events is possible, defined by the transitions between pairs of states of the process.

5.1.1 Transition matrices

Let us now examine the properties of Markov chains with a fixed, finite number of possible states: the state space.

Marginal and conditional probabilities

The *marginal distribution* $\boldsymbol{\pi}_t$ at time t gives the probabilities of being in each possible state at that time. The *initial distribution* $\boldsymbol{\pi}_0$ is the marginal distribution when the process starts. As we have seen (Section 1.1.4), these distributions do not uniquely define the stochastic process (unless the successive states are independent). However, they do describe the underlying marginal trend of the process; this may often be independent of the specificities of any particular observed case of the process.

On the other hand, the *conditional probability* $\pi_{j|i}$ of being in any state j at time t will depend on the state i at the previous time point (or more than one previous state if $M > 1$). For $M = 1$, these probabilities may be assembled into a square *transition matrix*, say $\mathbf{T}$, with columns indicating the present state and rows the previous state so that the rows each sum to unity. Because of this constraint on the row sums, this is called a *stochastic matrix*. (The adjective 'stochastic' here does not imply that the matrix is changing randomly over time!)

$\mathbf{T}$ is called a transition matrix because premultiplying it by the vector $\boldsymbol{\pi}_t$ of marginal probabilities of the different states at a given time point t will give the marginal vector for the next time period $t+1$:

$$\boldsymbol{\pi}_{t+1}^{\top} = \boldsymbol{\pi}_t^{\top}\mathbf{T} \tag{5.1}$$

Thus, the transition matrix $\mathbf{T}$ represents the pattern of change and the marginal probabilities $\boldsymbol{\pi}_t$ the underlying trend of the process.

The transition matrix of conditional probabilities of each state, given the previous one, may be varying over time: $\mathbf{T}_t$. It is, then, said to be *nonhomogeneous*. However, a matrix that changes arbitrarily at each time point will result in too complex a model. Thus, nonhomogeneity usually is the result of the conditional probabilities depending on time-varying covariates through some regression functions in the model. Almost all theoretical results, unfortunately, hold only for the simpler case of a homogeneous transition matrix.

Stationarity

Thus, the transition matrix is *homogeneous* if it is the same at all time points. However, because this matrix defines the *conditional* probabilities, this assumption is not sufficient to ensure that the process itself is stationary in the sense of Section 1.1.4. The initial conditions, given by $\boldsymbol{\pi}_0$, also must be known and take a specific form.

The stationary marginal distribution $\boldsymbol{\pi}^s$, if it exists, is given by the *steady-state equation*

$$\boldsymbol{\pi}^{s\top} = \boldsymbol{\pi}^{s\top}\mathbf{T} \tag{5.2}$$

This is the stationarity (Section 1.1.4) of a Markov chain process. It generally will arise after the process has run for a sufficiently long time. Note that the stationary distribution may not be unique. Thus, if $\mathbf{T}$ is the identity matrix, then $\boldsymbol{\pi}^s$ can be any distribution.

If the process starts with the stationary initial distribution corresponding to the given transition matrix, so that $\boldsymbol{\pi}_0 = \boldsymbol{\pi}^s$, then $\boldsymbol{\pi}_t = \boldsymbol{\pi}^s$ for all t and the complete series is stationary. On the other hand, if $\boldsymbol{\pi}^s \neq \boldsymbol{\pi}_0$ is independent of the initial distribution $\boldsymbol{\pi}_0$, the Markov chain is *ergodic* (Bartlett, 1955, p. 33; see Section 1.1.4 and further details below).

The existence of a unique stationary distribution implies several approaches to estimating parameters in a Markov chain.

(i) The initial conditions may be known *a priori*. Thus, the process(es) might necessarily start in some given state. Then, $\boldsymbol{\pi}_0$ will contain all zeros except for a 1 for that state.
(ii) If the series is stationary, the marginal distribution at the first time point will be $\boldsymbol{\pi}^s$, which can be obtained from $\mathbf{T}$ using Equation (5.2) with no additional parameters other than those in $\mathbf{T}$. This marginal distribution can, then, be included in the likelihood.
(iii) If the series is nonstationary, the parameters of the initial marginal distribution cannot be obtained from $\mathbf{T}$. For only one series, there really is no information available to estimate $\boldsymbol{\pi}_0$ and one must condition on the first response. However, if there are enough replicate series, the marginal distribution of the first event can be estimated.

Reversibility

If the sequence of states of a Markov chain is reversed, this is still a Markov chain, but with transition probabilities given as a combination of the stationary probabilities and the forward transition probabilities: $\pi^s_j\pi_{j|i}/\pi^s_i$. In addition, if $\pi_{i|j}$ and $\pi_{j|i}$ in the original matrix are related by

$$\pi_{j|i} = \frac{\pi^s_j\pi_{i|j}}{\pi^s_i} \tag{5.3}$$

then the process is *time reversible* (see Section 5.4.1).

n-step probabilities

Let us now see how a multistate process can pass from one state to another indirectly, after first transiting by some other arbitrary states. Represent the probability of change between two given states i and j, in say exactly n steps, by $\pi_{j|i}^{(n)}$.

If there exists an n such that $\pi_{j|i}^{(n)} > 0$, then state j can be *reached* or is *accessible* from state i. If, in addition, state i is accessible from j, then $\pi_{i|j}^{(n)} > 0$ and the two states *communicate*. A communicating set or *class* of states is one in which all pairs of states communicate with each other and none communicate with states outside the class. If all states in a Markov chain communicate, that is, there is only one class, the chain is *irreducible*.

The value of $\pi_{j|i}^{(n)}$ is given by the Chapman–Kolmogorov equation:

$$\pi_{j|i}^{(n)} = \sum_k \pi_{k|i}^{(m)} \pi_{j|k}^{(n-m)} \qquad 1 \leq m \leq n-1$$

or

$$\mathbf{T}^{(n)} = \mathbf{T}^{(n-m)}\mathbf{T}^{(m)} \qquad 1 \leq m \leq n-1 \tag{5.4}$$

where $\mathbf{T}^{(n)}$ is the n-step transition matrix. But $\mathbf{T}^{(1)} = \mathbf{T}$, so that this implies that $\mathbf{T}^{(n)} = \mathbf{T}^n$. Thus,

$$\boldsymbol{\pi}_n^\top = \boldsymbol{\pi}_0^\top \mathbf{T}^n \tag{5.5}$$

which also is implied directly by Equation (5.1).

Under suitable conditions, a Markov chain will have a *limiting distribution* so that, for large n, $\mathbf{T}^n$ approaches a matrix with all rows equal to $\boldsymbol{\pi}^s$. This means that, for such a chain, if the process runs long enough, it reaches an *equilibrium* situation independent of the initial distribution.

Multiplying a matrix by itself n times can be time consuming and can also involve substantial round-off error. An efficient way to calculate the power of such a matrix, especially if n is reasonably large, is by spectral decomposition. If $\mathbf{W}$ is a matrix containing the eigenvectors of $\mathbf{T}$ as columns and $\mathbf{D}$ is a diagonal matrix containing the corresponding eigenvalues, then

$$\mathbf{T} = \mathbf{W}\mathbf{D}\mathbf{W}^{-1} \tag{5.6}$$

and

$$\mathbf{T}^n = \mathbf{W}\mathbf{D}^n\mathbf{W}^{-1} \tag{5.7}$$

Because $\mathbf{D}$ is diagonal, the power can be calculated element-wise, a much simpler and more precise operation than multiplying the matrix by itself many times. Note that this works only if the eigenvalues are distinct, but generally this is the case for the numerical matrices arising from the data encountered in statistical modelling.

Aggregation

It is important to note that, if some states of a Markov chain (with more than two states) are combined, the result generally is not a Markov chain (at least, not of the same order). Thus, in any given situation, the states must be defined very carefully.

The reverse of this fact can be useful. It is often possible to obtain the Markov property by extending the number of states. Thus, suppose that a process is of higher order than first ($M > 1$). By defining, as new states, all possible ordered combinations of M observed states and looking at overlapping sets of M adjacent states, we can create a first-order Markov chain. Thus, if a second-order Markov chain has two possible states, 1 and 2, the four new states will be $(1,1)$, $(1,2)$, $(2,1)$, and $(2,2)$ and the new series of responses, (y_1, y_2), (y_2, y_3), (y_3, y_4), and so on, will form a first-order Markov chain.

The unit of time is much less important, as long as the time points are equally spaced. Thus, if the daily transition matrix of a first-order Markov chain is $\mathbf{T}$, weekly observations will still form a Markov chain, but with transition matrix $\mathbf{T}^7$.

More generally, great care must be taken in checking the properties of a Markov chain. If $\boldsymbol{\pi}^s$ and $\mathbf{T}$ are estimated from any reasonably long series, they will closely satisfy Equation (5.2) so that this is not an indication that the series follows a Markov chain (Cohen, 1968). Finch (1982) describes some of the difficulties in checking for a Markov chain when a number of replications is available.

5.1.2 Time spent in a state

In previous chapters, we concentrated on the time spent in a given state. Now let us see what the assumptions of a Markov chain imply about such durations.

Duration in a state

Three possible cases of durations in a state can occur, two of which are extreme:

(i) If $\pi_{i|i} = 0$, the process never stays more than one time period in the state i.

(ii) If $\pi_{i|i} = 1$ (implying that $\pi_{j|i} = 0$ for all $j \neq i$), then the state i is *absorbing*. If the process reaches such a state, it stays there forever after (compare with Section 3.1.1). A process with one or more absorbing states cannot be irreducible. A whole class of states also may be absorbing or *closed*; then, once the process enters the class, it moves among those states, never leaving the class. This occurs if $\pi_{j|i} = 0$ for all j outside the class and i in the class. (This implies that $\sum_j \pi_{j|i} = 1$ where the sum is over states in the class.) A closed class of states can be studied independently of any other states in the system.

(iii) If $0 < \pi_{i|i} < 1$, the time spent in a state, once it is entered, has a *geometric distribution*

$$f(t; \pi_{i|i}) = \pi_{i|i}^{t-1}(1 - \pi_{i|i}) \tag{5.8}$$

with mean duration time, $1/(1 - \pi_{i|i})$. This is the discrete-time analogue of the exponential distribution so that there is some similarity with a Poisson process.

As well, if the stationary marginal distribution $\boldsymbol{\pi}^s$ exists, it gives the long-run proportion of time spent in each state.

Although a Markov chain may appear to be similar to a discrete-time Poisson process, care must be taken. Consider, say, a binary point process in discrete-time following a Markov chain with states 0 and 1 where the latter might indicate a recurrent event. The time without the event, that is, the duration in the first state between recurrent events, is a series of 0s. The distribution of this duration is geometric, with parameter $\pi_{1|1}$, as if it were the discrete-time analogue of a Poisson process. However, this distribution, but with parameter $\pi_{2|2}$, also applies to second state, that is, to the duration of a sequence of 1s. Thus, there can be clustering of the recurrent events, as compared to the randomness of a Poisson process. The mean number of recurrent events occurring consecutively (the time spent in state 1), is $1/(1-\pi_{1|1})$, which can be considerably greater than unity if $\pi_{1|1}$ is large. On the other hand, if $\pi_{1|1}=0$, two recurrent events can never occur consecutively.

Recurrence of a state

Now let us look at when a process can be expected to visit a given state.

Suppose that the process starts in state j. The time taken to return for the first time to this state is called the *return time*. In contrast, the time to reach state j starting from any arbitrary state is called the *hitting time*; it is zero if the process starts in state j. The *first passage* or *recurrence time* is the hitting time $+1$. These are all stopping times (Section 4.1.1).

Periodicity A state has period m if $\pi^{(n)}_{i|i}=0$ for all n not divisible by m, with m the greatest such integer. Such a Markov chain can be in state i only at intervals of time m. All states in a communicating class will have the same period. A Markov chain with $m>2$ cannot be reversible (Wolff, 1989, p. 183). If $m=1$, the state is *aperiodic*. An irreducible, periodic Markov chain will not reach an equilibrium limiting distribution, whereas an irreducible, aperiodic one will. The transition matrix of the latter has one unit eigenvalue and all others less than one in absolute value. Thus, as n increases, $\mathbf{D}^n$ in Equation (5.7) approaches a matrix of zeros except for a one corresponding to the unit eigenvalue.

Each time that a Markov chain returns to its initial state, the future of the process has the same distribution as the first time it was there. This is a regeneration point (Section 1.1.4). Thus, the times taken successively to return to the same state are independent, identically distributed random variables. (See Grimmett and Stirzaker, 1992, p. 198.)

Recurrent and transient states The *first passage distribution* gives the probability of arriving in the state j *for the first time* after various intervals of time $n>0$ starting from state i:

$$\nu^{(n)}_{j|i}=\Pr(y_n=j,y_t\neq j,t=1,\ldots,n-1|y_0=i) \tag{5.9}$$

Note that $\nu^{(n)}_{j|i}\leq\pi^{(n)}_{j|i}$ because the latter can include second or later returns to state j.

The values $\nu_{i|i}^{(n)}$ are of particular interest because they give the *recurrence probabilities* of returning to the same state for the first time. If $\sum_n \nu_{i|i}^{(n)} = 1$, state i is *persistent* or *recurrent*; otherwise, it is *transient*. A process always will return to a recurrent state sometime in the future (if we wait long enough), in fact infinitely often, whereas it has a positive probability of never returning to a transient state. Thus, for the former,

$$\sum_{n=1}^{\infty} \pi_{i|i}^{(n)} = \infty \tag{5.10}$$

whereas it is finite for the latter.

The states of a communicating class are either all recurrent or all transient. If the class is closed, they are all recurrent. Thus, in an irreducible Markov chain, all states are recurrent.

For a recurrent state, $\nu_{i|i}^{(n)}$ is a probability distribution for which, among other things, the *mean recurrence time* can be calculated. This is equal to the reciprocal of the stationary marginal probability of that state: $1/\pi_i^s$, a special case of the elementary renewal theorem (Section 4.1.4). On the other hand, if we count recurrence from when the process leaves state i, the mean time is $(1-\pi_i^s)/[\pi_i^s(1-\pi_{i|i})]$.

A state that is aperiodic and recurrent with finite mean recurrence time is said to be ergodic (Section 1.1.4). If all states are ergodic, the chain is ergodic. An irreducible, aperiodic Markov chain with a finite number of states is always ergodic with a unique stationary distribution (Cox and Miller, 1965, p. 107). For a recurrent state, $\nu_{i|i}^{(n)}$ approaches the ratio of the period to the mean recurrence time (hence, the reciprocal of this time for aperiodic states) for large n.

Probability of absorption Suppose that a process has both recurrent and transient states. For simplicity, assume that the former are absorbing. By an appropriate ordering of the states, the transition matrix for such a process can be written as

$$\mathbf{T} = \begin{pmatrix} \mathbf{I} & \mathbf{0} \\ \mathbf{B} & \mathbf{Q} \end{pmatrix} \tag{5.11}$$

Then, $(\mathbf{I}-\mathbf{Q})^{-1}$ is called the *fundamental matrix*, where $\mathbf{I}$ is the appropriate identity matrix in each case. The probabilities of ending up in each absorbing state starting from each transient state are given by $(\mathbf{I}-\mathbf{Q})^{-1}\mathbf{B}$. The extension to absorbing classes of recurrent states is straightforward (Bremaud, 1999, pp. 154–161).

5.1.3 Random walks

One rather famous type of stochastic process is called a *random walk*, applied to how a drunk tries to find his way home.

General case

The most general form of a random walk is generated by the successive sums of a series of independent, identically-distributed random variables occurring over time:

$$S_t = \sum_{i=1}^{t} Y_i \tag{5.12}$$

Obviously, if $Y_t > 0$, the process S_t can only increase, but if Y_t can also take negative values, the sum moves in both directions.

There may be *boundary conditions* such that, if S_t reaches the edge of or goes outside an interval, say (b_l, b_u), the process either stops or returns in the opposite direction. The first is called an *absorbing barrier* and the second a *reflecting barrier*.

The renewal processes of Section 4.1.4 can be thought of as random walks where S_t is the total time elapsed and each $Y_t > 0$ is an interarrival time.

Application to Markov chains

Another special case of a random walk is important in the context of Markov chains. Suppose that, on a line, successive moves are allowed to the left or right with (unconditional) probabilities π_l and π_r, respectively. Thus, Y_t can only take the values 1 and -1 and S_t gives the position on the line at time t. (The famous application is to a drunk moving along a narrow pavement, one step at a time, with $\pi_l = \pi_r$.) If $\pi_l \neq \pi_r$, there is *drift* so that the process will tend to move in the direction of higher probability. If the possibility of no move is allowed so that $y_t = 0$ is also possible, then $\pi_l + \pi_r < 1$ and this is a form of *birth and death* process.

These special random walks may be reformulated as Markov chains where different points on the line are different states of the process. The transition probabilities will be zero except for movement between adjacent states. If the line is unbounded, the Markov chain will have an infinite number of states.

If $\pi_l + \pi_r = 1$, the probability of exactly m steps to the right in a total of n steps will be given by the binomial distribution,

$$f(m; \pi_r, n) = \binom{n}{m} \pi_r^m \pi_l^{n-m} \tag{5.13}$$

The position (state) after n steps equals the number of moves to the right minus the number to the left: $m - (n - m) = 2m - n$. Thus, the mean position is $(2\pi_r - 1)n$ steps from the starting position. The corresponding variance is $4\pi_r\pi_l n$, so that the variability increases with time, the number of steps (Berg, 1993, p. 13).

All of the above results require a homogeneous transition matrix. In most realistic modelling situations, they primarily will be of academic interest because models will involve transition probabilities depending on covariates. If some of the covariates are time-varying, the matrix will be nonhomogeneous. Such models with covariates can be fitted using logistic or log linear regression.

5.2 Binary point processes

Let us examine first what contributions Markov chains can make to the study of recurrent events (Chapter 4) in discrete time. As we know, when one of only two

Table 5.1. *June days with measurable precipitation (1) at Madison, Wisconsin, USA, 1961–1971. (Klotz, 1973)*

Year	
1961	10000 01101 01100 00010 01010 00000
1962	00110 00101 10000 01100 01000 00000
1963	00001 01110 00100 00010 00000 11000
1964	01000 00000 11011 01000 11000 00000
1965	10001 10000 00000 00001 01100 01000
1966	01100 11010 11001 00001 00000 11100
1967	00000 11011 11101 11010 00010 00110
1968	10000 00011 10011 00100 10111 11011
1969	11010 11000 11000 01100 00001 11010
1970	11000 00000 01000 11001 00000 10000
1971	10000 01000 10000 00111 01010 00000

possible states is recorded in each successive equal time interval, this may often indicate the presence or absence of some event. As we have seen, in contrast to the recurrent event models of Chapter 4, Markov chains allow clustering of such events.

Here, time is discrete so that dependencies among such binary responses can be modelled either by logistic regression or by creating a contingency table and applying standard log linear models. (In contrast, when there are more than two states, log linear models most often will be required.) Here, I shall illustrate the equivalence of the two approaches for a two-state series.

5.2.1 Transition matrices

In Chapter 4, I looked at binary point processes in continuous time, but with no dependencies on previous events. The important special case of a Markov chain with only two states will allow us to study one form of dependencies in such processes in discrete time. If there are no covariates, the data may be summarised as a 2×2 contingency table and the transition matrix calculated from it.

Rainfall Measurable amounts of precipitation have been recorded in Madison, Wisconsin, USA, for a long time. The results for June days between 1961 and 1971 are shown in Table 5.1. There are 221 days without the rain and 109 with it. Thus, if we assume stationarity, the (constant) marginal probabilities can be estimated to be, respectively, 0.67 and 0.33: it rained on about one-third of the days. The cumulative numbers of rainy days in June each year are plotted in Figure 5.1. We see that there is some divergence over the years in the amounts of rain during this month.

Next, we can examine the dependence among days by calculating the estimates of the parameters of a Markov chain. If we assume that the transition matrix is homogeneous, the same each year, we can aggregate the data to form a contingency

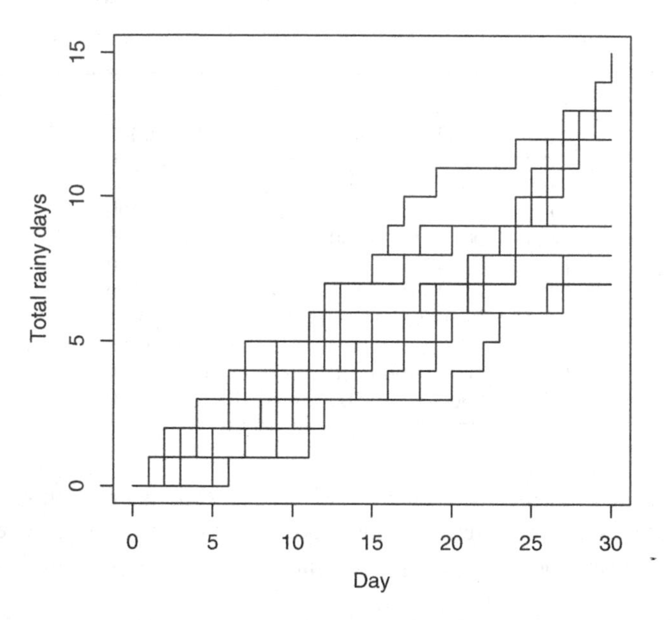

Fig. 5.1. The counting processes of rainy days in Madison each June from 1961 to 1971, from Table 5.1.

table (left), and obtain the corresponding estimated matrix of conditional transition probabilities (right), dependent on the state on the previous day:

	Day 1		Day 1	
Day 0	0	1	0	1
0	152	59	0.72	0.28
1	64	44	0.59	0.41

The two rows of the transition matrix are quite different, indicating that there may be dependence on the state occurring the previous day. If this matrix is a valid representation for all pairs of days in the series, the mean length of a run of days with rain would be $1/(1-0.41) = 1.67$ days, whereas without rain it would be $1/(1-0.72) = 3.58$ days. This indicates some clustering of rainy days.

We also can look at how the present state depends on that two days before:

	Day 2		Day 2	
Day 0	0	1	0	1
0	136	67	0.67	0.33
1	73	32	0.70	0.30

Here, the conditional probabilities are similar to the marginal ones. If there is rain one day, the probability is estimated to be 0.30 that there will be rain two days later,

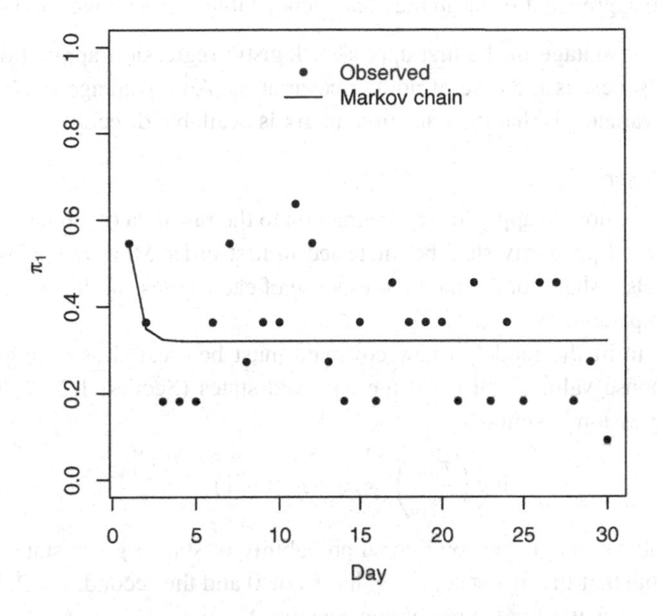

Fig. 5.2. The observed marginal probability of rain on each day of June over the 11 years and the corresponding probabilities predicted from the Markov chain for the data in Table 5.1.

similar to the probability of rain when there was none two days before. Notice that

$$\begin{pmatrix} 0.72 & 0.28 \\ 0.59 & 0.41 \end{pmatrix}^2 = \begin{pmatrix} 0.68 & 0.32 \\ 0.67 & 0.33 \end{pmatrix}$$

$$\doteq \begin{pmatrix} 0.67 & 0.33 \\ 0.70 & 0.30 \end{pmatrix}$$

The observed two-step transition matrix is similar to that obtained by squaring the observed one-step transition matrix. Both are close to the stationary matrix.

Because the square of the observed one-step transition matrix is close to the stationary matrix, the model quickly predicts a constant marginal probability of rain equal to 0.33. If we start with the observed marginal probability of rain for the first day of June over all years (0.55) and update for each successive day by multiplying by the estimated transition matrix, we soon find this constant probability, as can be seen in Figure 5.2.

5.2.2 Logistic regression

Although we have estimated the parameters of this simple Markov chain by a direct approach above, it useful to be able to fit the model in a more general context. This will allow us more easily to compare different models. Thus, these simple Markov chain models can be fitted by applying logistic regression. This can be done in two distinct ways:

(i) The binary series can be used directly.
(ii) The aggregated series in the contingency table given above can be used.

One major advantage of the first approach, logistic regression applied directly to the binary series, is the ease of adding covariates. An advantage of the second, without covariates, is that the transition matrix is available directly.

Binary time series

Let us first see how to apply logistic regression to the raw data of a binary series. In what follows, I primarily shall be interested in first-order Markov chains, so that, in the models, I shall not use the first response of each series. In this way, the AICs will be comparable.

In order to fit the model, a new covariate must be created as a lagged vector of the response values, that is, of the observed states (Section 1.2.3). Then, the logistic regression is simply

$$\log\left(\frac{\pi_{1|i}}{\pi_{2|i}}\right) = \beta_0 + \beta_1(i-1) \tag{5.14}$$

where, as above, $\pi_{j|i}$ is the conditional probability of state j given state i the time before. (Note that the first state, $i = 1$, has value 0 and the second, $i = 2$, has value 1, so that $i - 1$ is used in the regression equation.)

Rainfall For the rainfall data, the model with no dependence (corresponding to relying on the constant marginal probability for prediction) has an AIC of 201.7, compared to 200.0 for a first-order Markov chain (the conditional probabilities for dependence one day back, above). The logistic regression model is estimated to be

$$\log\left(\frac{\widehat{\pi_{1|i}}}{\widehat{\pi_{2|i}}}\right) = -0.95 + 0.57(i-1)$$

The positive sign of $\widehat{\beta}_1$ indicates that there is a higher probability of rain if there was some the day before.

For these data, few interesting covariates are available. We can, however, look at differences among the years and trends over the month (as in Sections 4.2.1 and 4.3 above for recurrent events). First, assume that the transition matrix is arbitrarily different each year. To fit this model, we add a factor variable for the year. The AIC rises to 204.1 without interaction and to 213.2 with an interaction between year and rain on the previous day. If instead, we look for a linear trend in the change in the transition matrix over the years, the AIC is 200.9. These results indicate that, over the 11 years, there are neither different amounts of rain nor varying effects of the previous day. A model with a linear trend over the 30 days of the month also has an AIC of 200.9, again a poorer fit than the homogeneous Markov chain. Thus, there is no indication that the transition matrix is not homogeneous over the years and over the month.

Higher order dependence It is also possible to check for dependence further back in time. For a second-order model, another lagged variable must be created, by displacing the response vector two time points. This also means losing two observations.

Rainfall Refitting the above models, plus the new one, each with two days removed per year, shows that present rainfall does not depend on that two days before. This is not surprising, given the results of our inspection of the transition matrices above. Note that these AICs are not comparable with those just given, because two days are lost each year.

Contingency table

As we have seen, if no covariates are present, the contingency is square, containing only four elements. If covariates are used, larger contingency tables will have to be created. Unfortunately, with this approach, it is generally not feasible to check in detail for nonhomogeneity, as we did above.

When logistic regression is applied to such a contingency table for a Markov chain, the table is split into two response vectors in the usual way. Here, these correspond to the two possible states at the present time point. Appropriate covariate vectors must then be constructed to describe the elements of these response vectors. In the simplest case, with a first-order chain and no covariates (except the previous state), each of the two response vectors contains only two elements, corresponding to the possible states at the previous time point so that the only necessary covariate vector will be that giving these states. Thus, in Equation (5.14), i now is this vector of previous states instead of a lagged variable.

Rainfall When logistic regression is used with the contingency table, the fitted regression equation is identical to that given above. Here, the AICs for independence and first-order dependence are respectively 9.0 and 7.3; the change in AIC is the same as above.

5.2.3 Log linear models

In contrast to logistic regression, log linear models for Markov chains usually are applied to contingency tables, not to the raw data of the time series. To fit a log linear model in most software, the Poisson distribution is used with the frequencies in the contingency table as the 'response' variable.

The independence model contains the two states:

$$\log(\mu_{ij}) = \beta_0 + \beta_1(i-1) + \beta_2(j-1) \tag{5.15}$$

For first-order dependence, the interaction between the states on two consecutive days is added to the independence model:

$$\log(\mu_{ijk}) = \beta_0 + \beta_1(i-1) + \beta_2(j-1) + \beta_3(i-1)(j-1) \tag{5.16}$$

To look at second-order dependence, a three-way table must be created. The independence model now contains the three states:

$$\log(\mu_{ijk}) = \beta_0 + \beta_1(i-1) + \beta_2(j-1) + \beta_3(k-1) \tag{5.17}$$

whereas, for first-order dependence, the interactions between the states on consecutive days is added:

$$\begin{aligned}\log(\mu_{ijk}) &= \beta_0 + \beta_1(i-1) + \beta_2(j-1) + \beta_3(k-1)\\ &\quad +\beta_4(j-1)(k-1) + \beta_5(i-1)(j-1)\end{aligned} \tag{5.18}$$

Finally, the second-order chain is the saturated model, with dependence on both previous days plus the interaction between them, thus introducing all possible interactions in the contingency table.

With a two-state process, the results will be the same as those obtained by logistic regression. On the other hand, when there are more than two states, i, j, and k in the above equations must be replaced by factor variables.

Rainfall For the independence model in Equation (5.15), the AIC is 17.8. Again, this and the following AIC are not comparable to the ones for the logistic models because the data are presented in a different way. For first-order dependence, the AIC is 16.2. As expected, the difference in AIC is the same as above (within roundoff error). In the estimated log linear model,

$$\log(\widehat{\mu_{ijk}}) = 5.02 - 0.87(i-1) - 0.95(j-1) + 0.57(i-1)(j-1)$$

the estimates of β_2 and β_3 are identical to the coefficients in the logistic regression given above.

5.3 Checking the assumptions

In order to use Markov chains, we often have to make rather strong assumptions, such as homogeneity and order one. However, when possible, it is important to check them. We already have done this to a certain extent above. Let us formalise these procedures and, at the same time, extend the methods outlined above to the analysis of series with more than two states.

The approach using logistic regression only is applicable if there are two states. As we have seen, this approach is simpler to apply for such binary responses; in addition, if the raw time series is used, it allows the direct inclusion of both continuous and categorical covariates in addition to the previous state(s).

If there are more than two states, usually it will be easiest to create contingency tables and apply log linear models, especially if no covariates are present or if all covariates are categorical. However, log linear models can also be applied to the raw data of a time series with more than two states using the technique described by Lindsey (1995a, pp. 76–82). Here I only shall apply log linear models to contingency tables, in the same way as I did for binary series above.

5.3.1 Homogeneous transitions

One simplifying assumption for a Markov chain is that the same pattern of change is occurring in each period. The transition matrix **T**, as calculated above, represents this pattern. As we have seen (Section 5.1.1), if it is the same over each period, the matrix is homogeneous. In this section, I shall look more closely at this aspect of a Markov chain, assuming the first-order assumption to hold. In the next section I shall look at the latter assumption. Recall that stationarity occurs when the marginal distribution of states remains constant over time. I shall look at that in Section 5.4.1.

Global homogeneity

In order to study homogeneity, it is useful to create a set of two-way contingency tables showing the changes of state between successive consecutive pairs of time periods. This only can be done if a sufficient number of replicate series is available. Then, the transition matrix for all periods combined, assuming homogeneity, can be obtained by aggregating all of these tables and calculating the conditional probabilities from the resulting table. The question, then, will be whether or not this global transition matrix is applicable to all of the tables in the series.

In order to model this set of two-way tables, we shall have three variables available: the state at the beginning of any period, the state at the end of any period, and the time periods themselves. If the transition matrix is homogeneous, the state at the end of a period will depend on that at the beginning of the period, but this dependence will not change with the time period. With this latter relationship omitted, we can write the model as

$$\log(\mu_{ijk}) = \mu + \alpha_i + \beta_j + \zeta_k + \delta_{ij} + \epsilon_{jk} \tag{5.19}$$

where i indexes the state at the beginning of a period, j that at the end, and k the period. (As in Section 2.3, appropriate constraints must be applied.) Thus, δ_{ij} represents the constant first-order Markov dependence over the time periods.

Voting One possible application of Markov chains is to voting behaviour. Here, I shall look at successive monthly expressions of intention to vote in the 1940 presidential elections in the USA, obtained from a panel of people in Erie County between May and October of that year, as given in Table 5.2. These data come from a survey of potential voters; only information on those (445) who responded to all six interviews is available. Unfortunately, the individual series of voting intentions for each person over the complete six months were not published.

In Section 5.2.1, the example involved a few reasonably long (30 observations) series of rain indicators for the months of June in a number of years. By assuming that all series had the same properties, I could aggregate them into contingency tables. Here, a large number of individuals is followed to provide short series (six observations), but with many replicates. Because of the large number of individual series involved, it is possible to aggregate to create a separate contingency table at each consecutive pair of time points, as can be seen in Table 5.2. Thus, the data used here consist of a set of five two-way tables corresponding to transition

Table 5.2. *Frequencies of one-step transitions for voting intentions in the 1940 USA presidential elections for Erie County, USA. (Goodman, 1962)*

		Party		
	Party	Democrat	Republican	Undecided
			June	
May	Democrat	106	7	15
	Republican	5	125	16
	Undecided	18	11	142
			July	
June	Democrat	109	6	14
	Republican	3	124	16
	Undecided	9	22	142
			August	
July	Democrat	111	6	4
	Republican	2	146	4
	Undecided	36	40	96
			September	
August	Democrat	140	4	5
	Republican	1	184	7
	Undecided	12	10	82
			October	
September	Democrat	146	2	5
	Republican	1	192	5
	Undecided	12	11	71

matrices for the five periods between successive interviews, yielding a three-way table over the five time periods.

Here, the very strong uncheckable assumption is that all series, that is, all people involved, had the same properties. This is probably less realistic than that the 11 months of June had the same properties.

The aggregated table, combining the subtables for all periods, and the conditional probabilities calculated from it are

	After			After		
Before	D	R	U	D	R	U
D	612	25	43	0.90	0.04	0.06
R	12	771	48	0.01	0.93	0.06
U	87	94	533	0.12	0.13	0.75

I now shall examine whether or not this is a suitable representation of change in all five periods.

Here, the three variables are: the voting intention at the beginning of any period (i), the voting intention at the end of any period (j), and the time periods themselves

(k), with respectively three, three, and five categories. In contrast to the binary case above, these will be factor variables.

If the transition matrix is homogeneous, the intention to vote at the end of a period will depend on that at the beginning of the period, but will be independent of the time period. The AIC is 179.3 for a homogeneous transition matrix, as compared to 1554.4 for independence and 152.5 for a nonhomogeneous matrix, that is, the saturated model with five distinct matrices. The assumption of homogeneity must be rejected decisively (as must independence!). The process of changing intentions to vote varies over the months of observation.

Local homogeneity

Checking homogeneity can be difficult if we only have a series for one individual. It is not possible to create a contingency table for each consecutive pair of time points. Instead, one will have to cut up the observations into several periods and assume homogeneity within each. For a collection of series, if global homogeneity is not reasonable, the same thing may be done to determine whether or not there is homogeneity within shorter periods of time.

Voting One important supplementary piece of information in the voting study is that the Democratic convention was held during the third period. Thus, it will be of interest to model separately the first two time periods and the last two, applying the above procedure successively to each. The AICs are 549.8, 60.8, and 63.1 before the convention and 703.9, 53.4, and 58.7 after it, for independence, first-order homogeneity, and nonhomogeneity, respectively. Thus, homogeneity of the transition matrix is now acceptable in both separate tables. The complete model for the data, then, has three different transition matrices: for the first two periods, for the third, and for the last two. The combined AIC is 145.0, better than that for the saturated model above, with five different transition matrices.

The transition matrices for the first and last periods are estimated, respectively, to be

	After			After		
Before	D	R	U	D	R	U
D	0.84	0.05	0.11	0.95	0.02	0.03
R	0.03	0.86	0.11	0.01	0.96	0.03
U	0.08	0.10	0.83	0.12	0.11	0.77

We may note, as might be expected, that the diagonal transition probabilities for the two major parties are considerably smaller before the convention than after, and the reverse for the undecided. In the first table (May–July), 86 per cent of those intending to vote for the Republicans and 84 per cent of those for the Democrats are estimated as not having changed their minds over a period of one month, whereas, after the convention (August–October), the percentages are 96 and 95, respectively.

Remember, however, that this analysis assumes that intentions at one point in time only depend on intentions one month before. This is the assumption of a *first-order* Markov chain that will be checked in the next section.

Table 5.3. *Two-step transitions for voting intentions in the 1940 USA presidential elections, Erie County, USA. (Goodman, 1962)*

		Time $t+2$		
Time $t+1$	Time t	Democrat	Republican	Undecided
	Democrat	435	9	22
Democrat	Republican	8	3	0
	Undecided	63	6	6
	Democrat	0	18	5
Republican	Republican	6	557	16
	Undecided	1	71	11
	Democrat	10	4	24
Undecided	Republican	5	17	21
	Undecided	54	62	346

5.3.2 Order

In order to examine whether a series of responses follows a first-order Markov chain or one of higher order, we require the details of changes for each series over successive periods, and not only over one period at a time, as in the previous section. We can, then, check whether or not the categories at the present time point are independent of those two time periods before, given the immediately preceding state. If so, the process is of first-order (or, more precisely, the second order is not required, although higher orders may be). If not, it is at least of second order. I applied such checks in Section 5.2.2 for two states; here, I shall look at a multistate process.

The states at the three times t (indexed by i), $t-1$ (indexed by j), and $t-2$ (indexed by k) will form a three-way table. To check whether or not the state at time t is independent of that at time $t-2$, the appropriate log linear model is the extension of Equation (5.18) to series with more than two states:

$$\log(\mu_{ijk}) = \mu + \alpha_i + \beta_j + \zeta_k + \delta_{ij} + \epsilon_{jk} \tag{5.20}$$

Voting Let us analyse further results from the same voting study as in the previous example. As mentioned above, individual sequences of states are not available. Ideally, we should at least have a table for three consecutive time periods. Unfortunately, only a collapsed table combining all sets of two-step transitions was published. These are aggregated over the four three-month periods (May–July, June–August, July–September, August–October), as given in Table 5.3. This table cannot be obtained from the one in the previous section which only covered two-month periods.

Because of the form of the data available, here I shall have to assume second-order homogeneity (most likely wrongly: we have seen that the data are not first-order homogeneous). If the raw data were available, I also would examine this assumption in a way similar to that of the preceding section. However, only the tables used in these two sections were published by Goodman (1962).

In summary, I here shall assume that the sequence of changes in voting intentions is identical over any consecutive three-month period (homogeneity) during the study. I shall, then, check whether current intentions in any month depend only on intentions in the previous month (first-order) or also on intentions two months before (second-order). Thus, here, I am obliged to ignore possible effects of the Democratic convention.

With an AIC of 212.6 as compared to 86.6 for the saturated model, this first-order assumption is rejected. Note, however, that part of the problem may arise from aggregating periods with nonhomogeneous transition matrices. Given homogeneity, present voting intentions depend on more than just those of the preceding time point. They depend, at least, on the two previous points. With sufficient data, the assumptions of higher order Markov chains could be checked in the same manner.

5.4 Structured transition matrices

If the state space of a Markov chain contains several states, estimation of the transition matrix can involve a fair number of parameters. In certain situations, the model can be simplified by imposing some informative structure on the probabilities in the matrix.

5.4.1 Reversibility and equilibrium

Let us look first at two possible symmetry characteristics of transition matrices:

(i) As we saw in Section 5.1.1, if a Markov chain is *reversible*, the probability of moving between a pair of states is the same in both directions in time. Rearranging Equation (5.3), we have

$$\pi_j \pi_{i|j} = \pi_i \pi_{j|i} \tag{5.21}$$

(ii) If, as in Equation (5.2), the Markov chain is in the *equilibrium* state, the marginal distribution π^s is stationary, not changing over time. (Recall that this is not the same as the homogeneity of the transition matrix, also studied in Section 5.3.1.)

In the analysis of contingency tables, reversibility is known as *quasi-symmetry* and equilibrium is known as *marginal homogeneity*. The combination of the two yields a *symmetry* model for the table. However, either is possible alone, as a weaker assumption:

- Quasi-symmetry implies that the table would be symmetric if it were not for the distorting effect of the changing marginal totals.
- Marginal homogeneity implies that the marginal totals are the same but the body of the table is not symmetric.

The latter is not a log linear model; a special algorithm must be used to fit it.

Table 5.4. *Results from Glass's study of British social mobility. PA: professional and high administrative; MS: managerial, executive, and high supervisory; IS: low inspectional and supervisory; SM: routine nonmanual and skilled manual; UM: semi- and unskilled manual. (Bishop* et al., *1975, p. 100, from Glass)*

	Son				
Father	PA	MS	IS	SM	UM
PA	50	45	8	18	8
MS	28	174	84	154	55
IS	11	78	110	223	96
SM	14	150	185	714	447
UM	3	42	72	320	411

The log linear model for reversibility (quasi-symmetry) is

$$\log(\mu_{ij}) = \mu_i + \delta_j + \alpha_{ij} \text{ with } \alpha_{ij} = \alpha_{ji} \tag{5.22}$$

Suppose, say, that the Markov chain has five states; then, the appropriate symmetric factor variable for fitting this model will be

$$\begin{matrix} K & A & B & C & D \\ A & L & E & F & G \\ B & E & M & H & I \\ C & F & H & N & J \\ D & G & I & J & O \end{matrix}$$

This variable has as many levels as there are possible paired combinations of states, plus one for each position on the diagonal. Each symmetric pair has the same level.

For complete symmetry, that is reversibility plus equilibrium, the *joint* probabilities in opposing cells across the diagonal are equal:

$$\pi_{ij} = \pi_{ji} \tag{5.23}$$

which is Equation (5.21) written in a different way. The parameters for the margins in Equation (5.22) are not fitted, so that the model becomes

$$\log(\mu_{ij}) = \alpha_{ij} \text{ with } \alpha_{ij} = \alpha_{ji} \tag{5.24}$$

Social mobility In the 1950s, Glass carried out a now classical study of British social mobility. One summary of his results is provided in Table 5.4. From this, the transition matrix is estimated to be

$$\begin{pmatrix} 0.39 & 0.35 & 0.06 & 0.14 & 0.06 \\ 0.06 & 0.35 & 0.17 & 0.31 & 0.11 \\ 0.02 & 0.15 & 0.21 & 0.43 & 0.19 \\ 0.01 & 0.10 & 0.12 & 0.47 & 0.30 \\ 0.00 & 0.05 & 0.08 & 0.38 & 0.48 \end{pmatrix}$$

and the marginal probabilities, (0.04, 0.14, 0.15, 0.43, 0.24) for the father and (0.03, 0.14, 0.13, 0.41, 0.29) for the son.

For this table, the model for independence has an AIC of 480.5 and the saturated model for a full Markov chain has 100.4, providing strong indication of dependence. This is not surprising because sons often tend to have the same class as their fathers. However, this model has 20 parameters. In contrast, the reversibility model, with six fewer parameters, fits somewhat better, having an AIC of 96.7. On the other hand, equilibrium fits more poorly: 112.8, as does the model with both: 109.1.

These results indicate that the (marginal) social structure of the categories has changed between father and son, but that, given this change, there may be equal movement in both directions between pairs of social categories. The estimates of the symmetry parameters in the reversibility model, with the first being baseline, are (0, −1.37, −0.81, −1.76, 0.93, 1.62, 0.61, 1.88, 1.12, 2.70), so that the movement between the first and last categories has the smallest probability and that between the last two categories has the highest.

5.4.2 Random walks

Let us now consider a random walk (Section 5.1.3) on a finite line segment with reflecting barriers and without drift. With five states (positions on the line), the appropriate factor variable for a log linear model will be

$$\begin{array}{ccccc}
\text{—} & \mathrm{A} & \text{—} & \text{—} & \text{—} \\
\mathrm{A} & \text{—} & \mathrm{A} & \text{—} & \text{—} \\
\text{—} & \mathrm{A} & \text{—} & \mathrm{A} & \text{—} \\
\text{—} & \text{—} & \mathrm{A} & \text{—} & \mathrm{A} \\
\text{—} & \text{—} & \text{—} & \mathrm{A} & \text{—}
\end{array}$$

However, observed processes often have nonzero values elsewhere in the contingency table besides these two minor diagonals. Thus, it is useful to generalise the random walk to allow no move and moves greater than one step. Such a process corresponds to a model for *diagonal symmetry* in a contingency table, with factor variable

$$\begin{array}{ccccc}
\mathrm{E} & \mathrm{A} & \mathrm{B} & \mathrm{C} & \mathrm{D} \\
\mathrm{A} & \mathrm{E} & \mathrm{A} & \mathrm{B} & \mathrm{C} \\
\mathrm{B} & \mathrm{A} & \mathrm{E} & \mathrm{A} & \mathrm{B} \\
\mathrm{C} & \mathrm{B} & \mathrm{A} & \mathrm{E} & \mathrm{A} \\
\mathrm{D} & \mathrm{C} & \mathrm{B} & \mathrm{A} & \mathrm{E}
\end{array}$$

If drift is allowed, this becomes

$$\begin{array}{ccccc}
\mathrm{I} & \mathrm{A} & \mathrm{B} & \mathrm{C} & \mathrm{D} \\
\mathrm{E} & \mathrm{I} & \mathrm{A} & \mathrm{B} & \mathrm{C} \\
\mathrm{F} & \mathrm{E} & \mathrm{I} & \mathrm{A} & \mathrm{B} \\
\mathrm{G} & \mathrm{F} & \mathrm{E} & \mathrm{I} & \mathrm{A} \\
\mathrm{H} & \mathrm{G} & \mathrm{F} & \mathrm{E} & \mathrm{I}
\end{array}$$

Table 5.5. *Initial and follow-up blood pressure, classified according to hypertension status. (Lawal and Upton, 1990, from Freeman)*

	Follow-up		
Initial	Normal	Borderline	Elevated
Normal	105	9	3
Borderline	10	12	1
Elevated	3	2	7

Blood pressure The blood pressures of 152 people were recorded at an initial baseline time and subsequently at a follow-up visit. The results were classified as normal, borderline, or elevated, as shown in Table 5.5. Let us examine whether or not the probability of moving among states is the same in either direction on this ordered scale.

The (conditional) transition matrix is estimated to be

$$\begin{pmatrix} 0.90 & 0.08 & 0.03 \\ 0.43 & 0.52 & 0.04 \\ 0.25 & 0.17 & 0.58 \end{pmatrix}$$

and the marginal probabilities, (0.77, 0.15, 0.08) at the initial time and (0.78, 0.15, 0.07) at the follow-up. The latter are very similar. The matrix of joint probabilities is

$$\begin{pmatrix} 0.69 & 0.06 & 0.02 \\ 0.07 & 0.08 & 0.01 \\ 0.02 & 0.01 & 0.05 \end{pmatrix}$$

This may have the form of a random walk.

For these data, the model for independence has an AIC of 47.7 and the saturated model for a full Markov chain has 25.7, a strong indication of dependence. This is not surprising because many people have the same classification at the two time points. The random walk without drift is a small improvement, with 24.2, better than that with drift, with 25.1. Thus, we can conclude that the probability of moving in opposite directions on the scale may be the same.

5.4.3 Mover–stayer model

Another special type of Markov chain model frequently has been applied in mobility studies. As we have seen in some of the preceding examples, dependence over time often results primarily from the large number of individuals staying in the same state between two successive observation points. In order to construct an appropriate model, let us assume that the population contains two subpopulations, called movers and stayers. The result, then, can be called a *mover–stayer* model.

In such a model, the off-diagonal individuals must belong to the mover subpopulation. However, those on the diagonal have the same state at the two observation points. These may be stayers, but they could also be individuals who move at some

Table 5.6. *Migrant behaviour in Great Britain between 1966 and 1971. (Fingleton, 1984, p. 142)*

	1971			
1966	Central Clydes.	Urban Lancs. and Yorks.	West Midlands	Greater London
Central Clydesdale	118	12	7	23
Urban Lancs. and Yorks.	14	2127	86	130
West Midlands	8	69	2548	107
Greater London	12	110	88	7712

other time, before or after the observations were made. Thus, the diagonal contains a mixture of the two subpopulations; without further information, we cannot separate them. Thus, there is a problem for estimation of the parameters in such a model. (This model is similar to the survival models of Section 3.4, for which the assumption was made that the people with censored duration times arose from a mixture of two unobservable subpopulations: those who would and would not have an event.)

If only observations at two points in time are available, a Markov chain cannot be fitted in these circumstances. Instead, let us make the further assumption that the two successive states of the movers are independent. We can check this on the individuals who actually moved, that is, ignoring the diagonal values. In terms of a log linear model for contingency tables, this is called *quasi-independence*.

Migration Data on migrant behaviour are available between 1966 and 1971 for moves among the four most important centres of population in the UK, given in Table 5.6. These data originated as a 10 per cent sample from the published migration reports of the 1971 British census; they describe the place of residence in 1966 and 1971 of a subgroup of migrants born in the New Commonwealth. The geographical locations are ordered from the north to the south of Great Britain.

The transition matrix is estimated to be

$$\begin{pmatrix} 0.74 & 0.08 & 0.04 & 0.14 \\ 0.01 & 0.90 & 0.04 & 0.06 \\ 0.00 & 0.03 & 0.93 & 0.04 \\ 0.00 & 0.01 & 0.01 & 0.97 \end{pmatrix}$$

and the marginal probabilities, (0.01, 0.18, 0.21, 0.60) in 1966 and (0.01, 0.18, 0.21, 0.61) in 1971, that is, virtually identical.

For these data, the model for independence has an AIC of 9999.5 and the saturated model for a full Markov chain has 66.5, indicating strong dependence. Let us see whether or not this dependence arises primarily from the stayers. The mover–stayer model, obtained by weighting out the diagonal of the table and refitting the model for independence, has an AIC of 62.6, an improvement. For the movers, those individuals who are susceptible to migrate (or at least those who did in the observation period), the new place of residence does not depend on the original residence, a rather surprising conclusion, given the varying distance among the regions.

We can use the quasi-independence model for the observed movers to predict the numbers of movers on the diagonal, those not happening to move in the observation period (by multiplying together the appropriate marginal probabilities). The estimates are, respectively, (1.6, 95.2, 60.3, 154.6) for the four regions, so that, by subtraction, the numbers of stayers are estimated to be (116.4, 2031.8, 2487.7, 7557.4). 92.6 per cent of the population is estimated as not being susceptible to migration.

Further reading

Derman *et al.* (1973, pp. 550–662) provide a useful elementary introduction to Markov chains. For more advanced treatments, see Bailey (1964, Ch. 5), Bremaud (1999), Doob (1953, Ch. 5), Feller (1950, Ch. 15), Grimmett and Stirzaker (1992, Ch. 6), Iosifescu (1980), or Karlin and Taylor (1975, Chs. 2–3). Berg (1993) provides many interesting applications of random walks in biology.

For a general introduction to categorical data analysis, the reader may like to consult Agresti (1990), Fingleton (1984), or Lindsey (1995a). The latter two have major sections devoted to applications in stochastic processes.

Exercises

5.1 The Six Cities study looked at the longitudinal effects of air pollution on health. Part of the data involved children in Steubenville, Ohio, USA, who were followed once a year from ages seven through ten. Each year, the wheezing status of the child was recorded, as shown in Table 5.7. Whether the mother smoked or not when the child was seven is also available, although this could have evolved over time.

- (a) How does the child's wheezing evolve over time?
- (b) Does it depend on the smoking status of the mother?
- (c) Are the transitions between the two states homogeneous over time?
- (d) What order of model is required?
- (e) Are the respondents a uniform group after the smoking status of the mother is taken into account?

5.2 A panel study was conducted to follow the interurban moves of people in the USA, classified by age and ownership of their home, as shown in Table 5.8.

- (a) Develop an appropriate model for these data.
- (b) Are the transitions between the two states homogeneous over time?
- (c) What order of model is required?
- (d) How does moving depend on age and ownership status?
- (e) Does the probability of moving depend on the number of previous moves?

Table 5.7. *Wheezing status of children in Steubenville, Ohio, USA, from ages seven through ten, with the mother's smoking status when the child was age seven. (Fitzmaurice and Laird, 1993)*

Age				Smoking	
7	8	9	10	No	Yes
No	No	No	No	237	118
No	No	No	Yes	10	6
No	No	Yes	No	15	8
No	No	Yes	Yes	4	2
No	Yes	No	No	16	11
No	Yes	No	Yes	2	1
No	Yes	Yes	No	7	6
No	Yes	Yes	Yes	3	4
Yes	No	No	No	24	7
Yes	No	No	Yes	3	3
Yes	No	Yes	No	3	3
Yes	No	Yes	Yes	2	1
Yes	Yes	No	No	6	4
Yes	Yes	No	Yes	2	2
Yes	Yes	Yes	No	5	4
Yes	Yes	Yes	Yes	11	7

5.3 A number of women in the USA were followed over five years, from 1967 to 1971, in the University of Michigan Panel Study of Income Dynamics. The sample consisted of white women who were married continuously to the same husband over the five-year period. Having worked in the year is defined as having earned any money during the year. The sample paths of labour force participation are given in Table 5.9. Here, there are two types of stable behaviour that might be called stayers.

(a) Study how the most recent employment record of each woman depends on her previous history.
(b) Are the transitions between the two states homogeneous over time?
(c) What order of model is required?
(d) Is there indication of heterogeneity among the women?

5.4 After the accident at the Three Mile Island nuclear power plant in the spring of 1979, mothers of young children living within ten miles of the plant were interviewed four times, in winter 1979, spring 1980, autumn 1980, and autumn 1982, to determine stress levels. The mothers were classified by distance from the plant, as shown in Table 5.10.

(a) Study how the level of stress changes over time.
(b) Is there any evidence that stress differs with distance from the plant?
(c) Notice that the interviews were not equally spaced in time. What difference does this make to the interpretation of your results?

Table 5.8. *Interurban moves over five two-year periods for a sample of people in Milwaukee, Wisconsin, USA, classified by age and ownership. M: person moved in the two-year period, S: he or she stayed in the same place. (Crouchley* et al.*, 1982, from Clark* et al.*)*

					Renters		Owners	
				Age	25–44	46–64	25–44	46–64
S	S	S	S	S	511	573	739	2385
S	S	S	S	M	222	125	308	222
S	S	S	M	S	146	103	294	232
S	S	S	M	M	89	30	87	17
S	S	M	S	S	90	77	317	343
S	S	M	S	M	43	24	51	22
S	S	M	M	S	27	16	62	19
S	S	M	M	M	28	6	38	5
S	M	S	S	S	52	65	250	250
S	M	S	S	M	17	20	48	14
S	M	S	M	S	26	19	60	25
S	M	S	M	M	8	4	10	3
S	M	M	S	S	8	9	54	21
S	M	M	S	M	11	3	18	1
S	M	M	M	S	10	3	21	1
S	M	M	M	M	4	1	8	2
M	S	S	S	S	41	29	134	229
M	S	S	S	M	16	15	23	10
M	S	S	M	S	19	13	36	25
M	S	S	M	M	2	4	1	0
M	S	M	S	S	11	10	69	24
M	S	M	S	M	11	2	15	3
M	S	M	M	S	1	9	13	2
M	S	M	M	M	2	2	2	0
M	M	S	S	S	7	5	40	18
M	M	S	S	M	4	2	9	2
M	M	S	M	S	8	1	15	3
M	M	S	M	M	1	0	5	0
M	M	M	S	S	8	1	22	7
M	M	M	S	M	3	2	7	2
M	M	M	M	S	5	0	9	2
M	M	M	M	M	6	3	5	0

5.5 The original British social mobility table, from which Table 5.4 was obtained, is given in Table 5.11.

(a) Does this table show the same dependence structure as the more condensed one in Table 5.4?

(b) Was it reasonable to combine the categories in the reduced table?

5.6 Table 5.12 shows the results of a Danish social mobility study.

(a) How are the transition probabilities structured?

(b) Compare your results to those for British social mobility.

Table 5.9. *Labour force participation of a panel of women in the USA, 1967–1971. (Heckman and Willis, 1977)*

				1971	
1970	1969	1968	1967	Yes	No
Yes	Yes	Yes	Yes	426	38
No	Yes	Yes	Yes	16	47
Yes	No	Yes	Yes	11	2
No	No	Yes	Yes	12	28
Yes	Yes	No	Yes	21	7
No	Yes	No	Yes	0	9
Yes	No	No	Yes	8	3
No	No	No	Yes	5	43
Yes	Yes	Yes	No	73	11
No	Yes	Yes	No	7	17
Yes	No	Yes	No	9	3
No	No	Yes	No	5	24
Yes	Yes	No	No	54	16
No	Yes	No	No	6	28
Yes	No	No	No	36	24
No	No	No	No	35	559

5.7 Results from a study of voting during the Swedish elections of 1964 and 1970 are shown in Table 5.13.

(a) Find an appropriate model to describe these data.
(b) Interpret the structure that you have found.

5.8 Frequency of company moves between British regions was recorded between 1972 and 1978, as shown in Table 5.14.

(a) Does the probability of a move depend on the geographical distance?
(b) How important is the role of stayers for these data?

5.9 Changes in votes were recorded for a sample of people in the 1964, 1968, and 1970 Swedish elections as shown in Table 5.15. These provide a different classification of the parties than that given in Table 5.13.

(a) Construct an appropriate Markov chain model for these data.
(b) Check for order and homogeneity.
(c) Can you take into account the unequal periods between elections?
(d) How do the results compare to those for Table 5.13?

5.10 Table 4.3 gave a series involving the numbers of accidents in a section of a mine in the UK.

(a) Analyse these data as a Markov chain applied to a binary point process where the states are presence or absence of accidents on each day.
(b) What order of model is required?

Table 5.10. *The stress levels, classed as low (L), medium (M), and high (H), of mothers of young children living within ten miles of the Three Mile Island nuclear power plant at four time points after the nuclear accident of 1979. (Conaway, 1989, from Fienberg* et al.*)*

			< 5 miles			> 5 miles		
Wave			Wave 4					
1	2	3	L	M	H	L	M	H
L	L	L	2	0	0	1	2	0
L	L	M	2	3	0	2	0	0
L	L	H	0	0	0	0	0	0
L	M	L	0	1	0	1	0	0
L	M	M	2	4	0	0	3	0
L	M	H	0	0	0	0	0	0
L	H	L	0	0	0	0	0	0
L	H	M	0	0	0	0	0	0
L	H	H	0	0	0	0	0	0
M	L	L	5	1	0	4	4	0
M	L	M	1	4	0	5	15	1
M	L	H	0	0	0	0	0	0
M	M	L	3	2	0	2	2	0
M	M	M	2	38	4	6	53	6
M	M	H	0	2	3	0	5	1
M	H	L	0	0	0	0	0	0
M	H	M	0	2	0	0	1	1
M	H	H	0	1	1	0	2	1
H	L	L	0	0	0	0	0	1
H	L	M	0	0	0	0	0	0
H	L	H	0	0	0	0	0	0
H	M	L	0	0	0	0	0	0
H	M	M	0	4	4	1	13	0
H	M	H	0	1	4	0	0	0
H	H	L	0	0	0	0	0	0
H	H	M	1	2	0	1	7	2
H	H	H	0	5	12	0	2	7

(c) Reanalyse the data as a three-state model where the states are no accidents, one accident, and more than one accident.

(d) Does this change the conclusions?

5.11 The numbers of patients arriving at an intensive care unit per day in a hospital within the Oxford, UK, Regional Hospital Board were recorded from February 1963 to March 1964, as shown in Table 5.16.

(a) Perform an analysis similar to that for the mine accidents in Exercise 5.10, where the two states are whether or not any patients arrive. Compare the results from logistic and log linear regression.

(b) Fit models where the states are the actual numbers of patients arriving and compare your results to those with a binary state.

5.12 Patients with acute spinal cord injury and bacteriuria (bacteria in their

Table 5.11. *The complete table for Glass's study of British social mobility. I: professional and high administration, II: managerial and executive, III: inspection, supervisory, etc. (high), IV: inspection, supervisory, etc. (low), Va: routine nonmanual, Vb: skilled manual, VI: semiskilled manual, VII: unskilled manual (Duncan, 1979, from Glass)*

	Son							
Father	I	II	III	IV	Va	Vb	VI	VII
I	50	19	26	8	7	11	6	2
II	16	40	34	18	11	20	8	3
III	12	35	65	66	35	88	23	21
IV	11	20	58	110	40	183	64	32
Va	2	8	12	23	25	46	28	12
Vb	12	28	102	162	90	554	230	177
VI	0	6	19	40	21	158	143	71
VII	0	3	14	32	15	126	91	106

Table 5.12. *The results of a Danish social mobility study. 1: professional and high administrative, 2: managerial, executive and high supervisory, 3: low inspectional and supervisory, 4: routine nonmanual and skilled manual, 5: semi- and unskilled manual. (Bishop* et al., *1975, p. 100, from Svalastoga)*

	Son				
Father	1	2	3	4	5
1	18	17	16	4	2
2	24	105	109	59	21
3	23	84	289	217	95
4	8	49	175	348	198
5	6	8	69	201	246

urine) were assigned randomly to one of two treatment groups and followed for up to 16 weeks. The first group (A) was treated for all episodes of urinary tract infection that occurred, whereas the second group (B) was only treated if two specific symptoms appeared. All patients entered the study with bacteriuria so that the first response is always positive. Having bacteriuria for a longer period did not mean necessarily that a patient was sicker.

The resulting binary series for the 36 patients in each group having at least four weeks of observation are shown in Table 5.17. (Joe, 1997, imputed 11 missing values.)

(a) Can these responses be described by a first-order Markov chain?
(b) Is there any evidence of difference between the two groups?
(c) Apparently, subjects in the first group systematically had bacteria removed, whereas those in the second group would retain the in-

Table 5.13. *The voting changes between the Swedish elections of 1964 and 1970. SD: Social Democrat, Con: Conservative. (Fingleton, 1984, p.138)*

	1970				
1964	Comm.	SD	Centre	People's	Con.
Communist	22	27	4	1	0
Social Dem.	16	861	57	30	8
Centre	4	26	248	14	7
People's	8	20	61	201	11
Conservative	0	4	31	32	140

Table 5.14. *Company moves between British regions, 1972–1978. SE: South East, EA: East Anglia, SW: South West, WM: West Midlands, EM: East Midlands, Y: Yorkshire, NW: North West, N: North, W: Wales, S: Scotland. (Upton and Fingleton, 1989, p. 6)*

	1978									
1972	SE	EA	SW	WM	EM	Y	NW	N	W	S
SE	958	119	70	16	95	48	39	46	89	41
EA	10	121	44	0	7	2	8	8	3	6
SW	9	11	162	5	2	2	9	6	25	6
WM	7	0	1	288	13	4	10	7	33	6
EM	10	9	5	6	332	16	4	6	6	4
Y	6	1	2	1	20	356	3	9	1	9
NW	8	1	10	6	10	15	331	12	21	8
N	0	2	4	1	5	4	5	119	1	3
W	5	0	3	1	0	0	0	2	154	1
S	1	1	2	0	1	2	4	5	4	269

fection for some time after acquiring it. What evidence do the data provide for this?

Table 5.15. *Voting in the 1964, 1968, and 1970 Swedish elections. SD: Social Democrat, C: Centre, P: People's, Con: Conservative. (Fingleton, 1984, p. 151)*

		1970			
1964	1968	SD	C	P	Con
SD	SD	812	27	16	5
	C	5	20	6	0
	P	2	3	4	0
	Con	3	3	4	2
C	SD	21	6	1	0
	C	3	216	6	2
	P	0	3	7	0
	Con	0	9	0	4
P	SD	15	2	8	0
	C	1	37	8	0
	P	1	17	157	4
	Con	0	2	12	6
Con	SD	2	0	0	1
	C	0	13	1	4
	P	0	3	17	1
	Con	0	12	11	126

Table 5.16. *Numbers of patients arriving at an intensive care unit per day from February 1963 to March 1964. (from Cox and Lewis, 1966, pp. 254–255; read across rows)*

0 0 0 2 0 0 0 1 0 0 1 0 0 0 0 1 0 2 0 2 1 0 0 0 2 0 0 1 1 0 1 0 0 0 2 0 1 0 0
0 0 0 1 1 1 0 0 1 0 1 0 3 0 0 0 2 0 1 0 0 2 0 0 0 1 0 0 2 1 1 0 0 0 1 1 0 0 0
0 0 0 1 0 0 0 0 1 1 0 0 0 0 1 0 1 1 0 0 0 1 0 1 2 0 1 0 1 1 1 0 1 2 1 1 0 0 1
0 1 0 2 0 1 0 1 0 0 0 0 2 0 1 2 0 0 1 0 2 0 0 2 1 0 2 2 1 1 0 1 1 0 2 0 0 0 0
0 1 1 1 0 1 3 0 0 0 0 0 0 1 1 0 1 1 0 0 2 0 1 0 0 1 1 2 0 0 0 3 0 0 0 1 0 3 0
0 0 0 0 1 0 1 0 0 1 0 0 0 0 0 0 1 0 1 0 0 1 0 1 1 0 2 1 0 1 0 1 1 0 1 0 0 1 0
1 2 0 0 1 1 0 0 1 1 2 0 0 1 0 1 0 0 0 1 1 0 0 0 0 0 1 3 0 1 1 0 0 1 0 0 0 1 1
2 0 0 2 1 1 1 1 2 0 1 1 0 0 3 1 1 1 1 0 0 1 2 1 0 1 1 0 1 2 1 1 4 0 2 0 2 1 1
0 1 1 1 2 1 1 1 0 0 2 1 1 2 2 2 0 0 1 0 0 1 0 1 1 0 1 1 3 0 0 2 1 1 0 2 1 0 1
0 1 1 1 0 0 1 1 0 0 0 1 1 1 0 0 0 0 2 1 1 0 0 0 2 1 0 0 0 0 1 3 1 1 0 0 1 1 1
1 0 0 0 1 0 1 0 1 0 0 0 1 1 0 0 0 0 0 0 2

Table 5.17. *Time series of infections (1) with bacteriuria for patients randomised to two treatment groups. (Joe, 1997, p. 353)*

Group A	Group B
1001010100101	111100011
10001	1111111111111111
1010010010101010	1111111111111001
101000101	1000111111110111
100011010111111	1111111111010000
1100100010101001	1111100111111111
100010101	1000011110111001
1000	111000111011
1001000000100010	1110111111111111
1010011	1110
1010110101	1011111111
1001010101	1011111111111111
100110	1111111100000110
10100011010100	100010
10010	1001111011011111
1000101100100100	10011
1010100110110100	11111100000
1000100010000000	1000111111111000
1010001010011001	1111001100011111
101000011011	1001111111111111
101000111011101	1000000000000000
1001000100	1100111111100111
1000101001110001	100100
1010001010010011	1111100000
1010101001001	100111110100
1000101010100010	1111111111
1010001001	1111111110111111
100010110101	1111101110111111
111000	110011
1001000010100010	1100000110010000
10000101001	100001111111111
110011011	100011110101001
100101100010001	1111110011111111
11010011001010	1111000101111111
10010010	1100000111
100100	1111000

6

Event histories

The stochastic processes that we have studied so far all involve data on the times between events. These can be called *event histories*. The simplest case occurred when there was one type of event and it was absorbing (Chapter 3). The classical examples are human mortality and machine breakdown. The situation is slightly more complex when the event is recurrent (Chapter 4). Examples include unemployment, sickness, and moving house. It is still more complex when individuals may change between distinct states (Chapter 5). Here, an example would be catching a disease, being hospitalised, recovering, or dying. However, up until now, I only have looked at this latter possibility in discrete time, through the use of Markov chains. We shall now see how to generalise those procedures to changes among several states in continuous time.

6.1 Theory

When an individual can change state at any point in time in an event history process, each event indicates a transition between states. However, by appropriate definition, it is possible to model almost any series of events, even recurrent events, as a set of transitions among states. I already have covered much of the basic theory in Section 4.1.

6.1.1 Diagrams

With several states and, perhaps, transitions only being possible among some of them, event history processes can be relatively complex. Generally, it is useful to clarify ideas by constructing a diagram for the states and permissible transitions between them. In this light, let us examine several special cases of particular importance:

- Mortality: two states of which the second is absorbing, the classical survival analysis covered in Chapter 3.
- Progression through a fixed series of irreversible states, such as disability: the example of moves through distinct states mentioned above, if death must be the final result.

  - Recurrent events (covered in Chapter 4), if appropriate, can be taken as a special case where the successive states are defined by the numbers of events already experienced.
- Alternance between two states, such as an alternating Poisson process having two different intensities.
- Competing risks: transition from one state to any one of several others.

Examples of diagrams for several of these common models are shown in Figure 6.1. Of course, several of these possibilities may need to be combined in order adequately to model an event history.

When a set of irreversible states forms a progressive series, such as in the advance of incurable diseases with a final, absorbing, death state, then there will be one less transition intensity than the number of states. However, when at least some of the state transitions are reversible, many more different intensity functions may need to be estimated and the problem becomes more complex. Further complications arise when any one of several states may be entered at a given time point, similar to those with competing risks in the simpler context of survival analysis.

6.1.2 Defining the states

Model construction depends heavily on how the series of states is defined. Generally, there will be no unique correct structure; as always, the model used will depend on the questions to be answered! However, where certain simplifying assumptions are possible, they will facilitate model building.

- As we have seen, a model is *progressive* if all states, except the first, have only one transition into them. Then, the current state defines what states were previously occupied and in what order, but not when the changes occurred.
- The transition probability to the next state is *Markovian* if it only depends on the present state and not on the previous history of the individual.

However, transition probabilities often may depend on the time in the state; thus, an extension is to allow them to depend on the time since the last event, a special case being the semi-Markov model (Section 6.1.4).

A multistate model should not have several direct transition routes from one state to another; instead several different states may be defined. For example, if subjects in the state of having a given disease may recover either by natural body defences or by medical treatment, either these can be defined as two different recovery states, as in the alternative outcomes model of Figure 6.1, or they should not be distinguished at all.

6.1.3 Continuous-time Markov chains

Markov chain models in discrete time (Chapter 5) describe processes that move from state to state, staying in each state a discrete length of time. In most cases, this duration follows a geometric distribution (Section 5.1.2). The continuous-time analogue will be the exponential distribution. Thus, the generalisation of

Progressive events

A B C

Recurrent events

0 1 2 3

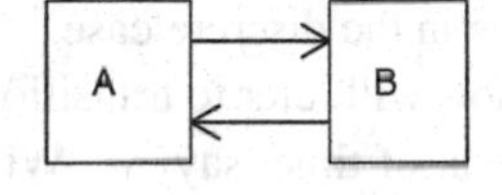

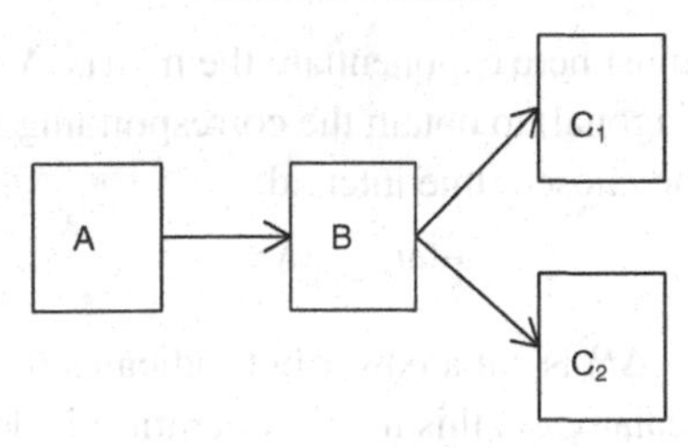

Fig. 6.1. Examples of diagrams for simple event histories.

Markov chains to continuous time will involve stays in each state following this latter distribution. Many results analogous to those for discrete time still hold, such as the existence of communicating classes, irreducibility, recurrence, and so on.

Differences with respect to discrete time

Let us assume that the process of interest must remain in each state i a strictly positive length of time (sometimes called a pure jump process). If the sojourn times in each state have independent exponential distributions, we shall have a continuous-time Markov chain. The mean times μ_i in the states, or the intensities $\lambda_i = 1/\mu_i$ of leaving them, will, in general, be different. If a state is absorbing, the mean duration will be infinite and $\lambda_i = 0$.

One of the simplest cases is the Poisson counting process (Section 4.1.1). This is the only renewal process that is a Markov chain (Grimmett and Stirzaker, 1992, pp. 229, 339).

In the case of discrete time (Chapter 5), we worked with a matrix $\mathbf{T}$ of conditional transition *probabilities* $\pi_{j|i}$. Here, it will be necessary to work with a matrix $\boldsymbol{\Lambda}$ of conditional transition *intensities* $\lambda_{j|i}$ of moving from state i to state $j \neq i$. In this matrix, the diagonal elements would correspond to staying in the same state, which does not make sense in continuous time. However, it is convenient to set them equal to $-\lambda_i$ where

$$\lambda_i = \sum_{j \neq i} \lambda_{j|i} \tag{6.1}$$

Then, the sum of each row will be zero, instead of unity for the matrix of conditional transition probabilities in the discrete case.

Transition probabilities now will refer to transitions occurring within some specified, arbitrarily fixed, interval of time, say Δt. With one intensity λ, from an exponential distribution, the probability is given by the survivor function in Equation (3.13):

$$S(\Delta t) = \mathrm{e}^{-\lambda \Delta t}$$

In a similar way, we must here exponentiate the matrix $\boldsymbol{\Lambda}$ of intensities (notice that $-\lambda_i$ appears on the diagonal) to obtain the corresponding matrix $\mathbf{T}$ of probabilities of transition within the chosen time interval:

$$\mathbf{T}^{(\Delta t)} = \mathrm{e}^{\boldsymbol{\Lambda} \Delta t} \tag{6.2}$$

where the superscript (Δt) is not a power but indicates the time interval.

In analogy to the scalar case, this matrix operation is defined by

$$\mathrm{e}^{\boldsymbol{\Lambda} \Delta t} = \mathbf{I} + \frac{\boldsymbol{\Lambda} \Delta t}{1!} + \frac{(\boldsymbol{\Lambda} \Delta t)^2}{2!} + \cdots \tag{6.3}$$

However, this is the definition, not necessarily the best procedure to calculate the exponential. A preferable way is by spectral decomposition, as for the power of a matrix in Equation (5.7). Again, the decomposition is given by Equation (5.6) and the exponential is then

$$\mathrm{e}^{\boldsymbol{\Lambda} \Delta t} = \mathbf{W} \mathrm{e}^{\mathbf{D} \Delta t} \mathbf{W}^{-1} \tag{6.4}$$

where $\mathbf{W}$ is a matrix with the eigenvectors of $\boldsymbol{\Lambda}$ as columns and $\mathbf{D}$ is a diagonal matrix containing the corresponding eigenvalues.

With the initial condition π_0, the marginal probabilities π_t now will be available, updated in the same way as in discrete time. The stationary marginal distribution of the chain, when it exists, also can be obtained from

$$\boldsymbol{\pi}^{s\top} = \boldsymbol{\pi}^{s\top}\mathbf{T}^{(\Delta t)} \tag{6.5}$$

analogous to Equation (5.2), but here also from

$$\boldsymbol{\pi}^{s\top}\boldsymbol{\Lambda} = \mathbf{0} \tag{6.6}$$

(Grimmett and Stirzaker, 1992, p. 244).

Embedded Markov chain

A second way to understand the functioning of a continuous-time Markov chain involves a corresponding discrete-time embedded Markov chain. This chain only will specify the moves from state to state, but not their times of occurrence. Let its probability transition matrix be $\mathbf{T}^b$ (the superscript is not a power; rather, it distinguishes this matrix from the probability transition matrix $\mathbf{T}^{(\Delta t)}$, obtained by matrix exponentiation above). This matrix will have $\pi^b_{i|i} = 0$ for all i, except absorbing states for which $\pi^b_{i|i} = 1$.

Then, the time spent in each state i will follow an exponential distribution with intensity λ_i, independently of what was the previous state and how long it spent there. In other words, the next state visited is independent of the amount of time spent in the previous state. The transition intensities $\lambda_{j|i}$ between states are given by

$$\lambda_{j|i} = \lambda_i \pi^b_{j|i} \quad i \neq j \tag{6.7}$$

Notice that the transition intensities $\lambda_{j|i}$ uniquely determine both the λ_i, from Equation (6.1), and the transition probabilities $\pi^b_{j|i} = \lambda_{j|i}/\lambda_i$, from Equation (6.7) (unless $\lambda_i = 0$ in which case $\pi^b_{i|i} = 1$; see Tijms, 1994, p. 132).

The mean recurrence time from entering state i to returning to it will be μ_i/π^{bs}_i and from exiting the state to returning to it will be $\mu_i(1 - \pi^{bs}_i)/\pi^{bs}_i$, where $\boldsymbol{\pi}^{bs}$ is the stationary marginal probability distribution of the embedded chain (generally different from $\boldsymbol{\pi}^s$). (See Iosifescu, 1980. p. 256 and compare these results with those in Section 5.1.2.)

6.1.4 Semi-Markov processes

The major constraining assumption of continuous-time Markov chains is that the intensity is constant while the process is in a given state. On the other hand, in the present context, the limitation of a renewal process (Section 4.1.4), with nonconstant intensity, is that only one type of event can recur. It would be useful to have models that combine this two approaches. We can imagine doing this in two ways:

(i) A continuous-time Markov chain can be generalised to a *semi-Markov process* by removing the assumption of constant intensity. Thus, the intensity will be allowed to vary over the time during which the process is in a given state.

(ii) An ordinary renewal process can be generalised to a *Markov renewal process* by removing the restriction that the same event is recurring all the time. In addition, the distributions of the durations between events (now transitions between states) need not be the same between all possible pairs of events. In the same way that a continuous-time Markov chain is a generalisation of a Poisson point process, this is a generalisation of a renewal process.

In fact, the final result is the same in both cases. Equation (4.11) already gave a general definition of this type of process.

In such a process, the intensity function depends on the time since the last event as well as on the two states between which the transition occurs. Thus, in contrast to a renewal process (Chapter 4) and to a Markov chain in continuous time (Section 6.1.3), here the intensity is a time-changing function in two ways: over time in a given state and between states.

An additional way to remove the constant intensity assumption is by letting the intensity be a function of time-varying covariates, another form of nonhomogeneous Poisson process within each state. A further extension is to introduce such covariates into a semi-Markov model.

6.2 Models for missing observations

In any study requiring observation over an extended period of time, major problems can arise if individuals disappear permanently before completion: nonrandom dropouts. An second problem may be that the series of some individuals cannot be observed at all required time points so that (some of) the exact times of change of state may be missing. This erratically missing information may also be nonrandom. If these phenomena are not modelled explicitly, the results may be misleading, as is always the case when inappropriate models are used.

6.2.1 Erratic and permanent missingness

In the context of multistate processes for event histories, one approach to modelling missingness is by including, in addition to the states of interest, one specific state for each type of missing observation. The two main possibilities are:

(i) a transient state for erratically missing observations or
(ii) an absorbing state for dropouts.

If it is not possible to determine whether or not missing observations are erratic or dropouts, one combined state may be appropriate. If there is no state exclusively for dropouts, the system may be closed because it will be possible to leave either the erratic or the combined erratic/dropout state.

As a simple case, consider a closed, three-state system with intensity transition matrix

$$\mathbf{\Lambda} = \begin{pmatrix} -\lambda_{2|1}-\lambda_{3|1} & \lambda_{2|1} & \lambda_{3|1} \\ \lambda_{1|2} & -\lambda_{1|2}-\lambda_{3|2} & \lambda_{3|2} \\ \lambda_{1|3} & \lambda_{2|3} & -\lambda_{1|3}-\lambda_{2|3} \end{pmatrix} \tag{6.8}$$

where the third state is for missing observations. Then, the corresponding probability transition matrix for the embedded discrete-time Markov chain will be

$$\mathbf{T}^b = \begin{pmatrix} 0 & \pi^b_{2|1} & \pi^b_{3|1} \\ \pi^b_{1|2} & 0 & \pi^b_{3|2} \\ \pi^b_{1|3} & \pi^b_{2|3} & 0 \end{pmatrix} \tag{6.9}$$

Thus, all states form one irreducible communicating class.

If the third state is for dropouts, the intensity transition matrix will be

$$\mathbf{\Lambda} = \begin{pmatrix} -\lambda_{2|1}-\lambda_{3|1} & \lambda_{2|1} & \lambda_{3|1} \\ \lambda_{1|2} & -\lambda_{1|2}-\lambda_{3|2} & \lambda_{3|2} \\ 0 & 0 & 0 \end{pmatrix} \tag{6.10}$$

and the corresponding embedded probability transition matrix

$$\mathbf{T}^b = \begin{pmatrix} 0 & \pi^b_{2|1} & \pi^b_{3|1} \\ \pi^b_{1|2} & 0 & \pi^b_{3|2} \\ 0 & 0 & 1 \end{pmatrix} \tag{6.11}$$

Of course, this approach to handling missing observations is not restricted to continuous-time Markov chains. It can also be used with semi-Markov processes.

Addiction In a clinical trial to evaluate the efficacy of buprenorphine for treating addiction, 162 patients were randomised to one of three groups: 8 mg of buprenorphine administered sublingually daily, 20 or 60 mg of methadone, each administered orally daily. Urine samples were collected three times a week, on Monday, Wednesday, and Friday, so that the observation times were unequally spaced. The samples were assayed for presence of opiates, with a positive assay indicating treatment failure; the first state will be for a negative assay and the second a positive one.

However, a third possibility at each given observation point was a missed visit for the sample. These missing observations clearly were nonrandom; hence, it is essential to model them. They could be sporadic or a permanent loss to follow-up. In fact, by the end of the study, the groups on buprenorphine and on the 60 mg dose of methadone had about the same amount of dropout, around 60 per cent, whereas the 20 mg dose group of methadone had about 80 per cent. However, at any given time, it was impossible to know whether a missing sample would be transient or permanent. Thus, only one missingness state will be used to model both cases.

The study ran for 17 weeks but only the data for the first three (methadone) or four (buprenorphine) weeks were published, as shown in Table 6.1. In the form presented, only transition frequencies between successive samples are available. The individual trajectories of the patients cannot be followed from these data. (The data

Table 6.1. *Transition frequencies among three states (1: negative; 2: positive; 3: missing) under three treatments in a clinical trial of buprenorphine. (Weng, 1994)*

		Transition								
Treatment	Days	11	12	13	21	22	23	31	32	33
Buprenorphine 8 mg	0–2	5	2	0	7	24	5	5	4	1
	2–4	14	2	1	11	14	5	2	2	2
	4–7	20	5	2	3	11	4	0	3	5
	7–9	17	3	3	3	14	2	2	1	8
	9–11	21	0	1	4	10	4	3	3	7
	11–14	24	4	0	2	9	2	0	3	9
	14–16	20	4	2	1	10	5	0	2	9
	16–18	18	2	1	5	7	4	0	4	12
	18–21	20	2	1	6	7	0	2	3	12
	21–23	23	3	2	2	8	2	0	1	12
	23–25	22	2	1	2	7	3	2	1	13
	25–28	20	2	4	4	6	0	1	2	14
Methadone 20 mg	0–2	2	0	0	15	23	8	1	6	0
	2–4	10	5	3	5	23	1	1	3	4
	4–7	14	1	1	2	24	5	1	3	4
	7–9	12	3	2	1	21	6	2	1	7
	9–11	8	5	2	5	16	4	2	0	13
	11–14	9	5	1	6	13	2	4	2	13
	14–16	3	7	9	2	13	5	1	5	10
	16–18	3	3	0	4	18	3	3	4	17
	18–21	7	3	0	5	15	5	2	1	17
Methadone 60 mg	0–2	5	1	0	10	28	5	1	3	1
	2–4	11	2	3	9	19	4	2	2	2
	4–7	15	4	3	1	16	6	1	3	5
	7–9	9	6	2	5	14	4	2	3	9
	9–11	10	3	3	7	13	3	1	4	10
	11–14	11	5	2	2	13	5	1	4	11
	14–16	10	2	2	7	14	1	1	4	13
	16–18	13	5	0	2	16	2	1	0	15
	18–21	11	3	2	4	14	3	2	2	13

are in a similar form to those for voting in Table 5.2.) Hence, only a continuous-time Markov chain, and not semi-Markov models, can be fitted. According to Weng (1994), the periods presented in the table correspond to time segments when the intensities were constant.

To model these data, we shall require the probability transition matrices for two and for three day intervals. We can obtain them by exponentiation of the matrix of intensity parameters being estimated. Then, because the data are aggregated, we can assume each row of the observed frequency matrix (for two or three days) to have a multinomial distribution with these probabilities, conditional on the previous state.

The main question is whether or not there are differences in transition rates among the states for the three treatment groups. A model with different transition matrices for the three groups has an AIC of 1375.7 compared to 1380.1 for the null model with the same matrix in all groups. The intensity transition matrices and one-day probability transition matrices are

$$\hat{\Lambda} = \begin{pmatrix} -0.096 & 0.067 & 0.029 \\ 0.145 & -0.262 & 0.117 \\ 0.047 & 0.130 & -0.177 \end{pmatrix}, \quad \hat{\mathbf{T}}^{(1)} = \begin{pmatrix} 0.913 & 0.058 & 0.029 \\ 0.124 & 0.780 & 0.096 \\ 0.049 & 0.106 & 0.845 \end{pmatrix}$$

for buprenorphine,

$$\hat{\Lambda} = \begin{pmatrix} -0.277 & 0.186 & 0.091 \\ 0.128 & -0.222 & 0.094 \\ 0.077 & 0.119 & -0.196 \end{pmatrix}, \quad \hat{\mathbf{T}}^{(1)} = \begin{pmatrix} 0.770 & 0.151 & 0.079 \\ 0.104 & 0.815 & 0.081 \\ 0.068 & 0.103 & 0.830 \end{pmatrix}$$

for the 20 mg dose of methadone, and

$$\hat{\Lambda} = \begin{pmatrix} -0.205 & 0.141 & 0.064 \\ 0.143 & -0.232 & 0.089 \\ 0.045 & 0.136 & -0.181 \end{pmatrix}, \quad \hat{\mathbf{T}}^{(1)} = \begin{pmatrix} 0.824 & 0.118 & 0.058 \\ 0.117 & 0.807 & 0.076 \\ 0.045 & 0.114 & 0.841 \end{pmatrix}$$

for the 60 mg dose of methadone. These estimated values are quite different from those given by Weng (1994), probably because he used least squares estimation.

These probability transition matrices are not those for the embedded discrete-time Markov chain; they are obtained by matrix exponentiation of the intensity transition matrix. The embedded matrices $\hat{\mathbf{T}}^b$ are, respectively,

$$\begin{pmatrix} 0 & 0.70 & 0.30 \\ 0.55 & 0 & 0.45 \\ 0.27 & 0.73 & 0 \end{pmatrix}, \begin{pmatrix} 0 & 0.67 & 0.33 \\ 0.58 & 0 & 0.42 \\ 0.39 & 0.61 & 0 \end{pmatrix}, \begin{pmatrix} 0 & 0.69 & 0.31 \\ 0.62 & 0 & 0.38 \\ 0.25 & 0.75 & 0 \end{pmatrix},$$

all fairly similar.

From the intensity matrices, we see that the exponential intensity $\widehat{\lambda_1}$ of moving out of the negative assay state is quite different for the three treatments, 0.096, 0.277, and 0.205, respectively. These correspond to mean sojourn times of 10.4, 3.6, and 4.9 days in the negative assay state. This indicates superiority of buprenorphine. The rates of movement out of the other two states are much more similar among the three treatments. Buprenorphine is also superior in inducing higher intensity of movement from the positive assay state to the negative than in the reverse direction. This does not occur with the two doses of methadone. (Notice, however, that this is not true of the embedded probability matrices.)

Closer inspection reveals no evidence of a difference between the transition matrices for the two doses of methadone. The model with the same transition matrix for the two, but different for buprenorphine, has an AIC of 1371.4. However, the saturated model has 1343.6, strong evidence that the transition matrices are not constant over time.

6.2.2 Trends in missingness

An approach similar to that used in Section 5.3.1 can be applied to check for homogeneity of intensity matrices. Here, I only shall consider log linear changes over time in the intensity of moves out of the missingness state. Thus, in the three-state matrix above, this can be modelled by

$$\log(\lambda_{i|3}) = \beta_{0i} + \beta_{1i}t \tag{6.12}$$

where t is the elapsed time.

Addiction For the buprenorphine data, let us make an additional assumption: that the change over time in the intensity of the moves from missingness back to the two other states is the same (that is, only one additional parameter per matrix, $\beta_1 = \beta_{11} = \beta_{12}$). With a different intensity matrix for each treatment, the AIC is reduced from 1375.7 to 1356.9. Combining the two methadone groups lowers it further, to 1351.7. The (log linear) slope parameters over time for change in the intensity of leaving the missingness state are estimated to be $\widehat{\beta}_1 = -0.0877$ for buprenorphine and $\widehat{\beta}_1 = -0.0963$ for the two doses of methadone: those on methadone stop returning to the study more quickly. Thus, most of the above lack of fit arises from this source, modelled by only two additional parameters. Of course, no firm conclusions can be drawn without the complete data for the 17 weeks.

6.3 Progressive states

One of the simplest cases of an event history has *progressive* states. This reduces very considerably the number of transition intensities that must be modelled, as compared to a complete transition matrix, like those in the previous section.

6.3.1 Constant intensities within states

When the states are progressive, a Markov chain is very similar to a Poisson point process. The difference is that the events are not recurrent, so that the process changes state at each event. Then, the intensity, constant between events, differs among states. The intensity matrix will have the form

$$\Lambda = \begin{pmatrix} -\lambda_{2|1} & \lambda_{2|1} & 0 & 0 & \cdots & 0 \\ 0 & -\lambda_{3|2} & \lambda_{3|2} & 0 & \cdots & 0 \\ 0 & 0 & -\lambda_{4|3} & \lambda_{4|3} & \cdots & 0 \\ 0 & 0 & 0 & -\lambda_{5|4} & \cdots & 0 \\ \vdots & \vdots & \vdots & \vdots & \ddots & \vdots \\ 0 & 0 & 0 & 0 & \cdots & 0 \end{pmatrix} \tag{6.13}$$

where, with a finite number of states, the last necessarily is absorbing. Here, from Equation (6.1), the transition intensity out of the state i is $\lambda_i = \lambda_{i+1|i}$ because only one type of transition is possible. The corresponding probability transition matrix for the embedded discrete-time Markov chain is

$$\mathbf{T}^b = \begin{pmatrix} 0 & 1 & 0 & 0 & \cdots & 0 \\ 0 & 0 & 1 & 0 & \cdots & 0 \\ 0 & 0 & 0 & 1 & \cdots & 0 \\ 0 & 0 & 0 & 0 & \cdots & 0 \\ \vdots & \vdots & \vdots & \vdots & \ddots & \vdots \\ 0 & 0 & 0 & 0 & \cdots & 1 \end{pmatrix} \tag{6.14}$$

In this model, the transition probabilities can be obtained explicitly. Thus, for three states, the matrix is

$$\mathbf{T}^{(t)} = \begin{pmatrix} e^{-\lambda_{2|1}t} & \frac{\lambda_{2|1}\left(e^{-\lambda_{3|2}t}-e^{-\lambda_{2|1}t}\right)}{\lambda_{2|1}-\lambda_{3|2}} & 1+\frac{\lambda_{3|2}e^{-\lambda_{2|1}t}-\lambda_{2|1}e^{-\lambda_{3|2}t}}{\lambda_{2|1}-\lambda_{3|2}} \\ 0 & e^{-\lambda_{3|2}t} & 1-e^{-\lambda_{3|2}t} \\ 0 & 0 & 1 \end{pmatrix} \tag{6.15}$$

Thus, numerical matrix exponentiation is not necessary here.

If $\lambda_i = \lambda$ is a constant for all states, we have an ordinary Poisson process (stopping after a fixed number of recurrent events). On the other hand, if the states in a progressive model simply indicate the number of previous recurrent events, we have a *pure birth process*. The special case given by Equation (4.8) was a *linear birth process*. There, $\lambda_i = i\lambda$, i being the number of recurrent events so far. This is a special case of Equation (4.26).

Myeloid leukæmia In Section 3.5.1, I applied survival models with time-varying covariates to data, from Klein *et al.* (1984), on the treatment of chronic myeloid leukæmia. Here, I shall take a different approach. Instead of concentrating on the absorbing death state, let us assume the state space to be ordered with three states: stable, blast, and death. Thus, two types of transition are possible instead of only one. This means that, as compared to that section, there are a few more events: 38 transition events instead of 30. The Kaplan–Meier curves for the times in the stable and blast states are plotted in the left graph of Figure 6.2.

The null model has the same constant intensity in both states (a Poisson process), with an AIC of 273.6. In a Markov chain model, with an exponential distribution of times between events, the constant intensity can change between states. Here, this improves the model only slightly, with an AIC of 272.3 for a different intensity of leaving each state. The transition intensity matrix among the three states is estimated to be

$$\Lambda = \begin{pmatrix} -0.0019 & 0.0019 & 0 \\ 0 & -0.0062 & 0.0062 \\ 0 & 0 & 0 \end{pmatrix}$$

We see that the transition intensity from the blast state to death is estimated to be more than three times as high as that from stable to blast, according to this model. However, because the AICs are similar, there is not strong evidence that the intensities actually are different.

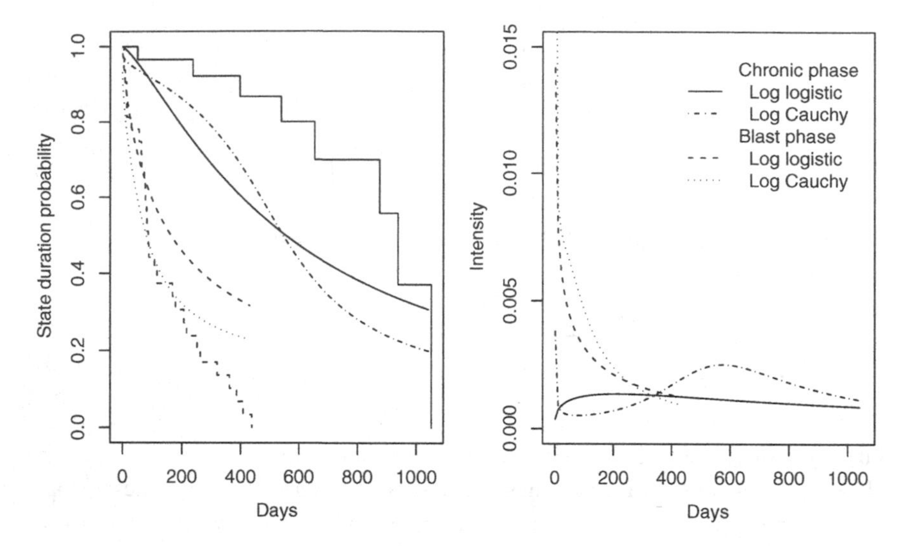

Fig. 6.2. The Kaplan–Meier curves (left) for the chronic and blast phases, ignoring ADA level, for the study of chronic myeloid leukæmia. The survivor curves (left) and intensity curves (right) for the log logistic and log Cauchy models are also plotted. These are for the mean levels of ADA (8.52 and 21.22×10^{-8} mol of inosine/h per million blood cells, respectively, for the chronic and blast phases).

6.3.2 Intensities depending on time

More realistic models will involve semi-Markov processes. Nonconstant intensity functions for the two transitions will replace the exponential intensity: special cases are the Weibull, gamma, log normal, log logistic, and log Cauchy. In all of these models, in contrast to the exponential model, the shape parameter may differ among states as well as the location parameter.

Myeloid leukæmia When these models are applied to the leukæmia data, the resulting AICs are given in the first line of each of the second and third panels in Table 6.2. None of the models provides an improvement over the simpler Markov chain model above. The risk of moving to the next state does not appear to depend on the time spent in the present state. However, we have not yet accounted for the level of ADA in the model, the centre of interest in this study.

6.3.3 Intensities depending on covariates

An extension to semi-Markov processes is to allow the intensity of transition between pairs of states to be a function of some time-varying covariates as well as of time itself.

Myeloid leukæmia For the chronic myeloid leukæmia data, the level of the enzyme, adenosine deaminase (ADA), in leukæmia cells, is such a covariate. I shall

Table 6.2. *AICs for various models fitted to the myeloid leukæmia data.*

	Exponential	Weibull	Gamma	Log normal	Log logistic	Log Cauchy
Null	273.6	273.3	273.3	272.7	273.4	275.5
Location depends on						
State	272.3	272.6	272.6	272.4	272.8	274.3
ADA	266.3	266.8	266.9	263.4	264.7	268.0
State + ADA	265.4	266.2	266.3	263.5	264.7	260.5
State ∗ ADA	266.4	267.2	267.3	264.4	265.6	261.5
Location and shape depend on						
State	—	273.3	273.2	273.0	273.5	275.2
ADA	—	267.8	253.9	261.6	253.1	253.3
State + ADA	—	268.1	254.1	254.0	252.4	253.9

handle this covariate in the same way as in Section 3.5.1. The results also are given in Table 6.2.

Here, we find improvements over the Markov chain model. The log logistic and log Cauchy distributions fit best, closely followed by the gamma. When both of the parameters of these distributions are allowed to vary with ADA, the difference between states is not necessary, except for the log normal and log logistic intensities. Thus, there is an effect of ADA, with no difference in that effect between the two states (no interaction). From the log logistic and log Cauchy models, respectively, the location coefficients (for duration, not intensity) are estimated to be -0.097 and -0.152: the risk of changing state increases with higher levels of ADA. The corresponding coefficients for the shape parameters are 0.039 and 0.080.

The estimated survivor and intensity curves are plotted in Figure 6.2 at mean levels of ADA. Recall, however, that these levels are actually changing over time for each subject; the Kaplan–Meier curves ignore the ADA level, whereas the model curves are for the mean level. Both models give similar results, with the risk of changing from the chronic to the blast phase staying fairly constant over time, *if the adenosine deaminase level were to remain constant*, which it does not. On the other hand, the risk of dying is estimated to be highest just after the transition to the blast phase and decreases sharply thereafter.

Further reading

Iosifescu (1980), Karlin and Taylor (1975), Ross (1989), Tijms (1994), and Wolff (1989) provide material both on continuous-time Markov chains and on semi-Markov processes. Anderson (1991) and Bremaud (1999) each give advanced coverage of continuous-time Markov chains.

Table 6.3. *Frequencies of transitions among three states of smoking behaviour (N: never smoked; C: current smoker; Q: quit) for sixth-grade children in Oxford county, Ontario, Canada, given educational material on smoking during the first two months, distinguished by risk status. (Kalbfleisch and Lawless, 1985)*

Risk	Years	NN	NC	NQ	CN	CC	CQ	QN	QC	QQ
Low	0.15	32	2	0	0	0	2	0	0	1
Low	0.75	30	1	1	0	0	2	0	2	1
Low	1.10	27	1	2	0	1	2	0	0	4
Low	1.90	25	1	1	0	0	2	0	0	8
High	0.15	61	1	2	0	8	8	0	1	7
High	0.75	59	1	1	0	7	3	0	3	14
High	1.10	56	2	1	0	8	3	0	2	16
High	1.90	51	2	3	0	6	6	0	0	20

Exercises

6.1 Two exercises in Chapter 5 involved unequally spaced times. Hence, your analyses there may have been incorrect. Redo your modelling for

(a) the study of stress after the accident at the Three Mile Island nuclear power plant in Table 5.10 (Exercise 5.4) and

(b) the changes in Swedish voting in Table 5.15 (Exercise 5.9).

Does the use of continuous-time Markov chains alter your conclusions in either case?

6.2 A panel of children entering grade six in the counties of Waterloo and Oxford, Ontario, Canada, was followed over almost two years during which time they were interviewed five times. At the start of the study, the children were divided into two groups: one acted as control and the other was given educational material on smoking during the first two months. The resulting transitions occurring between each of the five interviews, only for those receiving the educational material in Oxford county, are shown in Table 6.3. These are classified by the children's risk situations, as determined by their exposure to friends and relatives smoking.

(a) Which transitions among the three states are impossible?

(b) Develop an appropriate stochastic model.

(c) Is there any evidence of a difference between the two risk groups?

(d) Is homogeneity a reasonable assumption?

6.3 Leukæmia patients who had attained remission were followed to subsequent relapse and possible death. Entry to the study was staggered so that termination of the study resulted in censoring after various lengths of observation. The data for 15 patients are shown in Table 6.4.

(a) Does the same distribution adequately describe both times to relapse and to death?

(b) Do any of the parameters of the distribution of times to death depend on the time to relapse?

Table 6.4. *The times (days; asterisks indicate censoring) from remission to relapse and to death for 15 leukæmia patients. (Clayton, 1988, with corrections)*

Relapse	238	72	99	140	294	937	517	126
Death	571	179	100	204	67	21	159	123
Relapse	783	56	889*	246	951*	931*	146	
Death	63*	11	—	151	—	—	40	

6.4 A clinical trial was conducted to study intensification therapy combined with a high-dose combination chemotherapy in the treatment of acute non-lymphoblastic leukæmia in adults. Patients were assigned to one of three groups. All subjects received the same treatment for the first 186 days. Then, because the first group had been treated prior to the development of the intensification program, it could not be given this therapy. The second group was eligible to receive the therapy but did not, whereas the third group received it. Times, in days, from remission to relapse and from relapse to death were recorded, as shown in Table 6.5. Notice that treatment is time-varying because it only starts after six months.

(a) What influence does therapy have on remission and survival?

(b) Thirteen additional subjects were lost before the end of the 186 days. Their survival times were 24, 68*, 76, 85, 90, 95, 119*, 124, 151, 158*, 172, 182, and 182*, the four indicated by asterisks being censored. Can you incorporate this information into your model?

6.5 In the National Cooperative Gallstone Study carried out in the USA, an important point of interest was the safety of the drug chenodiol. A major concern was that, as the gallstones dissolve, they might pass into the biliary tree and lead to increased gallbladder symptoms. Progression of the disease is indicated by the occurrence of biliary tract pain, perhaps accompanied by other symptoms that might require surgical removal of the gallbladder, called cholecystectomy. Thus, in this study of the disease progression of floating gallstones, biliary pain might be experienced, and, if so, might be followed by cholecystectomy. If the former did not occur, the latter would also be censored.

A series of observations on 113 patients under two treatments, placebo and the drug chenodiol, are presented in Table 6.6. Although patients were assigned randomly to high dose (305), low dose (306), or placebo (305), the table only contains the results for high dose and placebo. Patients could be followed for as long as 28 months.

(a) Find an intensity model for these two events.

(b) Does treatment have an effect on the risk of having either of them?

6.6 Breast cancer patients in a London, England, hospital were being treated for spinal metastases. They were followed over a five-year period, with their ambulatory state being recorded before treatment began, and at zero,

Table 6.5. *The times (days; asterisks indicate censoring) from remission to relapse and from relapse to death for leukæmia patients in three treatment groups. (Matthews, 1988, from Glucksberg* et al.*)*

Prior treatment		No treatment		Therapy after randomisation	
186	197	57	201	380	380
82	201	191	261	277	436
128	226	152	266	501	529
166	247	197	283	571	604
111	272	264	343	396	626
230	304	223	349*	635*	635*
304	333	254	437	688	767
273	334	249	490	358	812
294	342	258	552	506	971
284	347	269	576	1028	1151
304	355	583*	583*	1229*	1229*
117	359	534	687	1145	1237
247	365	510	743	1323*	1323*
341	378	171	1115	1349*	1349*
209	442	642	1447*	1482*	1482*
395	506	1584*	1584*	1492*	1492*
332	517	1812*	1812*	1076	1501
393	546	697	1865*	1553*	1553*
518	620	1908*	1908*	1352	1597*
487	670			1604*	1604*
608	702			736	1648
518	806			1820*	1820*
516	871			1989*	1989*
270	946				
1160	1355				
955	1392				
2182*	2182*				
2270*	2270*				
2331*	2331*				

three, and six months, one, two, and five years. Their ambulatory status, defined by whether or not they could walk unaided, is shown in Table 6.7.

(a) What are the probabilities of transition between the two ambulatory states?

(b) Do they evolve over time?

(c) What is the risk of death?

Table 6.6. *The times (days; asterisks indicate censoring) to biliary pain (first column) followed by cholecystectomy (second column) for 113 patients under two treatments, placebo and chenodiol, for gallstones. (Wei and Lachin, 1984)*

Placebo				Chenodiol			
741*	741*	35	118	735*	735*	742*	742*
234	234	175	493	29	29	360*	360*
374	733*	481	733*	748*	748*	750	750*
184	491	738*	738*	671	671	360*	360*
735*	735*	744*	744*	147	147	360*	360*
740*	740*	380	761*	749	749	726*	726*
183	740*	106	735*	310*	310*	727*	727*
721*	721*	107	107	735*	735*	725*	725*
69	743*	49	49	757*	757*	725*	725*
61	62	727	727*	63	260	288	810*
742*	742*	733*	733*	101	744*	728*	728*
742*	742*	237	237	612	763*	730*	730*
700*	700*	237	730*	272	726*	360*	360*
27	59	363	727*	714*	714*	758*	758*
34	729*	35	733*	282	734*	600*	600*
28	497			615	615*	743*	743*
43	93			35	749*	743*	743*
92	357			728*	728*	733*	755*
98	742*			600*	600*	188	762*
163	163			612	730*	600*	600*
609	713*			735*	735*	613*	613*
736*	736*			32	32	341	341
736*	736*			600*	600*	96	770*
817*	817*			750*	750*	360*	360*
178	727			617	793*	743*	743*
806*	806*			829*	829*	721*	721*
790*	790*			360*	360*	726*	726*
280	737*			96	720*	363	582
728*	728*			355	355	324	324
908*	908*			733*	733*	518	518
728*	728*			189	360*	628	628
730*	730*			735*	735*	717*	717*
721*	721*			360*	360*		

Table 6.7. *The successive states (0: dead; 1: unable to walk; 2: able to walk) of 37 breast cancer patients over a five-year period. (de Stavola, 1988)*

	Months					
Initial	0	3	6	12	24	60
2	2	2	2	2	2	0
2	2	2	2	0		
1	1	0				
1	2	1	2	2	2	0
2	2	2	2	2	1	0
2	2	2	2	0		
2	2	1	2	2	0	
1	2	1	0			
2	2	2	2	2		
1	1	0				
1	1	0				
2	2	2	2	0		
1	1	0				
1	2	2	2	0		
1	1					
1	1	0				
2	2	2	2	0		
1	1	1	0			
2	2	0				
2	2	1	1	1	0	
1	1	0				
1	2	2	2	2	2	
1	2	0				
2	2	1	—	0		
2	2	2	2	2	0	
1	1	1	1	1	0	
1	2	1	1	1	0	
1	2	2	2	2	0	
1	2	1	1	0		
1	2	2	1	0		
1	2	1	1	1		
2	2	2	1	1	0	
1	2	1	1	0		
2	2	2	2	0		
2	2	2	0			
1	1	0				
1	1	0				

7

Dynamic models

Markov chains (Chapter 5), and extensions of them (Chapter 6), use previous information from a multistate stochastic process to improve prediction of the present state. These state dependence models condition directly on the previous state(s) in the regression function, so that the other regression parameters only can be interpreted as conditional on those specific previous states of each given process.

In contrast, the underlying or marginal trend of any process is independent of the conditional dependence on the past of a specific observed process. This trend always can be obtained by summing or integrating at each time point, but it may not always be easy or practical to do so. In a Markov chain, the underlying trend is represented by the sequence of marginal distributions π_t (see Figure5.2). This is obtained by matrix multiplication, a summation, at each time point. However, in other models, such a representation may not be as easy to obtain.

Dynamic models provide one alternative to the above approaches. They use previous information in various different ways such that they are not directly state dependent. One simple case was serial dependence, described in Section 1.2 and developed further below in Section 7.1. There, dependence is on the previous residual, an unobservable (hidden) quantity, instead of on the previous state. Dynamic models generally will allow a theoretical underlying profile of the process to be plotted more easily, although it may not necessarily be a marginal distribution. However, a prediction of the actual observed process, based conditionally on its previous history, also will be available.

In dynamic models, certain parameters in the distribution of responses are allowed to vary randomly over time. Here, we may distinguish two cases:

(i) These parameters may be able to take a different value at each time point, yielding an infinite number of possible hidden states, the simplest cases often being called *dynamic linear models*.
(ii) The parameters of the distribution only may be allowed to vary among a finite set of possibilities, called *hidden Markov models*.

Classical dynamic linear models, involving a continuous observed state space, will be presented in Section 11.1.

For some reason, both of these types of models are sometimes called *state-space models*. However, as we know, *any* stochastic process has a state space, and not just these specific ones. They might more appropriately be called *hidden* state-space models, because they have hidden states, as well as the observed ones.

Although some of the examples in this chapter will involve discrete time, all of the models apply also to observations in continuous time.

7.1 Serial dependence

In Section 1.2.4, I introduced a general serial dependence model. Let us examine more closely the interpretation of such models.

Suppose that a process has an underlying trend that can be described by changes over time in the location parameter, often the mean of a distribution, that is, by some appropriate regression function μ_t. However, any given observed process may wander about this function. For each realisation of the process, this allows better predictions to be made, by taking into account the previous history, here by using some μ_{it} specific to that realisation, instead of μ_t. Now, assume that this μ_{it} depends on the distance of the previous observed response $y_{i,t-1}$ from the underlying trend μ_{t-1} at that time point. This is the previous residual $y_{i,t-1} - \mu_{t-1}$, an unobservable random quantity. Here, it will be useful to write this dependence as

$$\mu_{it} = \mu_t + \rho^{\Delta t}(y_{i,t-1} - \mu_{t-1}) \tag{7.1}$$

with Δt the time between observations, to allow for cases in which they are unequally spaced, and $\mu_0 = y_{i,0}$ the initial condition. Here, the hidden state is the unobservable previous residual $y_{i,t-1} - \mu_{t-1}$.

Notice that, if $\mu = \mu_t$ is a constant for a given series, not varying over time, this equation reduces to

$$\mu_{it} = \mu(1 - \rho^{\Delta t}) + \rho^{\Delta t} y_{i,t-1} \tag{7.2}$$

a simple state dependence model (Section 1.2.3).

In contrast to state dependence models, the serial dependence of Equation (7.1) usually is not applicable to strictly categorical responses, because the operation $y_{it} - \mu_t$ is meaningless. It can, however, be useful when only a series of aggregate counts is available over time. For some types of counts, this equation can be used directly whereas, for others, adaptation may need to be made.

7.1.1 Single count responses

Application of serial dependence models to series of responses involving aggregate counts of one kind of event is usually straightforward. Equation (7.1) often can be used directly, although care must be taken that negative values of μ_{it} are not produced.

University enrolment Yale University, USA, was founded in 1701, but the enrolment record is not complete before 1796. Counts of students enrolled are available each year from 1796 to 1975, as shown in Table 7.1 and plotted in the top graph of

Table 7.1. *Student enrolment at Yale University, USA, 1796–1975. (Anscombe, 1981, p. 130; read across rows)*

115	123	168	195	217	217	242	233	200	222
204	196	183	228	255	305	313	328	350	352
298	333	349	376	412	407	481	473	459	470
454	501	474	496	502	469	485	536	514	572
570	564	561	608	574	550	537	559	542	588
584	522	517	531	555	558	604	594	605	619
598	565	578	641	649	599	617	632	644	682
709	699	724	736	755	809	904	955	1031	1051
1021	1039	1022	1003	1037	1042	1096	1092	1086	1075
1134	1245	1365	1477	1645	1784	1969	2202	2350	2415
2615	2645	2674	2684	2542	2712	2816	3142	3138	3806
3605	3433	3450	3312	3282	3229	3288	3272	3310	3267
3262	2006	2554	3306	3820	3930	4534	4461	5155	5316
5626	5457	5788	6184	5914	5815	5631	5475	5362	5493
5483	5637	5747	5744	5694	5454	5036	5080	4056	3363
8733	8991	9017	8519	7745	7688	7567	7555	7369	7353
7664	7488	7665	7793	8129	8221	8404	8333	8614	8539
8654	8666	8665	9385	9214	9231	9219	9427	9661	9721

Figure 7.1. We see that the available series is fairly flat at first, up to about 1885, then rises sharply in the twentieth century, with drops during the two world wars. The change in slope may indicate a modification in the process producing the responses. The enrolment mechanism may have changed at the end of the nineteenth century. In addition, there are important perturbations due to war.

For comparison, the changes in enrolment between years and the ratios of successive enrolments are also plotted in Figure 7.1. We see that the levels of both of these are more stable than the absolute numbers, although their variability changes during some periods.

Because enrolment in a given year occurs virtually simultaneously for all students, not over a meaningful period of time, this is one example of a state space involving counts that cannot be disaggregated into a sequence of single events.

For single counts, the simplest model with which to start is the Poisson distribution. To see clearly the major effects that serial dependence can produce, let us attempt to model this series using a constant mean (except for serial dependence), that is, with no regression function to describe the increase in numbers over time. (This implies that it will also be a state dependence model.) Here, Equation (7.1) can be used for the serial dependence.

This model has an AIC of 4939.2 compared with 264 079.2 for independence. The parameter estimates are $\widehat{\mu_t} = 109.3$ and $\hat{\rho} = 1.02$. Clearly, there is strong dependence among these responses. Because μ_t is constant and ρ is estimated to be approximately one, Equation (7.1) reduces to $\widehat{\mu_{1t}} \doteq y_{t-1}$. (This implies that the constant mean is redundant, and can be set to zero.)

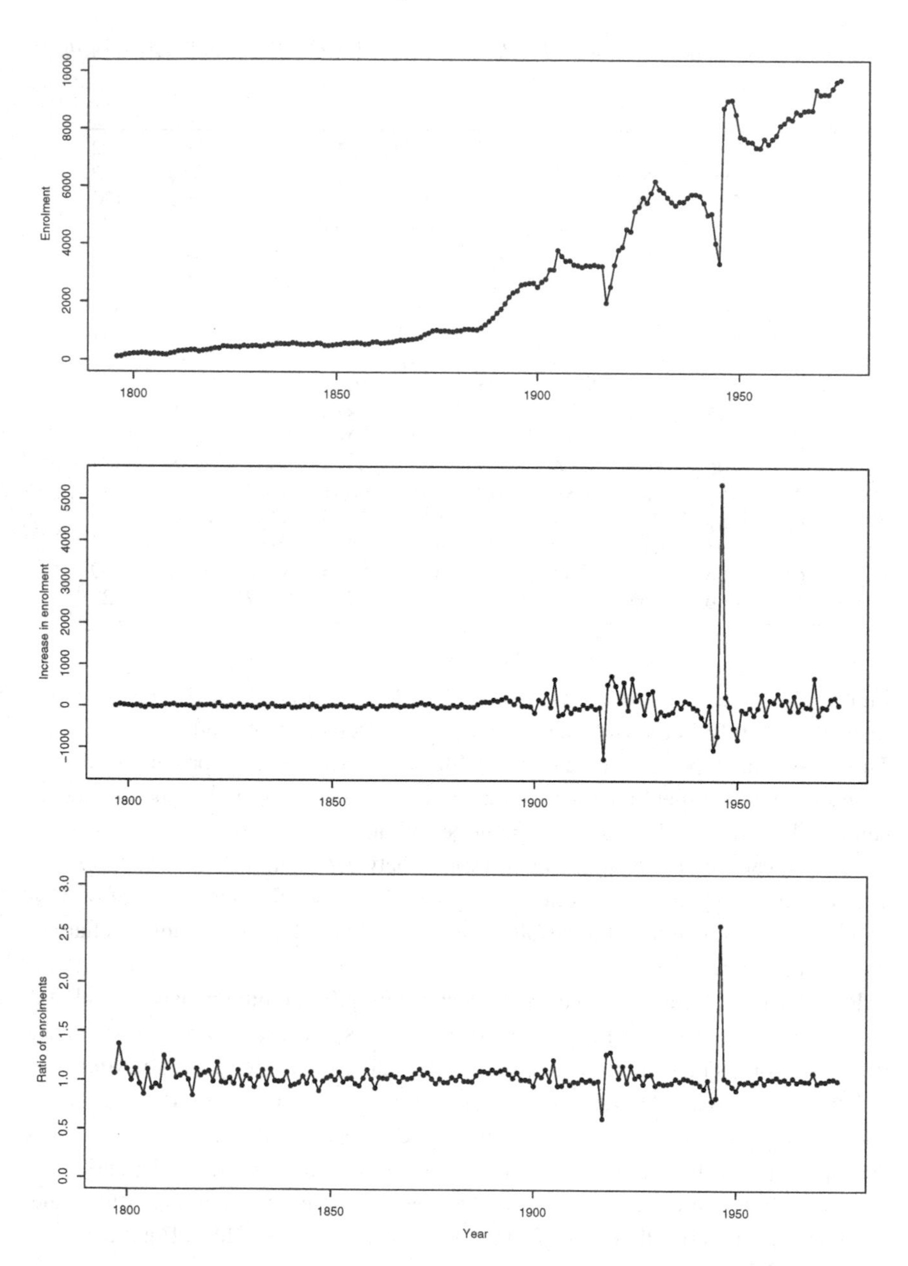

Fig. 7.1. Enrolment (top), changes in enrolment between years (middle), and ratio of enrolments in successive years (bottom) at Yale University from 1796 to 1975, from Table 7.1.

7.1.2 Binary count responses

When more than one type of event is counted, the standard serial dependence model given above usually will need to be modified. For two kinds of events, the re-

Table 7.2. *Numbers of successes out of 20 by 16 animals in a learning experiment under four different combinations of conditions. (Aickin, 1983, pp. 238–240)*

Not trained								Trained							
Light				Bell				Light				Bell			
6	1	2	1	1	0	2	1	2	0	0	3	0	0	2	0
8	6	7	6	0	0	2	1	0	0	0	4	0	0	0	3
3	16	5	15	2	0	9	0	0	10	0	4	4	7	3	2
6	17	13	19	3	0	4	0	7	17	10	6	4	5	0	0
6	17	19	17	16	0	1	4	18	17	15	8	15	3	0	0
5	8	19	19	12	0	9	4	19	19	18	15	10	4	7	0
18	18	18	17	17	0	15	5	19	19	19	17	11	6	14	3
18	17	19	19	18	0	16	7	18	19	15	15	13	2	11	2
17	18	17	19	15	14	17	13	18	19	19	14	19	11	15	8
19	18	20	20	16	15	17	17	20	19	19	16	20	12	18	12
19	19			17	18	19	15		20	18	19		6	18	15
18	20			20	16	19	15			19	18		13	17	18
20					18	20	19			20	19		14	20	17
					18		18				20		19		18
					17		20						18		18
					17								17		20
					19								18		
					19								20		
					19										
					20										

sponses will be binary. Then, the model might be adapted in the following way:

$$\pi_{it} = \pi_t + \rho^{\Delta t}(y_{i,t-1} - n_{i,t-1}\pi_{t-1}) \tag{7.3}$$

with n_{it} the total number of events in series i at time t and y_{it} the corresponding number of events of the first kind. (Even more care must be taken here that π_{it} does not take invalid values than for single counts above.) Here, the initial condition is $\pi_0 = y_{i0}/n_{i0}$ or $\pi_{i1} = \pi_1$, that is, the first residual is zero. As previously, π_t represents the common underlying profile for all individuals under the same conditions at time point t and π_{it} is the corresponding predicted individual value.

Animal learning In a learning experiment, 16 laboratory animals (of unstated species) were tested for their ability to learn using a 2×2 factorial experiment, the two factors being with or without training and with light or bell stimulus. Each animal was given a series of trials. At each trial, the animal was allowed 20 attempts to complete a task, as shown in Table 7.2. Trials for an animal stopped when a perfect score was reached. (No further information was provided.) The observed profiles are plotted in Figure 7.2. We see that many animals follow rather irregular trajectories, although a vaguely sigmoidal-shaped path is discernible.

These data have two levels of clustering: the 20 attempts within a trial and the series of trials of an animal. Here, in contrast to the Yale data, the individual events (the ordered results of the 20 attempts) making up the aggregated counts do have meaning. Knowing the order of successes and failures within a trial could

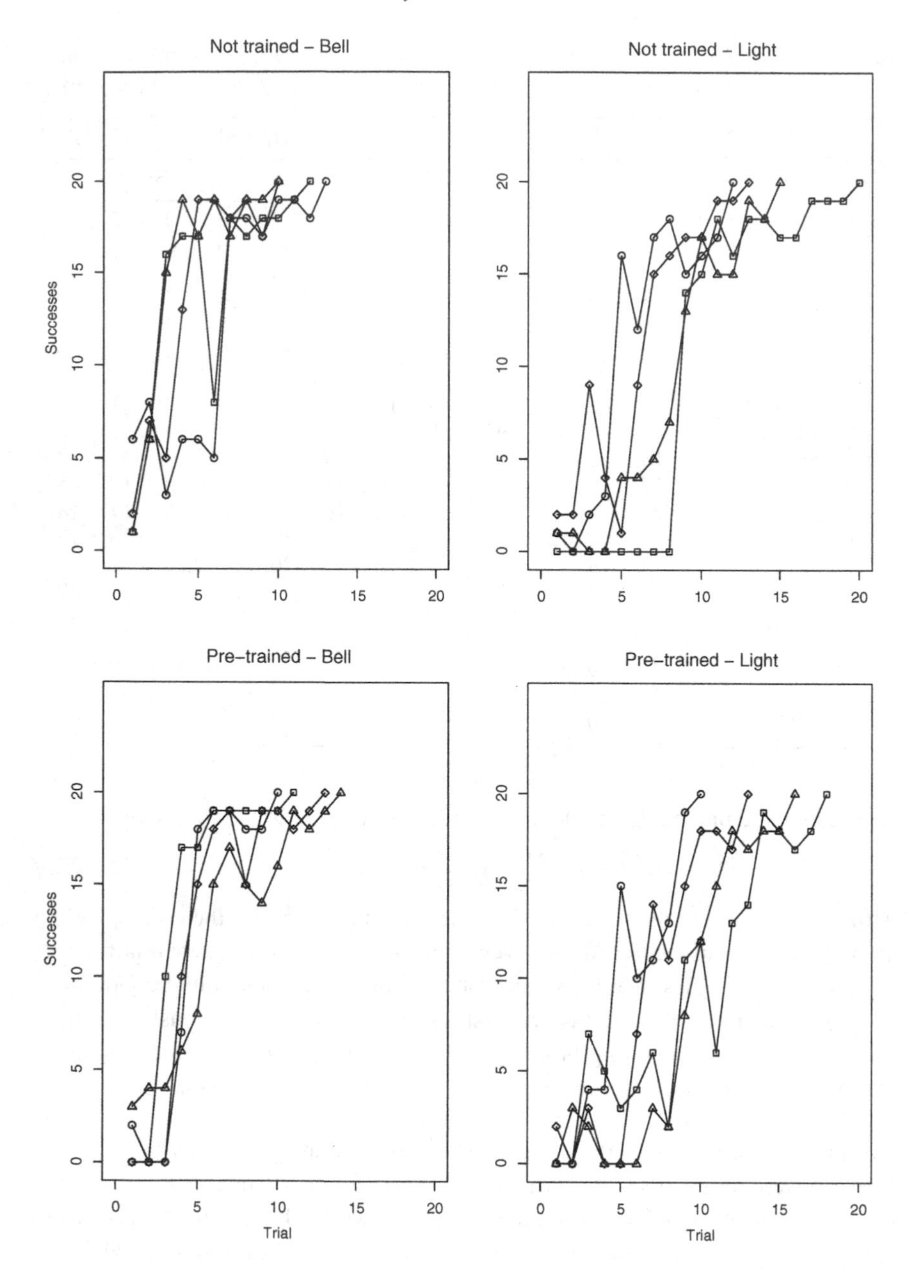

Fig. 7.2. The observed profiles for animals learning under the four conditions, from Table 7.2.

be very useful, but unfortunately we have no information about this. On the other hand, more elaborate modelling will be possible for the succession of counts of successes and failures for the set of trials of each animal.

At each time point (trial), there are 21 possible states, the number of successful

Table 7.3. *AICs for various binary models fitted to the animal learning experiment of Table 7.2. Cloglog: complementary log log link.*

	Independence			Serial dependence		
Link	Logit	Loglog	Cloglog	Logit	Loglog	Cloglog
Binomial	669.3	656.6	748.7	554.5	551.3	581.9
Beta-binomial	498.8	494.2	516.3	462.7	461.1	481.7
Multiplicative binomial	471.5	471.0	466.9	457.2	456.2	458.1
Double binomial	480.6	478.5	497.0	455.5	455.2	460.8

attempts. A Markov chain would require too large a number of parameters in the transition matrix to be practical. In addition, we shall be interested in the underlying profile of the learning curves under the four experimental conditions, independently of the short-term variability in the performance of a given animal.

Here, for binary count responses, the simplest models will be based on the binomial distribution. Various link functions are possible, such as the logit, $\log[\pi/(1-\pi)]$, the log log, $\log[-\log(\pi)]$, and the complementary log log, $\log[-\log(1-\pi)]$. From the graphs, we know that the probability of success is increasing over time. Each link function will yield a different form, and different interpretation, of the increasing curve over time (see Section 12.3 for more details).

In contrast to time and stimulus, the training never shows a significant influence on the responses. Thus, I shall only consider models for difference between the two stimuli and its interaction with time (the trial number). The AICs for these three link functions with the binomial distribution are given in the first line of Table 7.3. The log log link proves to be best. As with the Yale data, there is a clear need for serial correlation.

Unfortunately, for aggregate counts of events, the simple Poisson and binomial models rarely prove adequate. More complex models may prove to be necessary.

7.1.3 Overdispersion

When only aggregate counts of events are available, and not their timing, the standard Poisson or binomial models often do not fit well. This will arise when the aggregated events within each count are dependent. Then, overdispersion models (Sections 3.6.2 and 4.3.2) may be used to allow for this dependence.

For counts of one type of event, the overdispersed negative binomial distribution of Equation (4.36) can be used. For two types of events, the beta-binomial distribution was given in Equation (3.28). Two other possibilities for this latter type of response are the 'multiplicative' binomial distribution (Altham, 1978)

$$f(d|n;\pi,\kappa) = c_m(\pi,\kappa)\binom{n}{d}\pi^d(1-\pi)^{n-d}e^{\kappa d(n-d)} \tag{7.4}$$

and the double binomial distribution (Efron, 1986)

$$f(d|n;\pi,\kappa) = c_d(\pi,\kappa)\binom{n}{d}\frac{n^{n\kappa}\pi^{d(\kappa+1)}(1-\pi)^{(n-d)(\kappa+1)}}{d^{d\kappa}(n-d)^{(n-d)\kappa}} \tag{7.5}$$

where n is the total number of events, d is the number of the first kind of event, κ is a dispersion parameter, and $c_m(\pi,\kappa)$ and $c_d(\pi,\kappa)$ are intractable normalising constants.

University enrolment The negative binomial distribution without serial dependence has an AIC of 1618.9, whereas the corresponding Poisson model above had 264 079.2, showing that overdispersion plays an important role. When serial dependence is added, the AIC of the former reduces to 1190.0. The dependence parameter is estimated to be $\hat{\rho} = 1.03$. With constant mean $\widehat{\mu}_t = 112.2$, this again implies, from Equation (7.1), that $\widehat{\mu_{1t}} \doteq y_{t-1}$.

The recursive predicted values $\widehat{\mu_{1t}}$ for the negative binomial serial dependence model are plotted in the top graph of Figure 7.3. They follow the responses reasonably well although there is no explicit component in the model to account for the structural growth in numbers. However, adding an exponential regression on time reduces the AIC to 1181.6.

Animal learning Now, let us apply overdispersion models to the animal learning data. This will be similar to what I did in Section 3.6.2 for human mortality data. However, in contrast to those data and the Yale enrolment, here there is more than one series. Then, without serial dependence, these models should allow for the irregular aspects of the responses but will make no distinction between variability within and between animals. They will only account for the dependence among attempts within each trial.

Here, $n = 20$ and d is the number of successes in a trial in the overdispersed distributions given above. The AICs of various models are shown in the left half of Table 7.3. The multiplicative binomial model with a complementary log log link fits better than the others, although all the links are reasonably close for this distribution.

However, I have not yet made any allowance for dependence over time or for heterogeneity among the animals. For the serial dependence model of Equation (7.3), the double binomial distribution with a log log link is best. However, the model with logit link can be simplified by removing the trial–stimulus interaction, making it a slightly better model. The autoregression parameter is estimated to be $\hat{\rho} = 0.028$, and the intercepts to be -4.47 for the bell and -2.37 for the light, with common slope 0.57.

The underlying profile curve ($\widehat{\mu_{jt}}$, with j the group) and the predicted curves ($\widehat{\mu_{ijt}}$) for the first two animals in each group are shown in Figure 7.4. The individual predicted profiles attempt to follow the observed responses, but always overshoot when an extreme value previously has been recorded.

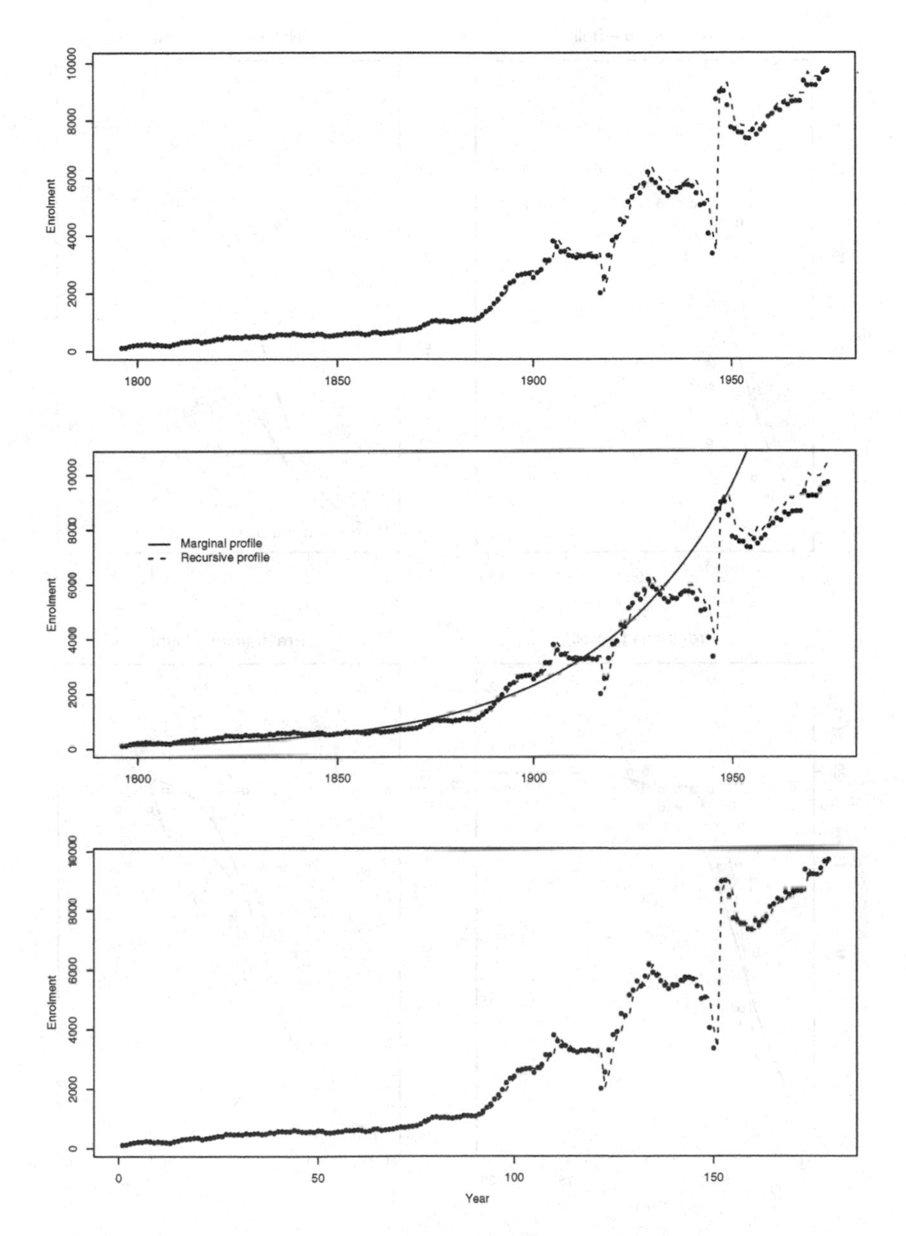

Fig. 7.3. The predicted enrolment from the serial dependence model without a regression on time (top), the serial dependence with mean and dispersion both depending exponentially on time (middle), and the gamma mixture dependence model (bottom). The origin in the bottom plot is 1795.

7.1.4 Changing variability over time

Overdispersion models introduce an extra parameter into the standard Poisson and binomial distributions in order to account for the large variability. In many circum-

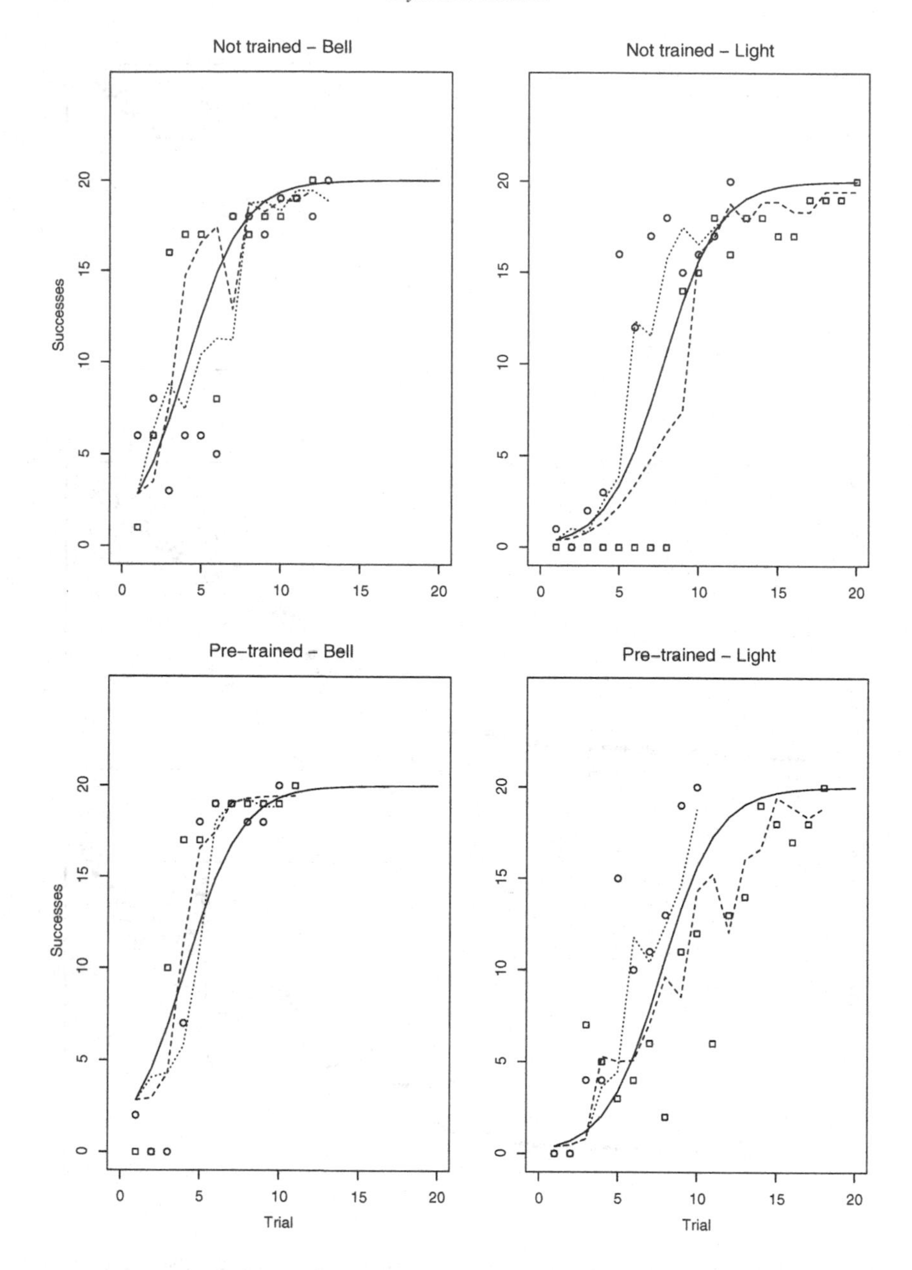

Fig. 7.4. The underlying profile (solid lines) from the double binomial serial dependence model with a logit link, along with the observed and predicted responses for the first (circles and dots) and second (squares and dashes) animals in each group, from Table 7.2.

stances, the variability may change over time so that it is necessary to model this parameter using a second regression function. In the context of normal distribution models, this is known as *heteroscedasticity* (Section 9.2.4).

University enrolment For the Yale enrolment data, the middle and bottom graphs of Figure 7.1 indicate that the variability may be changing over time. If I let the dispersion parameter of the negative binomial distribution also depend exponentially on time in the previous model above, the AIC is further reduced to 1158.4. On the other hand, if I now remove the exponential regression on time for the mean, it only rises to 1163.3, showing that, when the changing variability is accounted for, the exponential growth plays a relatively small role. The dispersion parameter is decreasing over time and the variance, which depends on the reciprocal of this parameter and on the mean, is increasing more quickly than the mean.

The model with both the mean and the dispersion depending exponentially on time is plotted in the middle graph of Figure 7.3. We see that, near the end of the series, the exponential regression for the mean pulls the recursive predictions above the observed values.

Models with a change in slope around 1900 do not fit as well as those with exponential regression on time just presented. Thus, there seems to be no indication of an abrupt change at that time. On the other hand, the exponential regression function climbs too steeply, so that other nonlinear functions should be investigated.

Changes in dispersion here were modelled as a deterministic function over time. They also can be handled dynamically, as another form of serial dependence; I shall reserve the main description of such models (called ARCH) for Section 10.4. However, I shall present one such dynamic model, specifically for changing dispersion in series of counts, in Section 7.4 below.

7.2 Hidden Markov models

In certain stochastic processes, an individual producing a series of responses may be subject to spells, that is, to having periods with somewhat different reactions. This may be modelled by assuming a sequence of changing unobserved states corresponding to these periods in time. Thus, in a binary point process, some of the recurrent events may be clustered, corresponding to periods of higher activity. The dynamic models to be described here provide one way to handle such data.

7.2.1 Theory

In a hidden Markov model, an individual is assumed to occupy an underlying, unobservable sequence of states that follow a Markov chain (Chapter 5). The hidden state at a given time point determines the probabilities of the various possible observed states. In contrast to serial dependence models, with their infinite number of possible hidden states, here the number is finite (and usually rather small).

In the original presentation of the model, Baum and Petrie (1966) only considered responses consisting of a finite number of observable states. They called their models probabilistic functions of Markov chains. However, the applications have proved to be much wider than this. As a simple case of such a stochastic process, each observed state might be generated by one of two Poisson processes, each having a different intensity. The process switches from the one to the other following a two-state hidden Markov chain, in this way generating dependence over time.

In the analysis of such series, three problems generally are distinguished. Given an observed sequence:

(i) How can the likelihood be calculated efficiently (the evaluation problem)?
(ii) How can the maximum likelihood estimates be obtained ('training' the model)?
(iii) How can the optimal hidden state sequence be chosen for a given model?

I shall look briefly at each of these in turn.

Hidden state space

Suppose that we assume the existence an unobservable, irreducible, homogeneous Markov chain with M states. This will give the conditional probabilities of changing among the hidden states. Often, we can assume that the marginal distribution is stationary. (If so, the latter can be calculated from the transition matrix using Equation (5.2) and, hence, does not introduce any new parameters.) The probability of the observed state y_t at time t will depend on the unobserved state m at that time. Thus, we are assuming that the sets of observed states in a given series are independent, given the hidden states.

Let us formulate this more rigorously. Suppose that the process under study is currently in some unobservable hidden state m. Assume that, under these conditions, the distribution of the response is $f(y_t|\psi_t = m, \mathcal{F}_{t-1}; \boldsymbol{\kappa}_m)$. Then, the unknown parameters $\boldsymbol{\kappa}_m$ generally will depend on covariates through some linear or nonlinear regression function, different for each hidden state m. This is called the *observation* or *measurement equation.* On the other hand, ψ_t will be a random parameter indicating the hidden state at time t and $\mathcal{F}_t$ the history of the process up until time t, as in Chapter 4. If m were known at each time point, this would be a standard regression model.

We now need to model how ψ_t changes over time, even though it is unobservable. Let us assume that it obeys a first-order Markov chain with transition matrix $\mathbf{T}$. This defines the *state transition equation*

$$\Pr(\psi_t = m|\psi_{t-1} = j, \mathcal{F}_{t-1}) = \pi_{m|j} \tag{7.6}$$

with corresponding marginal probabilities

$$\Pr(\psi_t = m|\mathcal{F}_t) = \pi_{tm} \tag{7.7}$$

(Often, the hidden Markov chain can be assumed to be stationary so that $\boldsymbol{\pi}_t = \boldsymbol{\pi}^s$; see Equation (5.2).)

We can now set the process in motion to make predictions. The probability of each hidden state at the next time point will be given by $\boldsymbol{\pi}_t^\top \mathbf{T}$ or

$$\Pr(\psi_{t+1} = m|\mathcal{F}_t) = \sum_{j=1}^{M} \pi_{tj}\pi_{m|j} \tag{7.8}$$

called the *one-step-ahead prediction* or *time update*. Next, we need to introduce the new information available from the observed response y_{t+1} at time $t+1$. To do this, we apply Bayes' formula, combining the time update with the observation equation. This yields the *filtering* or *observation update*

$$\Pr(\psi_{t+1}=m|\mathcal{F}_{t+1})=\frac{f(y_{t+1}|\psi_{t+1}=m,\mathcal{F}_t;\boldsymbol{\kappa}_m)\Pr(\psi_{t+1}=m|\mathcal{F}_t)}{\sum_j f(y_{t+1}|\psi_{t+1}=j,\mathcal{F}_t;\boldsymbol{\kappa}_j)\Pr(\psi_{t+1}=j|\mathcal{F}_t)} \tag{7.9}$$

Finally, we require the probability of the observed response, given that we do not know in what hidden state the process actually is. We can obtain this by summing, over all possible hidden states, the product of the observation equation for each hidden state times the probability of being in that state (the time update)

$$\begin{aligned} f(y_{t+1}|\mathcal{F}_t;\boldsymbol{\kappa}_1,\ldots,\boldsymbol{\kappa}_M,\mathbf{T}) &= \sum_{m=1}^{M} f(y_{t+1}|\psi_{t+1}=m,\mathcal{F}_t;\boldsymbol{\kappa}_m)\Pr(\psi_{t+1}=m|\mathcal{F}_t) \\ &= \sum_{m=1}^{M} f(y_{t+1}|\psi_{t+1}=m,\mathcal{F}_t;\boldsymbol{\kappa}_m)\sum_{j=1}^{M}\pi_{tj}\pi_{m|j} \end{aligned} \tag{7.10}$$

This is used to construct the likelihood function. (See Churchill, 1989, for a more complete description of these steps.)

This model will have $M\times(M-1)$ unknown parameters in the transition matrix (plus the parameters of the initial marginal distribution if the hidden chain is non-stationary), as well as M times the length of $\boldsymbol{\kappa}_m$ in the probability distributions. (More generally, $f(y_t|m;\boldsymbol{\kappa}_m)$ might be a completely different probability distribution in each state, with each $\boldsymbol{\kappa}_m$ a different length.)

Fitting a model

The probability of the observed response in Equation (7.10) is a form of mixture distribution. However, in contrast to the simple (static) mixture distributions often used, such as the finite mixtures in Section 3.4 or that yielding the negative binomial distribution in Section 4.3.2, this one becomes ever more complex at each time step, as the number of possible paths increases.

Evaluating the likelihood A direct evaluation of the probability of the responses in Equation (7.10) would involve about $2RM^R$ calculations for one series of length R. Fortunately, this probability can be rewritten in a recursive form over time involving about RM^2 calculations:

$$f(\mathbf{y};\boldsymbol{\kappa},\mathbf{T})=\boldsymbol{\pi}_0^\top\left(\prod_{t=1}^{R}\mathbf{T}\mathbf{F}_t\right)\mathbf{J} \tag{7.11}$$

where $\mathbf{F}_t$ is an $M\times M$ diagonal matrix containing, on the diagonal, the probabilities $f(y_t|\psi_t=m,\mathcal{F}_{t-1};\boldsymbol{\kappa}_m)$ of the observed responses, given the various possible states, and $\mathbf{J}$ is a column vector of 1s.

To calculate the likelihood function from this, the finite mixture distribution given by the hidden marginal probability times the observation probability for each state at time 1, say $a_m=\pi_{0m}f(y_1|m;\boldsymbol{\kappa}_m)$, is first calculated. At the second time point, the first step is to calculate the observation probability for each state multiplied by this quantity and by the hidden transition probabilities in the corresponding column of $\mathbf{T}$. These are summed yielding, say, $b_m = \sum_h a_h T_{hm} f(y_2|m; \boldsymbol{\kappa}_m)$. This is the new vector of forward recurrence probabilities. However, it should be divided by its average (to prevent underflow), yielding a new vector $\mathbf{a}$. This average is also cumulated as a correction to the likelihood function. These steps are repeated at each successive time point. Finally, the sum of these a_m at the last time point is the likelihood defined by Equation (7.11) except that it must be multiplied by the cumulative correction.

Maximising the likelihood Because the likelihood surface is complex, often with many local maxima, great care must be taken in exploring it. Thus, to obtain maximum likelihood estimates, various sets of initial values always should be tried. Unfortunately, no way has been found of guaranteeing that the global optimum has been attained.

A form of estimation-maximisation procedure, called the Baum–Welch algorithm (which preceded introduction of the EM algorithm), is often used to find the maximum likelihood estimates. Here, as for other models in this book, I shall used a numerical optimiser.

Optimal hidden state sequence At each step, the vector $\mathbf{a}$, normalised by dividing by its sum, gives the (filtered) conditional probabilities of being in the various possible hidden states, given the previous responses. The sequence of hidden states with maximum probability provides one solution to the problem of finding the optimal hidden state sequence. However, this does not take into account the dependence among the responses and may yield an impossible result. Thus, if certain transition probabilities are set to zero, some of the adjacent pairs of states in this sequence of highest probability may be impossible.

More sophisticated procedures use dynamic programming methods, generally the Viterbi algorithm (for details, see the references at the end of the chapter).

Continuous time

This model can be applied in continuous time by using a matrix of hidden transition intensities (Section 6.1.3), so that $\mathbf{\Lambda}$ has rows summing to zero (instead of one). Again, the matrix exponentiation of Equation (6.2) is used to obtain the transition probabilities $\mathbf{T}^{(\Delta t)} = \mathrm{e}^{\Delta t \mathbf{\Lambda}}$, where Δt is the time interval between observations.

Discretized hidden Poisson process

In Section 4.1.2, we saw that a homogeneous Poisson process can be discretized in time, in which case the numbers of recurrent events in each fixed time interval have a Poisson distribution. Let us now consider a further coarsening of the recorded responses, whereby we only have a binary indicator of whether or not one or more recurrent events occurred in each interval. Now, as above, assume that there is a different homogeneous Poisson process corresponding to each hidden state (Davison and Ramesh, 1996).

Again, we have a matrix $\mathbf{\Lambda}$ of transition intensities among the hidden states. Then, N_t will be the unobserved (hidden) counting process (Section 4.1.1) which is discretized, with intensity λ_m depending on the hidden state m. Thus, we have two sets of intensities: those between hidden states, and those for observable recurrent events, given the state. However, we only observe the binary time series, say y_i, indicating whether or not any recurrent events occurred in the interval from t_{i-1} to t_i, where the t_i are fixed time points.

Let $\mathbf{F}_t$ now be a diagonal matrix containing, not the probability distributions in given states as above, but the intensities λ_m corresponding to each hidden state. These may be a function of time and/or covariates, but are constant within each fixed time interval of observation. The probability transition matrix of moving between hidden states in time t is then

$$\mathbf{T}^{(t)} = \mathrm{e}^{t\mathbf{\Lambda}} \tag{7.12}$$

with that for no recurrent events ($y_t = 0$) occurring being

$$\mathbf{T}_0^{(t)} = \mathrm{e}^{t(\mathbf{\Lambda}-\mathbf{F}_t)} \tag{7.13}$$

using the Chapman–Kolmogorov equations (Section 5.1.1). Thus, the probability transition matrix for more than one recurrent event ($y_t = 1$) is

$$\mathbf{T}_1^{(t)} = \mathrm{e}^{t\mathbf{\Lambda}} - \mathrm{e}^{t(\mathbf{\Lambda}-\mathbf{F}_t)} \tag{7.14}$$

Finally,

$$f(\mathbf{y};\boldsymbol{\lambda},\mathbf{\Lambda}) = \boldsymbol{\pi}_0^\top \left(\prod_{t=1}^{R} \mathbf{T}_{y_t}^{(t)} \right) \mathbf{J} \tag{7.15}$$

gives the probability of the complete binary time series.

7.2.2 Clustered point process

One of the simplest applications of hidden Markov chains is to Bernoulli responses.

Locust activity Let us look at sequences of animal behaviour, that of locusts (*Locusta migratoria*), to determine whether or not they have spells of activity. Here, the investigation involves the effect of hunger on locomotory behaviour. To this end, 24 locusts, three days into the fifth larval stage, were placed individually in glass observation chambers. Even-numbered subjects were not fed for 5.5 h, whereas odd-numbered subjects received as much food as they could eat during this period. Within each group, individuals alternately were of the two sexes. During subsequent observation, neither food nor water was available. 161 observations, at 30 s intervals, were made on each animal. The data are given in MacDonald and Zucchini (1997, pp. 209–210). At each time point, every locust was classified either as locomoting (1) or not (0), the latter including quiescence and grooming.

I shall be interested in the evolution of locomotory behaviour over time, and whether it differs between the two treatment groups. There are 144 locomoting

Table 7.4. *AICs for models fitted to the locust data. The top panel contains models with the same intercept for all locusts and the middle panel has a different intercept for each treatment, whereas the bottom one has 24 different intercepts (for the hidden Markov models, these numbers are doubled, being different for each hidden state). 'Different trend' refers to treatment in the independence and Markov models and to hidden state in the hidden Markov models.*

	Independence	Markov	Hidden Markov
	Common intercept		
Null	2324.5	1869.0	1599.1
Trend	2263.2	1849.2	1591.6
	Treatment intercepts		
Null	2322.9	1869.2	1596.6
Same trend	2261.4	1849.1	1582.7
Different trend	2248.2	1844.4	1583.2
	Individual intercepts		
Null	1667.6	1577.6	1494.0
Same trend	1575.6	1521.6	1492.1
Different trend	1566.5	1513.5	1486.1

events in the fed group, but 973 in the unfed group, each in 1932 observation intervals. However, there is so much individual variability among the locusts that study of the differences between the two sexes will not be possible. I shall compare independence and ordinary Markov chain models (Chapter 5) with two-state hidden Markov models.

The results are summarised in Table 7.4. We see that a different intercept for each locust is required in all of the models. The ordinary Markov chain model is considerably worse than the hidden Markov model, although we shall see below that this must be nuanced. Allowing completely different parameters, that is, different transition matrices, in the hidden Markov model for the two treatment groups does not improve the model (AIC 1487.7). Thus, there is a clear time trend, different in the two treatment groups, but also state dependence.

Let us look more closely at the model with different hidden transition matrices. The slopes of the time trends, common to the two states, are estimated to be 0.0140 in the fed group and 0.0053 in the unfed group. Thus, locomotory behaviour increases faster in the former group, perhaps because it stays rather stable throughout the observation period for the unfed group. For the fed group, the probability of locomotion is small in both states, whereas, for the unfed group, there is a clear distinction between the two states, one of them indicating higher locomotory behaviour.

The two hidden transition matrices are estimated to be

$$\mathbf{T} = \begin{pmatrix} 0.978 & 0.022 \\ 0.073 & 0.927 \end{pmatrix}$$

and

$$\mathbf{T} = \begin{pmatrix} 0.980 & 0.020 \\ 0.028 & 0.972 \end{pmatrix}$$

in the fed and unfed groups, respectively, with corresponding stationary distributions (0.77,0.23) and (0.59,0.41). The large diagonal values indicate that the locusts tend to remain a relatively long time in the same state, a spell, inducing a dependence among consecutive observations. According to this model, the fed group stays about three-quarters of the time in the first state, whereas the unfed group spends about 60 per cent there.

The (filtered) conditional probabilities of being in the state with high probability of locomotion are plotted in Figure 7.5 for four selected locusts from each treatment group. We see the much higher probability of being in this state for unfed locusts. As with locust 1 in the fed group, many in this group did not move at all. On the other hand, locust 5 in the unfed group alternated frequently between locomoting and not. Although there is large variability among individual locusts, we have been able to detect substantial differences between the two treatment groups.

However, it turns out that the above model can be simplified further. For the fed group, a simple Markov chain fits better than the hidden one: the AIC for this group alone is 386.7 with 14 parameters as compared to 389.4 with 27 parameters for the hidden Markov model. Thus, with a Markov chain for the fed group and a hidden Markov chain for the unfed group, the AIC is 1485.0.

These data also can be interpreted as recording whether or not a locust made one or more movements in each observation interval so that a discretized Poisson process would be applicable. The null model with no covariates has a fit identical to the Bernoulli process. The other models give similar results to those above but fit slightly less well. The differences arise mainly from the Bernoulli processes using a logit link and the Poisson processes a log link.

A different type of application of hidden Markov models to point processes will be given in Section 8.5.2.

7.3 Overdispersed durations between recurrent events

We have studied recurrent events in some detail in Chapter 4. Here, I shall introduce some more complex models for such processes, using the dynamic approach.

7.3.1 Theory

Let us start with some arbitrary intensity function $\lambda(t)$ of interest for modelling a recurrent event process. From Equation (3.4), the corresponding integrated intensity will be

$$\Lambda(t) = \int_0^t \lambda(u) \mathrm{d}u$$

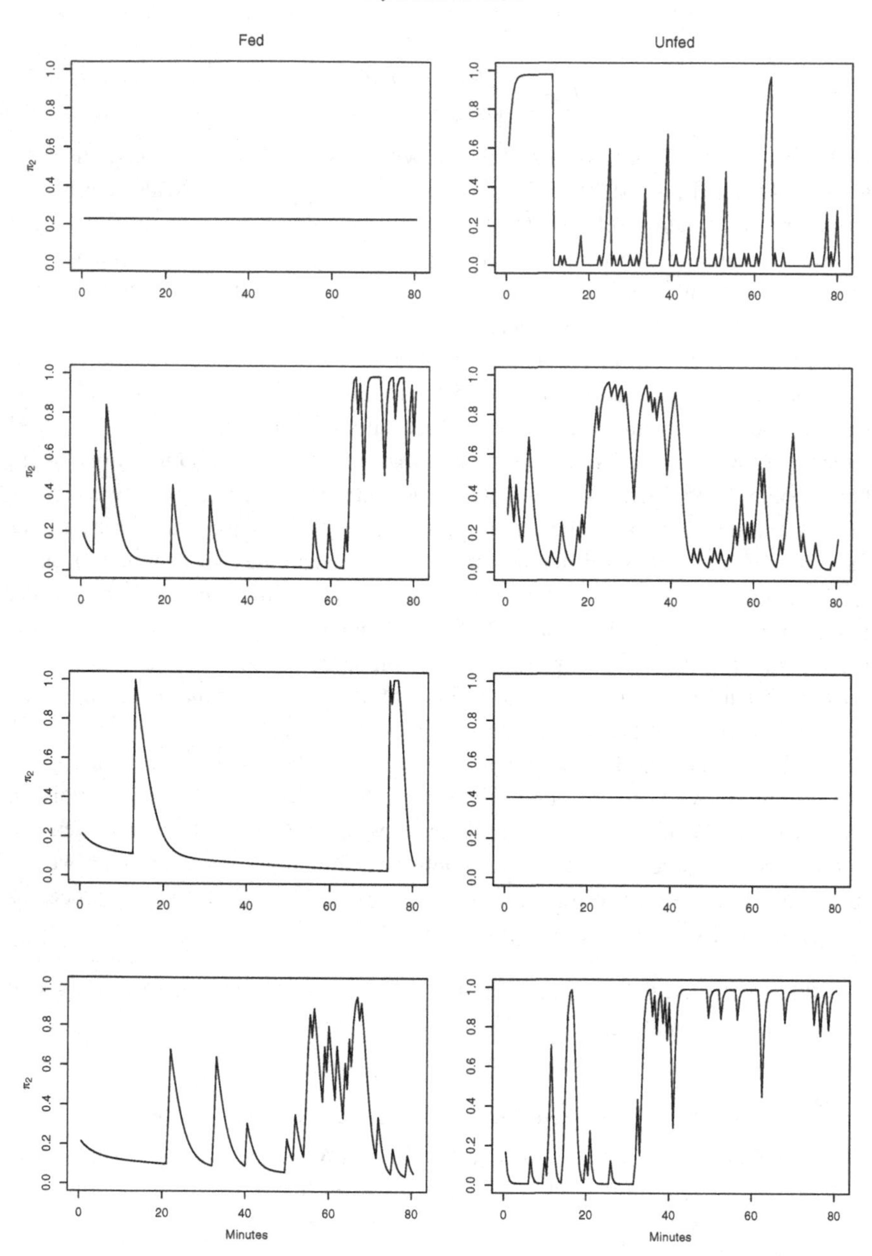

Fig. 7.5. The filtered probabilities of being in state 2, the state with generally higher probability of locomotion, for locusts 1, 2, 5, and 7 in each treatment group.

and the density and survivor functions

$$\begin{aligned} f_\lambda(t) &= \mathrm{e}^{-\Lambda(t)}\lambda(t) \\ S_\lambda(t) &= \mathrm{e}^{-\Lambda(t)} \end{aligned} \tag{7.16}$$

If $f(t)$ is a second, suitably chosen, density function, our goal will be to allow the unknown parameters in this latter density, somehow combined with $f_\lambda(t)$, to vary randomly over time in a dynamic model. In other words, the parameters in $f(t)$ will generate dependence structures over time, as a generalisation of the distribution $f_\lambda(t)$.

We can interpret $\Lambda(t)$ as a transformation of t, with $\lambda(t)$ its Jacobian. Let us apply this transformation to the responses in $f(t)$, in the same way that we could apply a logarithmic transformation to obtain a new distribution. Then,

$$\begin{aligned} f_m(t) &= f(\Lambda(t))\lambda(t) \\ &= f(\Lambda(t))\mathrm{e}^{\Lambda(t)} f_\lambda(t) \end{aligned} \tag{7.17}$$

will also be a probability density, the modification of the original density function $f_\lambda(t)$ that we are looking for. We see that $f(\Lambda(t))\mathrm{e}^{\Lambda(t)}$ modifies this function. The parameters in $f(t)$ will vary dynamically over time, so that the changes in the distribution $f_m(t)$ will primarily involve its *shape*.

Before proceeding to see how the parameters in $f(t)$ can vary dynamically over time, let us look in more detail at one especially simple choice for this distribution.

Gamma mixture

One particularly simple family of distributions $f_m(t)$ results when, for arbitrary $f_\lambda(t)$, we choose

$$f(t) = \frac{\alpha\beta^\alpha}{(\beta+t)^{\alpha+1}} \tag{7.18}$$

a special case of a Pareto distribution. Here, α and β will be the parameters that we shall (eventually) allow to change dynamically over time.

For any choice of $\lambda(t)$, this distribution yields

$$f_m(t) = \frac{\alpha\beta^\alpha}{[\beta+\Lambda(t)]^{\alpha+1}}\lambda(t) \tag{7.19}$$

The survivor function is

$$S_m(t) = \left[\frac{\beta}{\beta+\Lambda(t)}\right]^\alpha \tag{7.20}$$

so that it always will be explicitly available for any choice of $\Lambda(t)$. The corresponding intensity function is

$$\lambda_m(t) = \frac{\alpha\lambda(t)}{\beta+\Lambda(t)} \tag{7.21}$$

As well, if we take a suitable limit of Equation (7.19),

$$\lim_{\alpha=\beta\to\infty} f_m(t) = \mathrm{e}^{-\Lambda(t)}\lambda(t) \tag{7.22}$$

we have back the original probability density $f_\lambda(t)$, corresponding to the intensity function $\lambda(t)$.

Consider now a different justification for this particular choice of $f(t)$. Suppose that, instead of starting with a fixed intensity function $\lambda(t)$, we assume that the intensity is varying randomly (perhaps over time or over a set of different series). One way to do this is by introducing a random parameter, say ξ, which multiplies $\lambda(t)$, or equivalently $\Lambda(t)$. Now, let this parameter have a gamma distribution, as in Equation (2.7), but parametrised here as

$$p(\xi) = \frac{\beta^\alpha \xi^{\alpha-1} e^{-\beta\xi}}{\Gamma(\alpha)} \tag{7.23}$$

Now, the new survivor function will be given by $e^{-\Lambda(t)\xi}$, appropriately modified from Equation (7.16), multiplied by $p(\xi)$.

Because ξ is unobservable, it must be integrated out. Fortunately, this can be done explicitly:

$$\int_0^\infty e^{-\Lambda(t)\xi} \frac{\beta^\alpha \xi^{\alpha-1} e^{-\beta\xi}}{\Gamma(\alpha)} d\xi = \frac{\beta^\alpha}{[\beta + \Lambda(t)]^\alpha} \tag{7.24}$$

which is just the survivor function in Equation (7.20). As we have already seen in Section 4.3.2, an integral like this, modifying a density or survivor function, is called a *mixture distribution*. Then, Equation (7.22) can be interpreted as letting the variance of the gamma distribution go to zero while holding the mean equal to unity, so that the value $\xi = 1$ is known exactly and all responses have the same intensity $\lambda(t)$.

We saw in Equation (3.5) that an integrated intensity has a unit exponential distribution, so that its survivor function is $e^{-\Lambda(t)\xi}$, the modification of that in Equation (7.16) used above. The exponential distribution has, as its *conjugate*, the gamma distribution, which makes it possible to obtain this integral explicitly (see Section 14.1.2). Another explanation for this simple result is that Equation (7.24) is the Laplace transform $E(e^{-\Lambda(t)\xi})$ of a gamma distribution.

As we shall see, this choice of $f(t)$ in Equation (7.18) can yield useful results for any intensity function, interpretable in terms of a gamma mixing distribution.

Burr distribution As one possible choice for $f_\lambda(t)$, let us look at the Weibull distribution of Equation (2.15), with two parameters, and having the intensity function given in Equation (3.18),

$$\lambda(t) = \frac{\phi t^{\phi-1}}{\mu^\phi}$$

Then,

$$\Lambda(t) = \left(\frac{t}{\mu}\right)^\phi \tag{7.25}$$

The resulting probability density function will be

$$f_m(t) = \frac{\alpha\beta^\alpha \phi t^{\phi-1}}{\mu^\phi \left[\beta + \left(\frac{t}{\mu}\right)^\phi\right]^{\alpha+1}} \tag{7.26}$$

This is sometimes called a Burr distribution (Burr and Cislak, 1968). As might be expected, it is a form of Weibull mixture distribution, where the mixing distribution is a gamma distribution. The corresponding survivor function is

$$S_m(t) = \left[\frac{\beta}{\beta+\left(\frac{t}{\mu}\right)^{\phi}}\right]^{\alpha} \tag{7.27}$$

and the intensity function is

$$\lambda_m(t) = \frac{\alpha\phi t^{\phi-1}}{\mu^{\phi}\left[\beta+\left(\frac{t}{\mu}\right)^{\phi}\right]} \tag{7.28}$$

In contrast to the Weibull distribution with its monotone intensity (Section 4.4.1), for $\phi > 1$ this intensity function has a mode at $t = [\beta(\phi-1)]^{1/\phi}$. The mode of this density moves to the right and the tail becomes heavier as β increases relative to α. From Equation (7.22), as α and β increase together, $f_m(t)$ approaches a Weibull distribution.

Updating the parameters
A mixture distribution, such as that in Equation (7.24), yields a static model. To transform this into a dynamic model, we need to allow the parameters of $f(t)$ to vary randomly over time.

Thus, we must now explore how the intensities for successive durations in a stochastic process can be related through the parameters of $f(t)$. I shall continue with the Pareto distribution given in Equation (7.18), arising from a gamma mixing distribution. Thus, the parameters α and β will vary randomly over time in the sense that their present values will depend in some specific way on some (unobservable) function of the responses.

In this way, α and β will define the hidden states of the model. In a hidden Markov model, the distribution of responses differs in each hidden state; here, only certain parameters depend on the state. These parameters arise from the gamma mixing distribution, so that the *shape* of the mixture distribution is changing dynamically over time. This will be a form of *volatility model* (Chapter 10).

Frailty For α and β in Equation (7.19), particularly simple models can be constructed using the relationships

$$\begin{aligned} \alpha_{i+1} &= \alpha_i + 1 \\ \beta_{i+1} &= \beta_i + \Lambda(t_{i+1}) \end{aligned} \tag{7.29}$$

where i indexes the recurrent events and the initial values are set to $\alpha_0 = \beta_0 = \delta > 0$. Here, α_i counts the number of events whereas β_i cumulates the values of $\Lambda(t_i)$, the latter being larger for larger values of t_i.

By substitution of Equation (7.29), Equation (7.19) now becomes

$$f_m(t_i|t_0,\ldots,t_{i-1}) = \frac{\alpha_{i-1}\beta_{i-1}^{\alpha_{i-1}}}{\beta_i^{\alpha_i}}\lambda(t_i) \tag{7.30}$$

The complete multivariate distribution for a series of responses on an individual is built up in the usual way as an ordered product of the conditional distributions, at each point conditioning on the previous history. In this special case, successive ratios cancel, yielding a rather simple result:

$$\begin{aligned} f_m(t_1,\ldots,t_R) &= f_m(t_0)\prod_{i=1}^{R} f_m(t_i|t_0,\ldots,t_{i-1}) \\ &= \frac{\beta_0^{\alpha_0}}{\beta_R^{\alpha_R}}\prod \alpha_{i-1}\lambda(t_i) \end{aligned} \tag{7.31}$$

with the values at $t = 0$ suitably defined, as above.

Two special cases must be considered when analysing recurrent events. The discreteness of time in any empirical observation process implies that two or more events may occur in the same time interval. These ties can be handled by augmenting the counter accordingly:

$$\begin{aligned} \alpha_{i+1} &= \alpha_i + n_{i+1} \\ \beta_{i+1} &= \beta_i + \Lambda(t_{i+1}) \end{aligned} \tag{7.32}$$

where n_i is the number of ties in the interval i.

The second problem is right censoring. Generally, this will only occur for the last response in the series. Then, the event counter is not increased ($\alpha_i = \alpha_{i-1}$) and, as usual for duration distributions, the survivor function is used instead of the density:

$$S_m(t_i|t_1,\ldots,t_{i-1}) = \left[\frac{\beta_{i-1}}{\beta_i}\right]^{\alpha_i} \tag{7.33}$$

Because of the recursion in the updating equations, the above procedure allows all previous responses to influence the probability of the current one to the same extent. This is analogous to the classical normal distribution random effect model (see Chapter 14). However, here it is a random effect model with a gamma mixing distribution (for its application in the more classical context, see Section 14.4). Such models can be used to allow for unexplained (by covariates), time-constant heterogeneity among different series of recurrent events. In this context, in the medical literature, they are called *frailty* models; hence, here I shall call Equations (7.29) and (7.32) the *frailty update*.

Some of the same frailty models have been developed directly as random effect models, not as dynamic models. The Burr distribution is fairly well known because it has a tractable form. Other special cases include the repeated failure time model of Crowder (1985) and the frailty model of Aalen and Husebye (1991). Clayton (1988) uses a nonparametric intensity function to develop a proportional hazards frailty model.

Hidden Markov process Smith and Miller (1986) provide an interpretation of this model that is related more closely to that given for hidden Markov models above. They follow the same steps as in Section 7.2, taking ξ to be the random parameter and introducing two extra fixed parameters. For the moment, let us assume these to be set to unity: $\rho_1 = \rho_2 = 1$. Recall that here the random parameters, defining the hidden states, can take any positive value, in contrast to the finite set of hidden states of hidden Markov models.

Smith and Miller assume, as above, that the model of interest has intensity function $\lambda(t)$ multiplied by a random parameter ξ. In other words, the *observation or measurement equation* is here

$$f(t_i|\xi_i,\mathcal{F}_{i-1}) = \mathrm{e}^{-\Lambda(t_i)\xi_i}\lambda(t_i)\xi_i \tag{7.34}$$

where ξ_i is a random parameter having the gamma distribution of Equation (7.23):

$$p(\xi_i|\mathcal{F}_i) = \frac{\beta_i^{\alpha_i}\xi_i^{\alpha_i-1}\mathrm{e}^{-\beta_i\xi_i}}{\Gamma(\alpha_i)} \tag{7.35}$$

Then, they assume that this parameter ξ_i changes over time such that the ratio of successive values has a beta distribution with parameters $\rho_1\alpha_i$ and $(1-\rho_1)\alpha_i$, where α_i is the parameter of the above gamma distribution. This yields the *state transition equation*

$$p(\xi_{i+1}|\xi_i,\mathcal{F}_i) = \frac{\rho_2\Gamma(\alpha_i)(\rho_2\xi_{i+1})^{\rho_1\alpha_i-1}(\xi_i-\rho_2\xi_{i+1})^{(1-\rho_1)\alpha_i-1}}{\Gamma(\rho_1\alpha_i)\Gamma[(1-\rho_1)\alpha_i]\xi_i^{\alpha_i-1}} \tag{7.36}$$

The joint distribution $f(\xi_i,\xi_{i+1}|\mathcal{F}_i)$ can be obtained as the product of this equation and Equation (7.35). Integrating out ξ_i gives a gamma distribution for the *one-step-ahead prediction* or *time update*

$$p(\xi_{i+1}|\mathcal{F}_i) = \frac{(\rho_2\beta_i)^{\rho_1\alpha_i}\xi_{i+1}^{\rho_1\alpha_i-1}\mathrm{e}^{-\rho_2\beta_i\xi_{i+1}}}{\Gamma(\rho_1\alpha_i)} \tag{7.37}$$

Combining this with Equation (7.34) for t_{i+1} and applying Bayes' formula yields the *filtering* or *observation update*

$$p(\xi_{i+1}|\mathcal{F}_{i+1}) = \frac{[\rho_2\beta_i+\Lambda(t_{i+1})]^{\rho_1\alpha_i+1}\xi_{i+1}^{\rho_1\alpha_i}\mathrm{e}^{-[\rho_2\beta_i+\Lambda(t_{i+1})]\xi_{i+1}}}{\Gamma(\rho_1\alpha_i+1)} \tag{7.38}$$

Comparing this with Equation (7.35), we see that the updating equations are

$$\begin{aligned}\alpha_{i+1} &= \rho_1\alpha_i+1\\ \beta_{i+1} &= \rho_2\beta_i+\Lambda(t_{i+1})\end{aligned} \tag{7.39}$$

the same as Equations (7.29) when $\rho_1=\rho_2=1$.

The marginal distribution of t_i, given the previous history, $f(t_i|\mathcal{F}_{i-1})$, is obtained by combining Equations (7.34) and (7.37) (at observation i) and integrating out ξ_i. This yields Equation (7.19) from which the likelihood is constructed.

A more easily generalisable interpretation takes α_i and β_i, instead of ξ, as the random parameters. To show the Markov nature of the updating process for the hidden states, we can write Equation (7.29) in matrix form as

$$\boldsymbol{\psi}_{i+1}^\top = \boldsymbol{\psi}_i^\top\mathbf{T}+\mathbf{b}_{i+1}^\top+\boldsymbol{\epsilon}_{i+1}^\top \tag{7.40}$$

where $\boldsymbol{\psi}_i=(\alpha_i,\beta_i)^\top$ is the hidden state of the system at time t_i, $\mathbf{T}$ is an identity matrix, $\mathbf{b}_i=(1,0)^\top$ the deterministic input, and $\boldsymbol{\epsilon}_i=(0,\Lambda(t_i))^\top$ the random input. However, here $\boldsymbol{\psi}_i$ is not a vector of probabilities and, in general, $\mathbf{T}$ is not a stochastic matrix, so that this is not a Markov chain. (For further details about this representation, see Chapter 11 on classical dynamic linear models.)

Longitudinal dependence To model dependence in a stochastic process, we would prefer to have some way to allow the influence of previous states to diminish with their distance in time. This is the role that ρ_1 and ρ_2 above will play when they are no longer fixed at unity, but have $0 < \rho_1, \rho_2 < 1$. Thus, α and β will now be updated as in Equation (7.39), which can also be written in the form of Equation (7.40), but with

$$\mathbf{T} = \begin{pmatrix} \rho_1 & 0 \\ 0 & \rho_2 \end{pmatrix}$$

Often, the number of unknown parameters can be reduced by setting $\rho = \rho_1 = \rho_2$. This parameter has been called the discount because it reduces the influence of previous responses on the present probabilities. As above, I shall also set $\alpha_0 = \beta_0 = \delta$. The influence of previous values now reduces as ρ^i (recall that i is the number of events, not the time past). I shall call this the *Kalman update* because of its similarity to the classical dynamic linear models in Chapter 11. With this and the following updates, the product of conditional distributions no longer collapses to a simple form, as in Equation (7.31).

Further possibilities One modification to the previous relationships is not to discount the first update equation, that counting the recurrent events, so that $\rho_1 = 1$ and

$$\mathbf{T} = \begin{pmatrix} 1 & 0 \\ 0 & \rho \end{pmatrix}$$

or

$$\begin{aligned} \alpha_{i+1} &= \alpha_i + 1 \\ \beta_{i+1} &= \rho\beta_i + \Lambda(t_{i+1}) \end{aligned} \qquad (7.41)$$

I shall call this the *cumulated update*.

The following possibilities do not fall within the framework of the Smith and Miller (1986) interpretation described above, The first is to discount the event counting and not to update β_i from its previous value, with $\rho_2 = 0$:

$$\begin{aligned} \alpha_{i+1} &= \rho\alpha_i + 1 \\ \beta_{i+1} &= \delta + \Lambda(t_{i+1}) \end{aligned} \qquad (7.42)$$

Here,

$$\mathbf{T} = \begin{pmatrix} \rho & 0 \\ 0 & 0 \end{pmatrix}$$

and $\mathbf{b}_i = (1, \delta)^\top$, where δ is the initial value. I shall call this the *count update*. Here, the probability of the current response does not depend on the previous values, but only on the number of previous recurrent events. Hence, as the number of observations on a individual accumulates, α_i increases relative to β_i so that the probability of large values tends to diminish.

One major handicap of all of the previous updating procedures is that $\rho = 0$ does not imply independence. One possible alternative is to discount only the counter and $\Lambda(t_i)$, respectively, in the two update equations, and not the initial value, δ:

$$\begin{aligned} \alpha_{i+1} &= \rho\alpha_i + (1-\rho)\delta + 1 \\ \beta_{i+1} &= \rho\beta_i + (1-\rho)\delta + \Lambda(t_{i+1}) \end{aligned} \tag{7.43}$$

so that

$$\mathbf{T} = \begin{pmatrix} \rho & 0 \\ 0 & \rho \end{pmatrix}$$

and $\mathbf{b}_i = (1+(1-\rho)\delta, (1-\rho)\delta)^\top$. I shall call this the *event update* because ρ^i decreases with the number of previous events. With this method of updating, $\rho = 0$ indicates independence between successive responses.

An extension is to allow the discount in the previous update to decrease as a function of the time elapsed since the previous event:

$$\begin{aligned} \alpha_{i+1} &= \rho^{t_{i+1}}\alpha_i + (1-\rho^{t_{i+1}})\delta + 1 \\ \beta_{i+1} &= \rho^{t_{i+1}}\beta_i + (1-\rho^{t_{i+1}})\delta + \Lambda(t_{i+1}) \end{aligned} \tag{7.44}$$

I shall call this the *serial update* because ρ acts in a way more similar to serial dependence models (Sections 1.2.4 and 7.1). Again, $\rho = 0$ indicates independence.

One undesirable characteristic of these update procedures is that they produce nonstationary models. However, the last two update procedures easily can be modified to produce a model closer to first-order Markov. Thus, the serial update could become

$$\begin{aligned} \alpha_{i+1} &= \rho^{t_{i+1}}\alpha_i + (1-\rho^{t_{i+1}})\delta + 1 \\ \beta_{i+1} &= \delta + \rho^{t_{i+1}}\Lambda(t_i) + \Lambda(t_{i+1}) \end{aligned} \tag{7.45}$$

with

$$\mathbf{T} = \begin{pmatrix} \rho^{t_{i+1}} & 0 \\ 0 & 0 \end{pmatrix}$$

$\mathbf{b}_i = (1+(1-\rho^{t_i})\delta, \delta)^\top$, and $\boldsymbol{\epsilon}_i = (0, \Lambda(t_i) + \rho^{t_i}\Lambda(t_{i-1}))^\top$. I shall call this the *Markov update*.

None of the proposed update procedures is ideal. The dependence (discount) parameter in the Kalman and cumulated updates is difficult to interpret because $\rho = 0$ does not yield independence. With the count update, the present risk of an event is not modified by the lengths of times between previous events, but only by the number of previous events. This is similar to a birth process (Section 4.1.1). As well, most of the models are nonstationary.

7.3.2 Frailty

Clinical trials may involve series of observations on subjects who are rather heterogeneous, with inadequate covariates to account for these differences. Then, a frailty model may be appropriate.

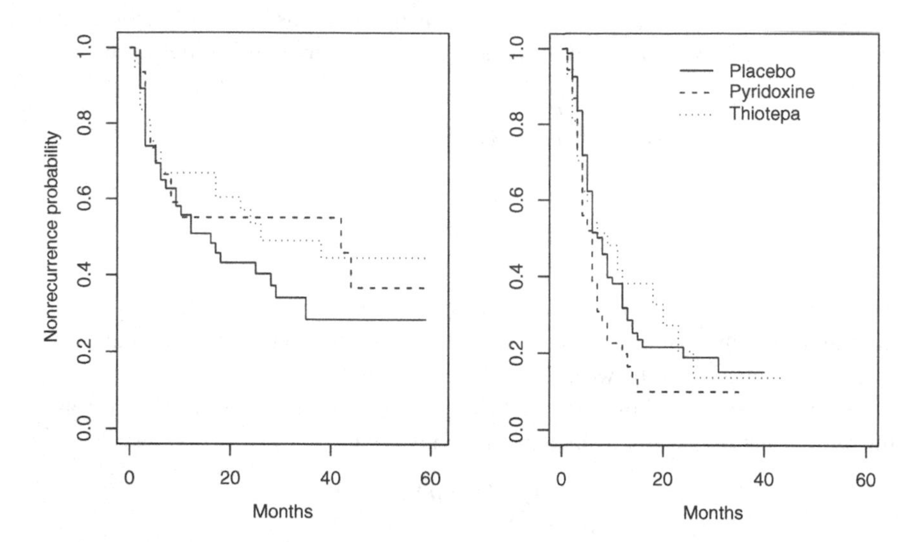

Fig. 7.6. The Kaplan–Meier curves for the first (left) and subsequent (right) recurrences of bladder cancer tumours.

Bladder cancer In a clinical trial conducted by the Veterans Administration Cooperative Urological Research Group in the USA, 116 patients with bladder cancer were randomised to one of three treatment groups: placebo, pyridoxine (vitamin B_6) pills, or periodic instillation of a chemotherapeutic agent, thiotepa, into the bladder (Andrews and Herzberg, 1985, pp. 254–259). When they entered the trial, all patients had superficial bladder tumours that were removed transurethrally. At each discovery of one or more new tumours at a visit, these were also removed. These times, in months, between recurrences of tumours were recorded for each patient. Thus, times are between visits when one or several tumours were detected, not between individual tumours. One might expect that the risk of further tumours might change depending on the specific history of each patient.

The Kaplan–Meier curves are shown in Figure 7.6, separately for the first and for subsequent recurrence times. Let us start with some simple models from Section 4.1.4. A Weibull renewal process has an AIC of 747.0 without treatment differences and 746.6 with them. The corresponding values for a renewal process based on the overdispersed gamma mixture, a Burr renewal process, are 698.5 and 698.9, respectively. However, neither the Kaplan–Meier curves nor these models take into account the individuality of the patients.

The frailty model of Equation (7.31) is meant to allow for individual differences among the patients. However, the AICs are 707.4 and 708.4, respectively, with the same number of parameters as the Burr renewal process. Thus, according to these models, the improvement comes from the overdispersion, not the heterogeneity among patients.

If I replace the Weibull distribution by the log normal as my choice for $f_\lambda(t)$,

Table 7.5. *AICs for various models fitted to the bladder tumour data of Andrews and Herzberg (1985, pp. 254–259).*

	Independent	Gamma–Weibull mixture			
	Weibull	Independence	Frailty	Markov	Count
Null	747.0	698.5	707.4	697.5	685.8
Treatment	746.6	698.9	708.6	699.2	685.9
Treatment + Birth	725.3	699.5	708.4	700.1	685.1
	Independent	Gamma–log normal mixture			
	log normal	Independence	Frailty	Markov	Count
Null	719.5	697.9	689.5	685.7	684.5
Treatment	720.4	698.0	691.2	686.5	684.6
Treatment + Birth	710.5	698.7	690.4	685.5	682.5

the fit improves. Thus, for the log normal renewal process, the respective AICs are 719.5 and 720.4. For the gamma–log normal mixture, they are 697.9 and 698.0; whereas, here, for the frailty model, they are 689.5 and 691.2. This now indicates that a model with heterogeneity among patients fits better. On the hand, there is no indication of difference due to treatment. These results are summarised in the first three columns of Table 7.5.

I shall continue with this example in the next section, and return to the application of frailty models briefly in Section 14.4.

7.3.3 Longitudinal dependence

When longitudinal measurements are made in a clinical trial, frailty models may be necessary, but only rarely do they prove to be adequate. The longitudinal dependencies among responses must (also) be taken into account.

Bladder cancer Let us now compare some models for longitudinal dependence with those for frailty above; the best fitting ones are presented in the last two columns of Table 7.5. Again, we see that the log normal intensities fit considerably better than the Weibull. In no case is there evidence of a treatment effect. On the other hand, these models clearly fit better than the frailty ones. For these responses, the dependence is best modelled by the nonstationary count update of Equations (7.42) rather than, say, by a Markov update from Equations (7.45).

One further possibility to allow for time dependence among the responses is a birth process (Section 4.1.1). This will have μ_t, a location parameter, depending directly on the number of previous recurrent events, in contrast to the updates in which the value of the parameter α, a dispersion parameter, depends on the number of previous events. Here, such birth models may be reasonable because there is a clear time origin when no tumours were present.

The results also are included in the table. The best model has the count update

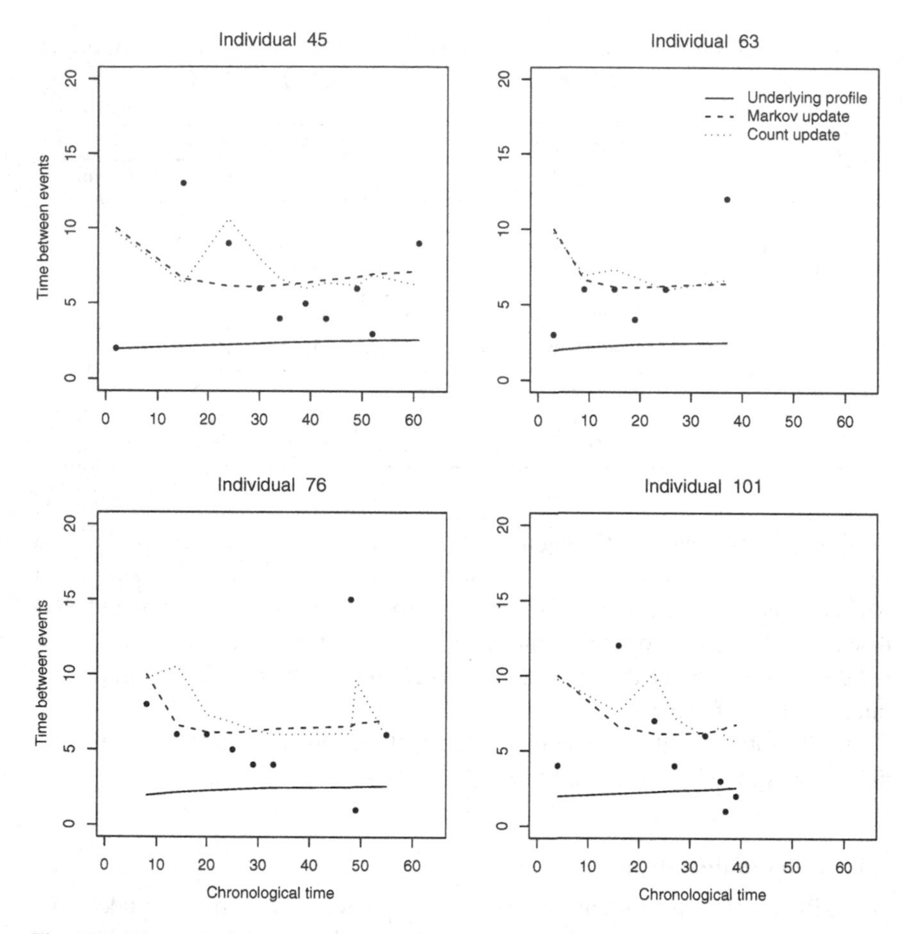

Fig. 7.7. The underlying marginal and individual profiles from the log normal mixture model with count and Markov updates and a birth process for selected individuals with recurrences of bladder cancer tumours.

with birth effect in a log normal intensity and gamma mixing distribution. However, the treatment effect is not required: when it is removed, the AIC is 682.6. It is interesting that the number of previous events appears in the model in two distinct ways: both as a birth model, changing the location parameter, and in the count update equation, changing the dispersion.

Profile plots for four selected subjects with high numbers of events are shown in Figure 7.7. We can see how the Markov update tries to adjust for the longer periods between recurrent events and often fails because the following period is much shorter.

7.4 Overdispersed series of counts

We saw in Equation (7.32) that several recurrent events might occur between observation points. Suppose now that the study is, in fact, designed in this way:

observations only are made at fixed time points, the numbers of recurrent events in these fixed time intervals being recorded. I analysed point processes in this way in Section 4.3. The dynamic models of the previous section also can be modified to accommodate this type of study.

7.4.1 Theory

As we saw in Section 4.3.2, if the intensity parameter λ of the Poisson distribution of Equation (3.26) is allowed to vary stochastically following the gamma distribution as parametrised in Equation (7.23), then the resulting mixture distribution is negative binomial with probability

$$\Pr(N=n;\alpha,\beta)=\frac{\Gamma(\alpha+n)}{\Gamma(\alpha)n!}\left(\frac{\Delta t}{\beta+\Delta t}\right)^{n}\left(\frac{\beta}{\beta+\Delta t}\right)^{\alpha} \tag{7.46}$$

a slightly different parametrisation than that in Equation (4.36). Then, if a sequence of overdispersed counts follows a negative binomial process, the probability of no events in an observation interval Δt is given by

$$\begin{aligned}\Pr(N=0;\alpha,\beta) &= \left(\tfrac{\beta}{\beta+\Delta t}\right)^{\alpha} \\ &= \Pr(T>\Delta t;\alpha,\beta)\end{aligned} \tag{7.47}$$

This survivor function is identical to that arising from the special case of the Pareto distribution given above in Equation (7.20). The average intensity over the interval Δt is α/β.

The same type of reasoning as in Section 7.3.1, but with Δt fixed, here gives, from Equation (7.46),

$$\Pr(n_i|n_1,\cdots,n_{i-1}) = \frac{\Gamma(\alpha_i)}{\Gamma(\alpha_{i-1})n_i!}\frac{\Lambda(\Delta t_i)^{n_i}\beta_{i-1}^{\alpha_{i-1}}}{\beta_i^{\alpha_i}} \tag{7.48}$$

the equivalent of the frailty update model of Equation (7.30) with multivariate distribution corresponding to Equation (7.31):

$$\Pr(n_1,\cdots,n_R) = \frac{\Gamma(\alpha_R)\beta_0^{\alpha_0}}{\Gamma(\delta)\beta_R^{\alpha_R}}\prod_{i=1}^{R}\frac{\Lambda(\Delta t_i)^{n_i}}{n_i!} \tag{7.49}$$

a form of negative multinomial distribution. As in Section 7.3.1, other updating equations also can be used. As for those models, the hidden states are defined by the random parameters α_i and β_i.

These models have certain limitations. When any longitudinal dependence update is used, a time origin must be specified. In addition, the dispersion parameter is used to construct the dependence in these models, so that it cannot depend directly on covariates, such as time, as was done in Section 7.1.4 above.

University enrolment As we saw in Section 7.1.4, the counts of student enrolment at Yale University from 1796 to 1975 have dispersion changing over time. Let us attempt to apply these models with dynamically changing dispersion. In them, the time origin is important; here, I shall set it to 1795.

Again, the count update of Equation (7.42) provides the best fit, with an AIC of 1244.6, slightly better than the negative binomial serial dependence model of Section 7.1. The recursive predictions are plotted in the bottom graph of Figure 7.3, very similar to the corresponding graph for serial dependence (but note that the time origin is set to zero). If the location parameter is allowed to depend exponentially on time, the AIC is only reduced to 1243.0.

However, the model fitted above in Section 7.1.4, with dispersion changing systematically over time, had an AIC of 1158.4. Thus, we may conclude that, for these data, this is a better model than one with dynamic changes.

As mentioned above, the dispersion parameter cannot be made to depend directly on time as it can in serial dependence models. Thus, these models are competitive with the latter models only if the dispersion is changing randomly, and not systematically.

Further reading

Durbin and Koopman (2001), Harvey (1989), and West and Harrison (1989) provide general presentations of dynamic models. The serial dependence and dynamic gamma mixture models are developed in Lindsey (1999a). Karr (1991) provides information, scattered throughout the text, about dynamic models for point processes.

For an introduction to hidden Markov models, see Guttorp (1995) or MacDonald and Zucchini (1997). Elliott *et al.* (1995) provide a more advanced presentation from an engineering point of view. These models are being used widely, for example in speech processing (Rabiner, 1989; Juang and Rabiner, 1991), molecular biological sequence analysis of nucleic acids in DNA and of amino acids in proteins (Churchill, 1989; Durbin *et al.*, 1998), ion channel kinetics in physiology (Kein *et al.*, 1997), econometrics (Hamilton, 1990), and panel studies in the social sciences (Bye and Schechter, 1986).

Exercises

7.1 Can you find some better nonlinear regression function to replace the exponential curve for the Yale enrolment data?

7.2 Obtain the enrolment history of your university, if possible since it was founded.

(a) Develop an appropriate model for your data.
(b) Are there any special historical events that influenced it, like the two world wars for Yale?
(c) Explain the differences between your model and those used for Yale above.

7.3 An endangered bird that has received much attention in North America is the whooping crane (*Grus americana*). These are large birds about 1.5 m tall with a 2.1 m wingspan and a relatively long life. They only reproduce

Table 7.6. *The whooping crane (*Grus americana*) population in Aransas, Texas, USA, from 1938 to 1989. Seshradi (1999, pp. 296–297; read across rows)*

18	22	26	16	19	21	18	22	25	31	30	34	31	25
21	24	21	28	24	26	32	33	36	39	32	33	42	44
43	48	50	56	57	59	51	49	49	57	69	72	75	76
78	73	73	75	86	97	110	134	138	146				

Table 7.7. *Monthly numbers of male deaths from bronchitis, emphysema, and asthma in the UK, 1974–1979. (Diggle, 1990, p. 238, from Appleton)*

Month	1974	1975	1976	1977	1978	1979
J	2134	2103	2020	2240	2019	2263
F	1863	2137	2750	1634	2284	1820
M	1877	2153	2283	1722	1942	1846
A	1877	1833	1479	1801	1423	1531
M	1492	1403	1189	1246	1340	1215
J	1249	1288	1160	1162	1187	1075
J	1280	1186	1113	1087	1098	1056
A	1131	1133	970	1013	1004	975
S	1209	1053	999	959	970	940
O	1492	1347	1208	1179	1140	1081
N	1621	1545	1467	1229	1110	1294
D	1846	2066	2059	1655	1812	1341

from age five, usually laying two eggs a year, of which only one often survives. They are a migratory species that winters in the southern USA. The winter population in Aransas National Wildlife Refuge in Texas, USA, has been followed closely since 1938. The numbers until 1989 are shown in Table 7.6.

(a) Develop a model to describe these data.

(b) What is your prediction for the population size in 2000? Find the observed size to check your result.

7.4 The reported total numbers of male deaths from bronchitis, emphysema, and asthma in the UK each month from 1974 to 1979 are presented in Table 7.7. Notice the particularly high values in the winter of 1975–1976.

(a) Fit a serial dependence model to these data. Clearly, you will require a seasonal effect.

(b) Is there any evidence of an increasing or decreasing trend over time?

(c) Give reasons why the dynamic overdispersed count models of Section 7.4 cannot be applied to these data.

7.5 In a study of fœtal lamb activity during the last two-thirds of gestation, body movements were recorded using ultrasound. The amount of activity

Table 7.8. *Numbers of movements of a fœtal lamb in 240 five-second periods. (Leroux and Puterman, 1992; read across rows)*

0 0 0 0 0 1 0 1 0 0 0 0 0 0 1 0 1 0 0 0 0 2 2 0 0 0 0 1 0 0 1 1 0 0 1
1 1 0 0 1 0 0 0 0 0 0 0 0 0 0 0 0 0 0 0 0 0 0 2 0 1 0 0 0 0 0 0 0 0 0
0 0 0 0 0 0 0 0 1 0 0 0 0 0 7 3 2 3 2 4 0 0 0 0 1 0 0 0 0 0 0 0 1 0 2
0 0 0 0 1 0 0 0 0 1 0 0 0 0 0 0 0 0 0 0 0 0 0 0 1 0 0 0 0 0 2 1 0 0 1
0 0 0 1 0 1 1 0 0 0 1 0 0 1 0 0 0 1 2 0 0 0 1 0 1 1 0 1 0 0 2 0 1 2 1
1 2 1 0 1 1 0 0 1 1 0 0 0 1 1 1 0 4 0 0 2 0 0 0 0 0 0 0 0 0 0 0 0 0 0
0 0 0 0 0 0 0 0 0 0 0 0 0 0 0 0 0 0 0 0 0 0 0 0 0 0 0 0 0 0

may change because of physical factors, such as reduction in the volume of amniotic fluid and space in the uterus, or through development of the central nervous system. The record for a series of 240 consecutive five-second intervals is shown in Table 7.8. For long-term development of the lamb, several such series spread over a longer time period would be required.

(a) Plot the data in appropriate ways.
(b) Is there evidence of clustering of recurrent events?

7.6 Apply the dynamic overdispersed duration models of Section 7.3 to the times between vehicles passing in Table 4.1.

(a) Do the models indicate dependence in the data?
(b) If so, which update performs best?
(c) There appears to be clustering in the graphs of Figure 4.8. Hidden Markov models can be applied either to the interarrival times or to the counts of events. Do either of these provide any additional information about these data?

7.7 Repeat the preceding exercise for the mine accident data in Table 4.3.

7.8 Check for heterogeneity ('frailty') among series for the following data:

(a) The replacement of valve seats in diesel engines in Table 4.7;
(b) The migrating motor complex phase in fasting individuals in Table 4.8;
(c) The failures of air-conditioning equipment in aircraft in Table 4.9;
(d) The recurrence of infections in kidney patients in Table 4.10.

8

More complex dependencies

In previous chapters, we have studied series of events occurring to one or more individuals observed over time. In Chapters 2 and 3, I looked at simple durations (survival times) where only one transition event to an absorbing state occurs. In Chapter 4, I extended this to series of duration times between recurrent events. In Chapters 5 and 6, each event resulted in a switch from one of a relatively limited number of states to another, with possible dependence on the previous state(s). We also have seen that dependence may be induced indirectly by postulating the existence of hidden states influencing the process, as in Chapter 7. It is now time to consider a few other, more complex, possibilities.

8.1 Birth processes

In Chapter 5 and Section 6.1.3, I looked at state dependence for categorical state spaces. This gives rise to Markov chain models. Another simple way to introduce dependence for such state spaces is by means of some form of birth process (Section 1.2.5; see also the example in Section 7.3.3). Instead of conditioning on the previous state(s), one conditions on the number(s) of previous times the series was in some given state(s). For binary point processes, this is just the number of previous occurrences of the recurrent event. In contrast to Markov chains, this will necessarily result in a nonstationary process.

8.1.1 Birth or contagion

Because birth or contagion processes are nonstationary, special care must be taken in choosing the time origin. It is important to begin observation before the first event occurs or at the point in time when it happens, or at least to know how many recurrent events previously had occurred in the process before recording of new events begins. Otherwise, the number of previous events usually is not well defined and the model may be meaningless. When modelling intensities in continuous time using Equation (4.8), it generally is most sensible to begin at the first event, because before that time $N_t = 0$ implying that $\lambda(t|N_t)$ is also zero. On the other hand, this is not necessary if Equation (4.26) is used.

Spermarche in boys A seven-year longitudinal study of spermarche in boys was begun in the spring of 1975 in Edinburgh, Scotland. Two classes of a municipal primary school participated; 40 of 42 boys and their parents gave informed consent. Approximately every three months, a 24 h urine sample was collected by each boy. These were analysed for the presence of sperm cells with results as shown in Table 8.1.

The ages of entry to and exit from the study varied. Entry age always was well before the first recurrent event so that no problem arises in counting the number of events. There is also no reason to suspect that the decision to withdraw was influenced by the sperm content of the urine so that it should be independent of the process under study. The analyses often were performed quite some time after the sample was taken and, in any case, the results were not revealed to the boys.

Age of entry might be used as a covariate in the models. Other possible covariates are time and the age of first sperm. We would expect to use the latter as an interaction with other covariates because the stochastic process should change once sperm start to appear.

Sperm samples were taken approximately every three to four months, but the exact times were not published. Hence, I have divided up the time between age of entry and of exit into equal periods. (This could have been made slightly more precise by also using the ages of first sperm, which were published.) The cumulated numbers of positive sperm samples n_{it} are plotted as counting processes in Figure 8.1. We see from Table 8.1 and this figure that sperm first appear between ages of about 12 and almost 16 years. Using a birth model, we can study how the probability of a positive sample increases with the number of previous such samples.

Here, I shall modify Equation (1.9) in order to use logistic regression:

$$\log\left(\frac{\pi_{it}}{1-\pi_{it}}\right) = \rho n_{it} + \boldsymbol{\beta}^\top \mathbf{x}_{it} \tag{8.1}$$

where π_{it} is the probability of a positive sample for boy i (being counted) at time t. In this formulation, n_{it} may be zero.

The null model, with no covariates, has an AIC of 439.7. Introduction of the birth process reduces it to 382.0. However, a simple time trend, by itself, gives 362.0 and the birth process is no longer necessary. Nevertheless, if we add an interaction between the two, the AIC is 336.5. We have not yet directly taken into account the age of first sperm. When this is used as an interaction with time (without the birth process), we discover that time is necessary only before first sperm. The AIC is 324.5. Reintroducing the birth process, we obtain a final AIC of 322.1. The estimated regression function is

$$\log\left(\frac{\pi_{it}}{1-\pi_{it}}\right) = -17.2 + 0.093 n_{it} + 16.7 I(n_{it} > 0) + 1.14 t I(n_{it} = 0)$$

where $I(\cdot)$ is the indicator function. Thus, the third term acts only after the sperm first appear and the fourth (time) only before. The probability increases with age, given by the coefficient of $tI(N_{it} = 0)$, until the first positive sample. It, then,

Table 8.1. *Presence of sperm cells in the urine samples of boys. (Jørgensen* et al., *1991)*

Age			
Entry	First sperm	Exit	
10.3	13.4	16.7	0 0 0 0 0 0 0 0 0 0 1 1 0 0 0 0 1 1 1 0 0
10.0	12.1	17.0	0 0 0 0 0 0 0 0 1 0 0 1 1 0 1 0 0 1 0 1 0 0 0 0 0 1 1
9.8	12.1	16.4	0 0 0 0 0 0 0 0 1 0 1 1 0 1 1 1 1 1 1 0 0 1 1 0 1
10.6	13.5	17.7	0 0 0 0 0 0 0 0 0 0 0 1 1 0 0 0 1 0 0 0 0
9.3	12.5	16.3	0 0 0 0 0 0 0 0 0 0 0 0 1 1 0 0 0 0 1 0 0 0 0 0 0 0 0
9.2	13.9	16.2	0 0 0 0 0 0 0 0 0 0 0 0 0 0 0 0 0 1 0 0 0 0 0 0 0
9.6	15.1	16.7	0 0 0 0 0 0 0 0 0 0 0 0 0 0 0 0 0 0 0 1 0 0 0 1
9.2	—	12.2	0 0 0 0 0 0 0 0 0 0 0 0
9.7	—	12.1	0 0 0 0 0 0 0 0 0
9.6	12.7	16.4	0 0 0 0 0 0 0 0 0 0 0 0 1 0 1 1 1 1 1 0 0 1 1 0 1
9.6	12.5	16.7	0 0 0 0 0 0 0 0 0 0 1 0 0 1 0 1 0 0 1 1 1
9.3	15.7	16.0	0 0 0 0 0 0 0 0 0 0 0 0 0 0 0 0 0 0 0 0 0 0 1 1
9.6	—	12.0	0 0 0 0 0 0 0 0 0
9.4	12.6	13.1	0 0 0 0 0 0 0 0 0 0 1 1 1 1
10.5	12.6	17.5	0 0 0 0 0 0 0 1 0 1 1 1 1 1 1 1 1 0 0 1 0 0 1 1
10.5	13.5	14.1	0 0 0 0 0 0 0 0 0 0 1 0 0
9.9	14.3	16.8	0 0 0 0 0 0 0 0 0 0 0 0 0 0 0 1 0 0 0 0 0 1 0 1
9.3	15.3	16.2	0 0 0 0 0 0 0 0 0 0 0 0 0 0 0 0 0 0 0 0 0 1 1 1
10.4	13.5	17.3	0 0 0 0 0 0 0 0 1 1 0 1 0 1 1 0 1 0 1 1 1
9.8	12.9	16.7	0 0 0 0 0 0 0 0 0 0 0 1 1 1 1 0 1 1 1 1 0 1 1 0 1 0 0
10.8	14.2	17.3	0 0 0 0 0 0 0 0 0 0 0 0 1 0 0 1 1 1 0 1
10.9	13.3	17.8	0 0 0 0 0 0 0 0 1 1 1 1 0 1 1 1 1 1 0 1 1 0 0
10.6	—	13.8	0 0 0 0 0 0 0 0 0 0 0
10.6	14.3	16.3	0 0 0 0 0 0 0 0 0 0 0 0 0 1 0 0 0 1 0 0 0
10.5	12.9	17.4	0 0 0 0 0 0 0 0 1 0 1 1 1 1 0 0 0 1 1 0 0 1 1 1 1
11.0	—	12.4	0 0 0 0 0 0
8.7	—	12.3	0 0 0 0 0 0 0 0 0 0 0 0 0 0
10.9	—	14.5	0 0 0 0 0 0 0 0 0 0 0 0 0
11.0	14.6	17.5	0 0 0 0 0 0 0 0 0 0 0 0 1 1 1 1 1 1 1 1 1 1 0 1
10.8	14.1	17.6	0 0 0 0 0 0 0 0 0 0 0 1 1 0 0 1 0 0 0 0 0 0
11.3	14.4	18.2	0 0 0 0 0 0 0 0 0 0 0 1 1 0 1 1 0 0 1 0 0 0 0 0
11.4	13.8	18.3	0 0 0 0 0 0 0 1 0 0 0 1 0 0 0 1 1 1 0 0 1 0 1
11.3	13.7	17.8	0 0 0 0 0 0 0 1 1 1 0 1 0 0 0 1 1 1 0 1 1
11.2	13.5	15.7	0 0 0 0 0 0 0 0 0 1 0 0 0 0 0 0 0 0
11.3	14.5	16.3	0 0 0 0 0 0 0 0 0 0 0 1 0 1 1 0 0 0
11.2	14.3	17.2	0 0 0 0 0 0 0 0 0 0 0 1 0 0 1 0 1 1 1 1 1 1 0
11.6	13.9	14.7	0 0 0 0 0 1 0 0 0
11.8	14.1	17.9	0 0 0 0 1 0 1 0 1 0 1 1 1 1 0 0 0 0
11.4	13.3	18.2	0 0 0 0 1 1 1 0 1 0 0 0 0 0 1 1 1 1 1 0 0
11.5	14.0	17.9	0 0 0 0 0 0 0 1 1 0 0 0 0 0 0 0 1 1 0 1 0

jumps to a much higher probability, given by the coefficient of $I(N_{it} > 0)$ and continues to increase with the number of previous positive samples, the coefficient of the birth process n_{it}, but no longer directly with time t.

This model makes biological sense. However, it was derived empirically from these data, so that it will require checking with other data.

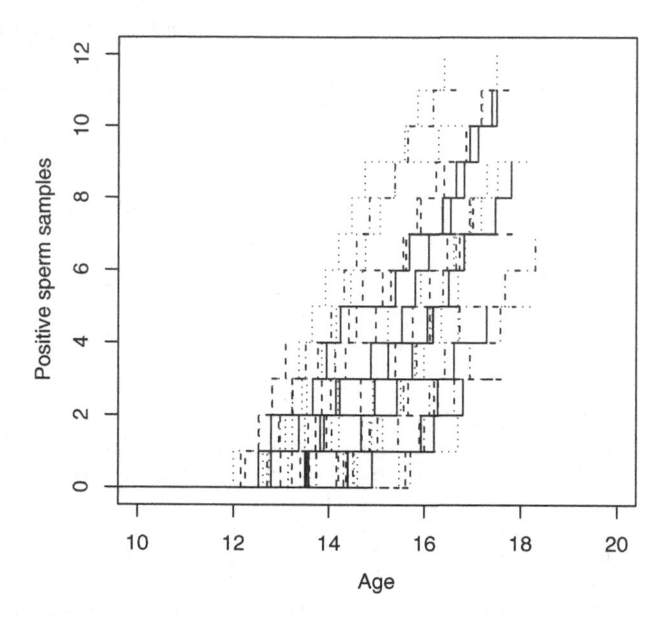

Fig. 8.1. The cumulated numbers of positive sperm samples in young boys, from Table 8.1.

8.1.2 Learning models

Birth models are most appropriate for recurrent series of events. However, in some circumstances, they also make sense when more than one type of event is possible. A generalisation uses the cumulative counts of each of several types of previous events. Conditioning on these may provide an indication of how well an individual learns. Thus, studying such a process may indicate which types of event experiences most influence present behaviour. The birth or contagion process of the previous section is the special case where only one type of event is being recorded and its probability depends on cumulative number of previous such events. In contrast, here the models will count separately the numbers of several different kinds of recurring events.

Dogs learning In the Solomon–Wynne experiment, dogs learned to avoid a shock. Each dog was kept in a compartment with a floor through which an intense shock could be applied. The lights were turned out and a barrier raised; ten seconds later, the shock occurred. Thus, the dog had ten seconds, after the lights went out, to jump the barrier and avoid the shock. Each of 30 dogs was subjected to 25 such trials. The results, in Table 8.2, were recorded as a shock trial ($y_{ik} = 0$), when the dog remained in the compartment, or an avoidance trial ($y_{ik} = 1$), when it jumped out before the shock. The cumulated numbers of avoidances are plotted as counting processes in Figure 8.2.

I shall assume that a dog learns to avoid shocks by its experience from previ-

Table 8.2. *The Solomon–Wynne dog experiment with 25 trials for each dog on a line. Avoidance of shock is indicated by 1. (Kalbfleisch, 1985, pp. 83–88)*

00101	01111	11111	11111	11111
00000	00100	00001	11111	11111
00000	11011	00110	10111	11111
01100	11110	10101	11111	11111
00000	00011	11111	11111	11111
00000	01111	00101	11111	11111
00000	10000	00111	11111	11111
00000	00110	01111	11111	11111
00000	10101	10100	01111	10110
00001	00110	10111	11111	11111
00000	00000	11111	10111	11111
00000	11111	00111	11111	11111
00011	01001	11111	11111	11111
00001	01101	11111	11111	11111
00010	11011	11111	11111	11111
00000	00111	11111	11111	11111
01010	00101	11101	11111	11111
00001	01011	11101	11111	11111
01000	01000	11111	11111	11111
00001	10101	10101	11111	11111
00011	11101	11111	11111	11111
00101	01111	11111	10011	11111
00000	00111	11111	11111	11111
00000	00011	10100	01101	11111
00000	01011	11010	11111	11111
00101	11011	01111	11111	11111
00001	01111	11111	11111	11111
00010	10111	01011	11111	11111
00001	10011	10101	01011	11111
00001	11111	01011	11111	11111

ous trials. The question is: from what aspects of the trials? The probability of avoidance may depend both on the number of previous shocks and on the number of previous successful avoidances, perhaps in different ways. Let π_{it} be the conditional probability of a shock at trial t ($t = 0, \cdots, 24$), for any dog i, given each particular dog's reactions on previous trials. If n_{it} is the number of avoidances by dog i before its trial t, then $t - n_{it}$ will be the number of previous shocks.

The standard approach of a statistician would be to use a logistic regression function to relate the binary probabilities to the numbers of previous events, as I did in the previous example. Here, instead, let us construct a mechanistic model by postulating that the probabilities have the following relationship. Assume that the ratio of successive probabilities over time are given by

$$\frac{\pi_{it}}{\pi_{i,t-1}} = \begin{cases} \kappa & \text{if } y_{i,t-1} = 1 \\ \upsilon & \text{if } y_{i,t-1} = 0 \end{cases} \tag{8.2}$$

Thus, the probability of shock will change by a factor of κ if there was an avoidance

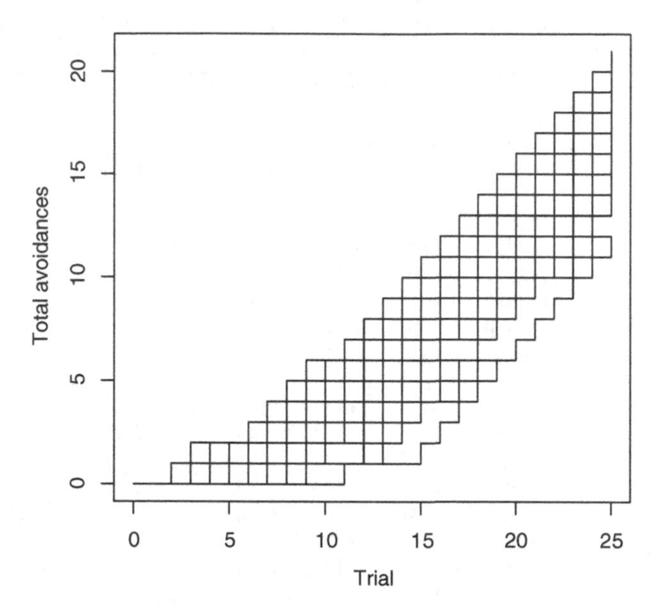

Fig. 8.2. The cumulated numbers of avoidances of shock by 30 dogs each on 25 trials, from Table 8.2.

at the previous trial or υ if there was a shock. This model can be rewritten as

$$\pi_{it} = \kappa^{n_{it}} \upsilon^{t-n_{it}} \tag{8.3}$$

or

$$\log(\pi_{it}) = \alpha n_{it} + \beta(t - n_{it}) \tag{8.4}$$

where $\alpha = \log(\kappa)$ and $\beta = \log(\upsilon)$. Formulated in this way, the parameters in the model have an immediate physical interpretation.

The parameter estimates are $\hat{\alpha} = -0.236$ and $\hat{\beta} = -0.0793$, so that $\hat{\kappa} = 0.790$ and $\hat{\upsilon} = 0.924$. This indicates that an avoidance is more effective than a shock in reducing the probability of future shock. Because $\hat{\alpha} \doteq 3\hat{\beta}$ or $\hat{\kappa} \doteq \hat{\upsilon}^3$, a dog learns about as much by one avoidance as by three shocks.

It is reassuring to note that this mechanistic model for binary responses, with a nonstandard log link function, but a clear theoretical interpretation, fits better than the corresponding *ad hoc* traditional models with a canonical logit link. The above model has an AIC of 278.1; that with the logit link replacing the log in Equation (8.4) has 304.9. Adding an intercept parameter to the latter model gives 283.2.

8.1.3 Overdispersion

In certain circumstances, it is not possible to record the times between events, but only the total numbers of events. As we know (Sections 4.1.4 and 4.3.2), this will lead to over- or underdispersed counts, because the process will not generally be Poisson. In certain situations, it may be possible, nevertheless, to develop models based on intensity functions for the associated stochastic processes suspected of producing the events.

Let us assume, here, that the system producing the events can be modelled as a birth process (Section 1.2.5), whereby the probability of an event depends, in some way, on the number of previous events:

$$\Pr(N_{t+\Delta t} = n+1 | N_t = n) = \lambda_n \Delta t + o(\Delta t) \tag{8.5}$$

where λ_n is some function of the previous number of events.

The resulting overdispersion model will depend on how this function of n is specified. For general count data, it can be modelled directly. As we already know, setting λ_n to a constant, independent of n, yields a Poisson process. A negative binomial distribution will be generated from

$$\begin{aligned} \lambda_n &= \beta_0 + \beta_1 n \\ &= (\alpha + n)\log(1+\upsilon) \end{aligned} \tag{8.6}$$

using the parametrisation of Equation (4.36). Other possibilities include exponential or logistic changes with the previous number of events,

$$\lambda_n = \mathrm{e}^{\beta_0 + \beta_1 n} \tag{8.7}$$

$$\lambda_n = \frac{\beta_2}{1 + \mathrm{e}^{\beta_0 + \beta_1 n}} \tag{8.8}$$

Indeed, any discrete distribution can be represented in some such way.

For a general counting process, n can take any non-negative integral value. However, for binomial-type processes, there will be some maximum, say M, such that $\lambda_M = 0$. This can be accommodated by including a factor of $M - n$ in the above models for λ_n. Thus, the constant λ_n of the Poisson process above becomes

$$\lambda_n = M - n \tag{8.9}$$

yielding a binomial distribution. In certain circumstances, the total possible number M might influence the intensity in other ways, such as

$$\lambda_n = \frac{(M-n)\mathrm{e}^{\beta_2 + \beta_3 M}}{1 + \mathrm{e}^{\beta_0 + \beta_1 n}} \tag{8.10}$$

Once a model for λ_n has been selected, the marginal probabilities $\pi_n(t) = \Pr(N_t = n)$ can be obtained from the Kolmogorov forward differential equations

$$\begin{aligned} \frac{\mathrm{d}\pi_0(t)}{\mathrm{d}t} &= -\lambda_0 \pi_0(t) \\ \frac{\mathrm{d}\pi_n(t)}{\mathrm{d}t} &= -\lambda_n \pi_n(t) + \lambda_{n-1}\pi_{n-1}(t) \end{aligned} \tag{8.11}$$

The process must start in state 1, so that the initial conditions are given by the marginal distribution $\pi(0) = (1,0,0,\ldots)^\top$. The matrix of transition intensities

(Section 6.1.3) corresponding to these equations is

$$\mathbf{T} = \begin{pmatrix} -\lambda_0 & \lambda_0 & 0 & \cdots & 0 \\ 0 & -\lambda_1 & \lambda_1 & \cdots & 0 \\ \vdots & \vdots & \vdots & & \vdots \\ 0 & 0 & 0 & \cdots & -\lambda_n \end{pmatrix} \tag{8.12}$$

which must be matrix exponentiated numerically. Multiplying the result by the initial marginal distribution shows that the first row of the probability transition matrix provides the required marginal probabilities $\pi_n(t = 1)$ (see Faddy, 1997; Faddy and Fenlon, 1999).

Nematodes Biological methods are increasingly being used to replace chemical pesticides in the control of insect pests. Parasite–host systems include nematodes such as those studied here in the control of the mushroom sciarid fly *Lycoriella solani*. Once a nematode finds and penetrates the larva of a fly, it kills it by releasing toxin-producing bacteria. It appears, however, that the first such invading nematode on a larva encounters more resistance than subsequent ones. Thus, the probability of a given nematode succeeding may depend on the number already present, a situation that might appropriately be described by a birth model.

Experiments were carried out with three strains of the nematode *Steinernema feltiæ*. The first of the strains used (Nemasys) has been marketed as a biological control agent for a number of years; the other two (Sus11 and Sus94) come from laboratory populations and may also potentially be useful. In this study, individual larvæ of the fly were challenged by varying numbers of the nematode. However, only the total numbers successfully penetrating the larvæ were recorded, not the successes and failures of individual nematodes, with results as shown in Table 8.3.

The results for the various models described above, separately for each strain, are summarised in Table 8.4. The binomial distribution fits relatively poorly in all cases. For Nemasys, the exponential form is sufficient, but for Sus94, the logistic form is somewhat better, and for Sus11, the results are influenced by the total number of challengers.

For Nemasys, the estimates are $\widehat{\beta_0} = -1.17$ and $\widehat{\beta_1} = 0.25$ using Equation (8.7) (multiplied by $M - n$); for Sus94, they are $\widehat{\beta_0} = 0.60$, $\widehat{\beta_1} = -0.61$, and $\widehat{\beta_2} = 0.72$ using Equation (8.8) (again multiplied by $M - n$); for Sus11, they are $\widehat{\beta_0} = 2.07$, $\widehat{\beta_1} = -0.87$, $\widehat{\beta_2} = 0.66$, and $\widehat{\beta_3} = -0.062$ using Equation (8.10).

The changing intensities λ_n and the values from the original equations, that is the series of $\lambda_n/(M - n)$, are both plotted in Figure 8.3. We see that the maximum intensity occurs early for Sus94 and late for Nemasys, with Sus11 in between. Fitting a common model (logistic with total) to all three strains yields an AIC of 1674.9 as compared to 1641.6 for the combination of three different models indicated by Table 8.4. Thus, there is strong evidence of different reactions of the various strains of attacking nematodes to those already present in a larva.

Table 8.3. *Numbers of nematodes (*Steinernema feltiæ*) of three different strains found in larvæ of the mushroom sciarid fly (*Lycoriella solani*) with various initial numbers (M) of challengers. (Faddy and Fenlon, 1999)*

Challengers (M)	Number of nematodes 0	1	2	3	4	5	6	7	8	9	10
					Nemasys						
10	1	8	12	11	11	6	9	6	6	2	0
7	9	14	27	15	6	3	1	0			
4	28	18	17	7	3						
2	44	26	6								
1	158	60									
					Sus94						
10	4	11	15	10	10	11	8	3	0	0	0
7	12	21	17	12	7	5	0	0			
4	32	22	15	6	0						
2	35	17	2								
1	165	59									
					Sus11						
10	21	13	11	11	9	4	2	2	1	0	0
7	34	15	13	1	2	3	1	1			
4	35	19	12	3	2						
2	45	26	3								
1	186	40									

Table 8.4. *AICs for various birth models fitted to the nematode data of Table 8.3.*

λ_n	Nemasys	Sus94	Sus11
Binomial distribution	627.8	560.3	555.8
Exponential form	590.9	544.1	514.4
Logistic form	591.5	543.0	512.6
Logistic total form	592.2	544.0	507.7

8.2 Autoregression

In Section 7.1, I looked at one way of constructing models incorporating serial dependence for responses involving counts of recurrent events. Special features of the exponential family can be used to construct another, rather special, type of autoregressive dependence, here for the durations between events. Such models have sometimes been called *continuous-state* Markov chains.

8.2.1 Theory

The simpler models in this class are very restrictive. It is not clear whether they actually can represent any real phenomenon. However, more flexible models are also available.

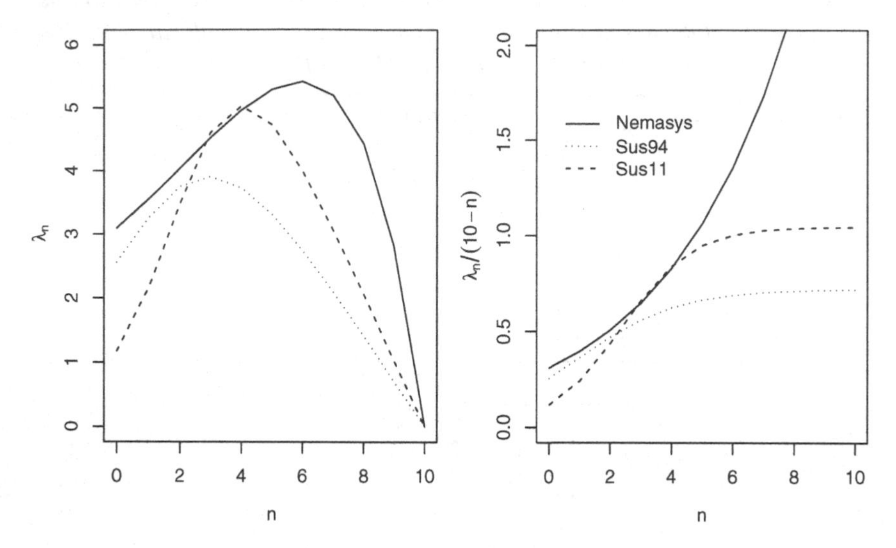

Fig. 8.3. The series of intensities λ_n and $\lambda_n/(10-n)$ from the best model for each strain of nematode for the data with $M = 10$ from Table 8.3.

Conditional exponential family

The conditional exponential family of Markov processes (Feigin, 1981; Küchler and Sørensen 1997, p. 60) involves density functions having the general form

$$f(y_t;\theta) = \exp[\theta d(y_t, y_{t-1}) - b(\theta, y_{t-1}) + c(y_t, y_{t-1})] \tag{8.13}$$

From the characteristics of the exponential family, this implies that

$$\mathrm{E}[d(Y_t, Y_{t-1})] = \frac{\mathrm{d}b(\theta, y_{t-1})}{\mathrm{d}\theta} \tag{8.14}$$

As a specific example, let us consider the gamma distribution in Equation (2.7). In the context of recurrent events, the response Y_t will be the time between successive events. Let us define new parameters $\rho = \mu_t/\phi_t$ and $\phi_t = y_{t-1}$, implying that $\mu_t = \rho y_{t-1}$. The parameter ρ is called the *autocorrelation* (see Section 9.1.2). Because μ_t and $\phi_t = y_{t-1}$ must both be greater than zero, in this model it is always positive.

This set-up yields

$$f(y_t|y_{t-1};\rho) = \frac{y_t^{y_{t-1}-1}\mathrm{e}^{-\frac{y_t}{\rho}}}{\rho^{y_{t-1}}\Gamma(y_{t-1})} \tag{8.15}$$

Here, $\theta = -\frac{1}{\rho}$ and

$$\begin{aligned} b(\theta, y_{t-1}) &= y_{t-1}\log(\rho) \\ &= y_{t-1}\log\left(-\tfrac{1}{\theta}\right) \end{aligned} \tag{8.16}$$

Then, we can check that indeed

$$\begin{aligned}\mu_t &= \mathrm{E}(Y_t|y_{t-1}) \\ &= -\frac{y_{t-1}}{\theta} \\ &= \rho y_{t-1}\end{aligned} \tag{8.17}$$

The autoregression relationship given by Equation (8.17) may be modelled in several ways. With a log link,

$$\log(\mu_{it}) = \log(y_{i,t-1}) + \log(\rho_i) \tag{8.18}$$

Here, the log lagged response is an offset, because it is not multiplied by an unknown parameter, and the linear model will be in terms of ρ_i. This essentially implies that only changes in the autoregression coefficient, depending on covariates, can influence the mean. In most situations, this will be of little use.

A more promising alternative is to use the identity link in an additive location model, such as

$$\mu_{it} = \rho_i y_{i,t-1} + \boldsymbol{\beta}_i^{\mathrm{T}} \mathbf{x}_{it} \tag{8.19}$$

The first response in a series, corresponding to the initial conditions, might be modelled by fitting a separate gamma distribution, with two distinct parameters, different from those for the rest of the series.

Recurrent tumours Let us return to the data on recurrences of tumours due to bladder cancer that I analysed in Sections 7.3.2 and 7.3.3. Models based on the gamma distribution and log link, without any dependence over time, have AICs of 750.9 and 750.2, respectively, without and with treatment effect.

For the usual state dependence model (Section 1.2.3), involving direct dependence on the previous time interval (a gamma distribution with constant dispersion parameter), the AICs are 727.7 and 728.7, showing an improvement on the independence model. A standard autoregression model (Section 1.2.4) with log link and log lagged recurrence time has AICs of 728.5 and 728.7. None of these models indicate a treatment effect.

In contrast, the conditional exponential model for the gamma distribution has AICs 1012.3 and 1001.2, respectively! Thus, this model fits very badly. This arises because of the strong constraint that the dispersion parameter must equal the previous time interval between recurrent events.

Exponential dispersion family

The conditional exponential family can be generalised by taking a member of the exponential dispersion family (Jørgensen, 1986; 1987; 1992), although the latter models originally were not presented in this way. Let us denote this family

$$f(\mathbf{y};\boldsymbol{\theta}) = \exp\{\phi[\mathbf{y}^T\boldsymbol{\theta} - b(\boldsymbol{\theta})] + c(\phi,\mathbf{y})\}$$

by $\mathrm{ED}^*(\boldsymbol{\theta},\phi) = \mathrm{ED}(\boldsymbol{\mu},\sigma^2)$, where $\boldsymbol{\mu} = \mathrm{E}(\mathbf{Y})$ and $\sigma^2 = 1/\phi$. This family has the convolution property for $\sum_{i=1}^n \mathbf{Y}_i/n$:

$$\mathrm{ED}^*(\boldsymbol{\theta},\phi_1) * \cdots * \mathrm{ED}^*(\boldsymbol{\theta},\phi_R) = \mathrm{ED}^*(\boldsymbol{\theta},\phi_1 + \cdots + \phi_R)$$

where $\phi\mathbf{Y} \sim \text{ED}^*(\boldsymbol{\theta}, \phi)$ when $\mathbf{Y} \sim \text{ED}(\boldsymbol{\mu}, \sigma^2)$. This can be used to model a process with stationary independent increments (Section 4.1.1).

Thus, the gamma process defined by Equation (2.7) will be denoted here by $Y \sim \text{Ga}(\mu, 1/\phi)$. Then, $\phi Y \sim \text{Ga}(\phi\mu, 1/\phi)$ so that

$$\phi\Delta Y_t \sim \text{Ga}\left[\phi\mu\Delta t, \frac{1}{\phi\Delta t}\right] \tag{8.20}$$

or

$$\phi Y_t \sim \text{Ga}\left[\phi\mu t, \frac{1}{\phi t}\right] \tag{8.21}$$

With $Y_t = t$, $\rho = \mu$, and $\phi = 1$, this is just the conditional gamma model of Equations (8.15) and (8.17), but derived in a different way. Thus, whereas that model had the dispersion parameter *equal* to the time interval between events, the present model is more flexible, having the dispersion parameter *proportional* to the interval.

Other members of this family include the Poisson process (Section 4.1.2) and Brownian motion, a Wiener diffusion process, based on the normal distribution (Section 10.1).

Recurrent tumours When this exponential dispersion model is fitted to the tumour data, the AIC is 723.4 both with and without treatment effect, an improvement on all of the autoregression models above, with constant dispersion parameter. The placebo and thiotepa groups have similar average recurrence times so that the model can be simplified, with an AIC of 722.8. If the number of previous recurrences is taken into account, a type of birth model, this is reduced to 721.2. The model can be simplified further by removing the direct dependence of the mean on the previous time so that only the dispersion parameter depends on it, with an AIC of 720.5. Finally, the model can be improved if the placebo is allowed to affect the first recurrence time differently than the two treatments, whereas pyridoxine affects subsequent times differently than the other two: the AIC becomes 718.9.

In summary, this model has the location regression function with a log link, a different intercept for the first and for subsequent time intervals, a placebo effect before the first event, an effect of pyridoxine after the first event, and a dependence on the number of previous events. The log dispersion parameter is constant until the first recurrent event and then becomes proportional to the previous time interval.

The mean time to first recurrence under placebo is estimated to be 30.4 mon, whereas it is 55.0 mon under pyridoxine and thiotepa. For subsequent recurrences, the mean time under pyridoxine is 11.1 mon, whereas it is 16.1 mon with placebo and thiotepa. The recurrence time reduces about 0.1 mon for each additional tumour. It is difficult to plot the theoretical survivor or intensity curves because of the dependence of the dispersion parameter on the previous time.

Note, however, that this gamma exponential dispersion model cannot compete with the dynamic models fitted in Section 7.3.3.

8.3 Marked point processes

Models for point processes of recurrent events were covered in Chapter 4. Often, however, each recurrent event is associated with some other response variable, called a *mark*. This should also be included in the model, resulting in a *marked point process*.

8.3.1 Theory

The mark recorded at each event usually will be either a count or some continuous measurement. Many types of processes usefully can be treated as marked point processes with almost any kind of mark. Thus, continuous-time Markov chains (Section 6.1.3) and semi-Markov processes (Section 6.1.4) could be analysed as point processes for which the mark is the type of transition.

Often, the presence of the mark is the only indication that an event has occurred. In these cases, a zero recorded value of the mark is impossible because the occurrence of the event is only indicated by the presence of a nonzero value.

As mentioned in Section 4.1.1, one important application of marked point processes arises when events can occur simultaneously. Thus, if the mark is the number of events occurring at each point, it often will be useful to distinguish an *occurrence function* from the intensity function. The former describes the rate for the points whereas the latter includes the multiplicity, referring to the recurrent events at these points. For such processes, the simplest preliminary approach may sometimes be first to concentrate on modelling the points, ignoring multiplicity, and then to handle the latter.

In analogy to a counting process (Section 4.1.1), we may wish to consider the *mark–accumulator process*. This will give the cumulative sum of the marks up until each point in time. It will be a piece-wise constant function of time, as is a counting process. However, instead of the jumps being unity, they will depend on the size of the mark. As long as zero is not allowed as a mark, the mark–accumulator process will contain all of the information in a marked point process.

In the most general marked point processes, the successive marks may not be independent of each other nor of the point process that determines their occurrence times.

Compound Poisson processes

One important simple case occurs when the point process is a Poisson process (Section 4.1.2), possibly nonhomogeneous, and the marks are independent, all having the same distribution. This is called a *compound Poisson process*.

Let us assume that the marks Y_i are independent and identically distributed variables and that N_t forms a Poisson process independent of these values. Then, the mark–accumulator process

$$Y_t^s = \sum_{i=1}^{n_t} Y_i \qquad (8.22)$$

is a compound Poisson process. This process retains the characteristic of an ordinary Poisson process of having stationary, independent increments. If the values of Y_i are discrete, the process can be modelled as a Markov chain in continuous time. (See Section 6.1.3 and Karlin and Taylor, 1981, pp. 427–428.)

Finite mixture models

One possible approach to modelling marked point processes, at least in discrete time, is to use a finite mixture distribution whereby the nonevent points are a mixture of the probability of no event and the probability of an event times the probability of recording zero for the mark:

$$f(Y=0;\pi,\boldsymbol{\kappa}) = 1-\pi+\pi f(Y=0|Z=1;\boldsymbol{\kappa}) \tag{8.23}$$

where Z is the event indicator, $\pi = \Pr(Z=1)$ is the probability of an event, and y is the mark observed when the event occurs. This is another application of the type of finite mixture model discussed in Section 3.4.

The Markov property can, then, be introduced by making π depend on whether or not an event occurred at the previous time point by means of a logistic regression, yielding a first-order, two-state Markov chain (Section 5.2.1). In addition, the conditional probability of the mark, $f(y|Z=1;\boldsymbol{\kappa})$, may depend on the previous occurrence of an event in a similar way.

8.3.2 Markov chains with marks

Marked point processes often are used to model weather phenomena. Thus, a point process can indicate whether or not precipitation occurs, with the mark being the amount.

Weather The Weather Service in the USA maintains a large number of precipitation monitors. For measuring the precipitation, a day is defined as being from 8:00 to the same time the next day. The recording instruments only can detect a minimum of 0.01 in of precipitation so that the raw data are left censored and

$$\Pr(Y=0|Z=1;\boldsymbol{\kappa}) = F(Y<0.01|Z=1;\boldsymbol{\kappa}) \tag{8.24}$$

for some cumulative distribution function $F(\cdot)$. However, the published data available, as described below, only are recorded to the nearest inch, so that I shall have to assume censoring at 0.5 in.

Guttorp (1995, p. 17) provides data for the month of January in the years 1948 to 1983 for the precipitation monitor located at Snoqualmie Falls in the foothills of the Cascade Mountains in western Washington state, USA. The observation for 31 December 1947, needed for conditioning in the Markov chain, is unavailable; I shall assume, arbitrarily, that precipitation occurred on that day, as it did on the following 12 days. The cumulative numbers of days with precipitation and the cumulative total precipitation, the mark–accumulator process, for January for each year are plotted in Figure 8.4. We can see that there is more variability in the amount of rainfall than in the number of wet days.

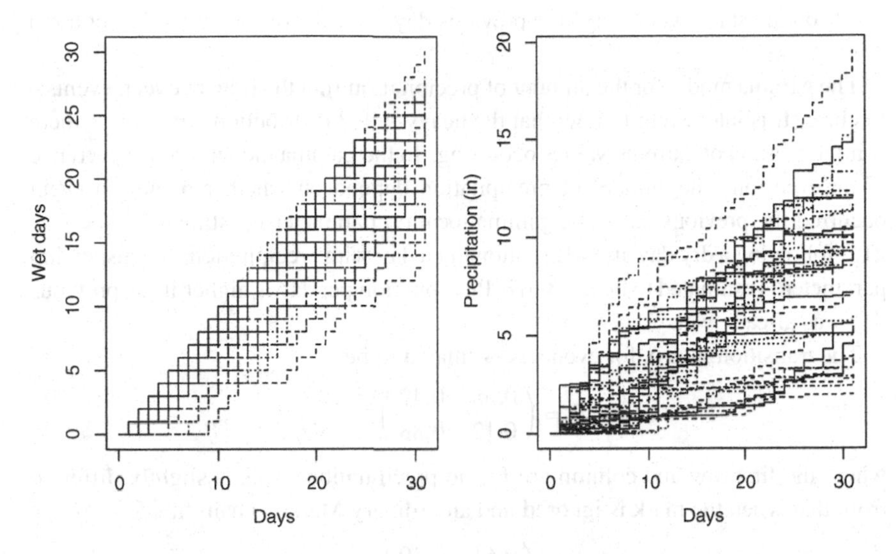

Fig. 8.4. The cumulative numbers of days with precipitation and the cumulative amount of precipitation (in) for the data from Snoqualmie Falls.

Table 8.5. *AICs for models fitted to the Snoqualmie Falls precipitation data.*

	Independence		Markov dependence	
		Mixture	Event	Event and mark
Gamma	4941.3	4893.0	4790.0	4777.3
Weibull	5012.3	4896.3	4793.3	4780.6
Inverse Gauss	5411.3	5009.7	4906.7	4902.9
Pareto	5349.4	4912.8	4809.8	4795.2
Log normal	5126.7	4939.6	4836.6	4826.9
Log logistic	5150.3	4953.5	4850.5	4839.8
Log Laplace	5192.1	4985.7	4882.7	4872.3
Log Cauchy	5186.1	5045.5	4942.4	4930.9

First, I shall consider the amounts of precipitation alone, as an independent and identically distributed series (with the zeros handled as left-censored). The results for various models are shown in the first column of Table 8.5. Next, I can add the assumption that some of the days have true zeros, the finite mixture, with the results in the second column of the table. All distributions show considerable improvement in fit. In both of these approaches, the gamma distribution fits best.

The next step is to introduce dependence among the successive responses into the finite mixture model. Suppose first that the probability of a wet or dry day depends on the state the previous day, a first-order Markov chain. The results are given in column three of Table 8.5. This is the discrete-time analogue of the compound Poisson process, with a geometric distribution describing the times between the recurrent events (Section 5.1.2). The models all improve further. Finally, assume that the mean amount of precipitation, not just its presence or absence, depends on the state (wet or dry) the previous day, with the results in the last column of the table.

The gamma model for the amount of precipitation (not the time between events!) fits best. It is interesting to note that the heavy-tailed distributions fit poorly: there is no indication of extreme values occurring. In the gamma model, both occurrence of an event and the amount of precipitation depend on whether or not an event occurred the previous day. The gamma location parameter is estimated to be $\hat{\mu} = 21.2$ following a dry day and 41.5 following a day with precipitation; the dispersion parameter is estimated to be $\hat{\kappa} = 0.67$. Precipitation per day is higher if the previous day was already wet.

The transition matrix for events is estimated to be

$$\mathbf{T} = \begin{pmatrix} 0.58 & 0.42 \\ 0.12 & 0.88 \end{pmatrix}$$

where the first row and column are for no precipitation. This is slightly different from that when the mark is ignored and an ordinary Markov chain fitted:

$$\mathbf{T} = \begin{pmatrix} 0.61 & 0.39 \\ 0.16 & 0.84 \end{pmatrix}$$

Thus, like days have a high probability of following each other, especially when precipitation is occurring. The mean length of a dry period is estimated to be $1/(1-0.58) = 2.4$ days and that of a wet period $1/(1-0.88) = 8.0$ days.

The Markov chain fitted here for recurrent precipitation events is rather crude. It is the discrete-time equivalent of an exponential distribution, with constant probability of an event over time. Thus, a more sophisticated model would allow for dependence on time since the previous event, both within wet and dry periods, such as an alternating semi-Markov process (Section 6.1.4).

8.4 Doubly stochastic processes

In modelling a point process, one may need to assume that the intensity function varies stochastically over time in some way. If this function only depends on the history of the process itself, it is said to be a *self-exciting process*. Simple cases include renewal processes (Section 4.1.4), where the intensity depends on when the previous recurrent event occurred, and alternating events (Section 6.1.1) with two different constant intensities, depending on the current state of the process.

A different situation occurs when the intensity function varies stochastically for reasons not depending on the previous history of the process under study. This is called a *doubly stochastic process* (Cox, 1955). We already have seen cases of such a processes in Chapter 7, such as the hidden Markov model for a binary series in Section 7.2.2.

8.4.1 Theory

The simplest case of a doubly stochastic process is the doubly stochastic Poisson process. This is conditionally a Poisson process, given the intensity process. Two especially simple possibilities can arise (Cox and Isham, 1980, p. 71):

(i) If the intensity is constant for each given series, only varying randomly among them, we have a *mixed Poisson process* in analogy with other mixture models.

(ii) The intensity may be constant during given intervals of time but vary randomly among the intervals.

  (a) In the simplest situation, the intervals are fixed, perhaps of equal length, and are known.

  (b) If the timing of the intervals is unknown, an important special case occurs when only two intensities alternate randomly in time: the *alternating Poisson process* (we just saw above that this is also a self-exciting process).

In Section 7.2.1, I presented a more elaborate example whereby two or more Poisson processes stochastically alternated according to a hidden Markov model, but where only the presence or absence of recurrent events was recorded in fixed time intervals.

However, more complex doubly stochastic processes are also possible, such as a doubly stochastic renewal process. I shall consider one here.

Gamma mixture of Weibull distributions

Suppose that we have a renewal process whereby the distribution of interarrival times is Weibull, as in Equation (2.15). Assume, however, that the location parameter of this distribution has a gamma distribution, as in Equation (4.35). We know, from Section 7.1.3, that the result is a Burr distribution, as in Equation (7.26). In that section, I allowed the parameters from the gamma distribution to change dynamically over time. Then, because the intensity of a Weibull process depends on its location parameter, it will be varying stochastically over time. Thus, the result was one form of doubly stochastic model.

Here, we can, more simply, allow one of these parameters to vary deterministically with time and still obtain a doubly stochastic model. Let us rewrite the Burr distributions as

$$f(y;\mu,\kappa,\nu) = \frac{\nu\kappa\left(\frac{y}{\mu}\right)^{\kappa-1}}{\mu^{\kappa}\left[1+\left(\frac{y}{\mu}\right)^{\kappa}\right]^{\nu+1}} \tag{8.25}$$

The parameter ν comes from the gamma mixing distribution; if it is allowed to depend on time, the distribution of the Weibull intensity will be varying stochastically over time because the gamma distribution is changing. Thus, I need only introduce a regression function on time for ν into the Burr distribution in order to obtain a doubly stochastic process. This will be a doubly stochastic renewal process; if $\kappa = 1$, it is still another type of doubly stochastic Poisson process.

Table 8.6. *Intervals between accidents causing fatalities in the mines in the seven divisions of the National Coal Board of Great Britain in 1950 (first column) and corresponding number of accidents (second column). The first and last values of each series are censored. (Maguire* et al.*, 1952)*

Division 1	Division 2		3	4	5	6	7	
16 1	4 1	3 1	12 1	16 1	21 1	7 1	17 1	1 1
3 1	9 2	2 1	1 1	8 1	2 1	9 1	2 1	15 1
1 2	11 1	3 1	7 1	24 1	15 1	10 1	4 1	2 1
7 1	1 1	4 1	1 1	6 1	1 1	4 1	1 1	5 1
7 1	12 3	3 2	4 1	5 1	5 1	2 1	2 2	3 1
20 1	1 1	5 2	5 1	14 1	1 1	6 1	6 1	14 1
6 1	2 1	2 1	9 1	2 1	9 1	9 1	1 1	2 1
1 1	26 1	13 1	6 1	11 1	1 2	14 2	4 1	5 1
20 1	2 1	2 1	5 1	8 1	17 2	7 2	2 3	24 2
5 1	4 1	3 2	8 1	16 1	1 1	5 1	14 1	1 2
10 1	4 1	3 2	2 1	8 1	24 1	1 1	2 1	8 0
4 1	6 2	9 2	13 1	98 1	14 1	14 1	10 1	
6 1	5 3	7 0	8 1	1 1	4 1	2 1	2 1	
19 1	2 1		1 1	3 1	9 1	9 1	5 1	
1 1	5 1		1 1	16 1	20 1	8 1	9 1	
9 1	13 1		4 1	9 0	14 1	2 1	6 1	
16 2	4 2		3 1		1 1	5 1	13 1	
2 1	3 1		6 1		1 1	10 1	8 1	
2 1	4 1		11 1		44 1	12 1	2 1	
2 1	4 1		11 1		4 1	3 1	1 1	
2 1	15 2		6 2		5 1	20 1	7 1	
6 1	5 1		13 1		1 1	7 1	4 1	
14 1	8 1		2 1		13 1	12 2	1 2	
10 1	3 1		13 1		6 1	23 1	5 1	
2 1	3 1		7 1		9 1	13 1	3 1	
2 1	1 2		5 1		3 0	19 1	6 1	
2 1	1 2		11 3			12 0	6 2	
7 1	5 1		3 1				1 1	
3 1	3 1		8 1				7 1	
15 1	1 1		5 1				3 1	
13 1	4 1		1 1				5 1	
4 1	2 1		13 1				2 4	
8 0	8 1		18 1				2 1	
	4 1		17 1				1 2	
	1 1		5 0				1 1	

Mine accidents Consider the intervals, in days, between accidents causing fatalities recorded for the mines in the seven divisions of the National Coal Board of Great Britain over a 245 day period in 1950, shown in Table 8.6. In each division, the first value begins at the start of the period, not at an accident, and the last value stops at the end of observation, not at an accident, so that both are censored. On a number of days, more than one accident occurred.

The question is whether or not there are differences among the divisions in the accident rate. The cumulative numbers of days with accidents and the cumulative

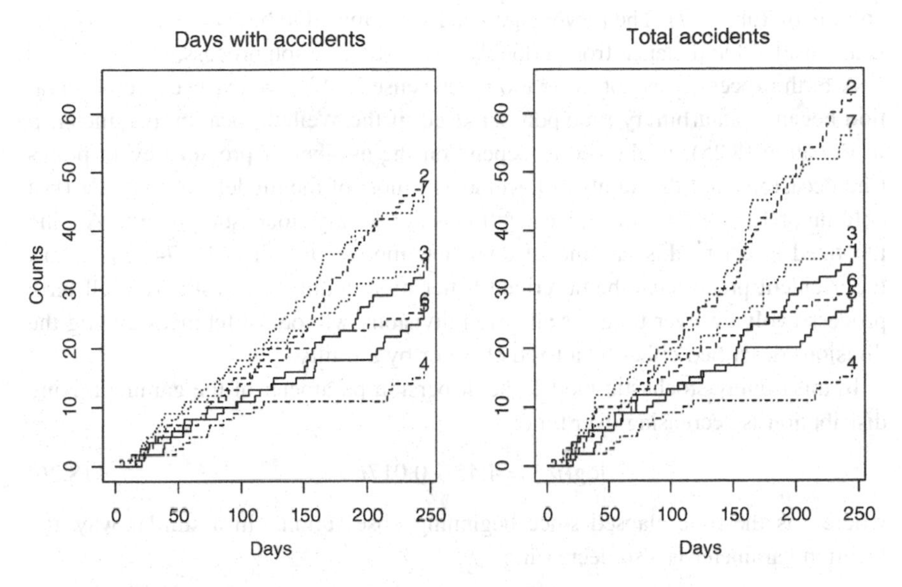

Fig. 8.5. The cumulative numbers of days with accidents (left) and the cumulative numbers of accidents (right) for the mine data of Table 8.6.

Table 8.7. *AICs for various models fitted to the mine accident data of Table 8.6.*

		Birth
Weibull	770.3	769.1
Fixed effect	756.8	757.7
Frailty	760.1	761.1
Burr	762.8	761.8
Fixed effect	753.1	754.1
Doubly stochastic	763.2	739.9

numbers of accidents are shown in Figure 8.5. We can see a very considerable divergence in the rate at which accidents accumulate in the seven divisions.

A Weibull renewal process (Section 4.1.4), ignoring any dependence among recurrent events in the same division, has an AIC of 770.3. This model can be modified in various ways in order to attempt to describe the differences among the divisions. A fixed effect difference among divisions improves the model considerably, as can be seen in the first column of Table 8.7. On the other hand, a gamma frailty model (Section 7.3), although an improvement over independence, does not fit as well as the fixed effect model. The Burr distribution, an overdispersed Weibull distribution, fits almost as well as the Weibull frailty model, although it does not take directly into account differences among divisions. Again, introducing a fixed effect improves the model.

None of the models so far take into account possible changes over time. However, allowing ν of the gamma mixing distribution to depend log linearly on time, a doubly stochastic process, does not improve the model (the last value in the first column of Table 8.7). The power parameter is estimated to be $\hat{\kappa} = 1.42$, so that this is a considerable distance from a doubly stochastic Poisson process.

A birth process does not make too much sense in this context because observation began at an arbitrary time point. Indeed, if the Weibull location parameter, μ in Equation (8.25), is allowed to depend on the number of previous event points (the occasions, not the number of accidents), most of the models fit no better (last column of Table 8.7). The only exception is the doubly stochastic process. Adding the fixed effect to this last model does not improve the fit (AIC 744.5). Thus, the accident process can be described better by a doubly stochastic Weibull birth process evolving over time for all seven divisions without differences among the divisions described either by a fixed effect or by a frailty.

In this doubly stochastic model, the dispersion parameter of the gamma mixing distribution is decreasing over time:

$$\log(\nu_t) = 4.45 - 0.017t \tag{8.26}$$

where t is the time elapsed since beginning observation. In a similar way, the location parameter is also decreasing:

$$\log(\mu_{it}) = 5.12 - 0.082n_{it} \tag{8.27}$$

where n_{it} is the number of previous accidents that occurred in the division i during the period of observation. Letting the Weibull power parameter depend either on time or on the number of events does not improve the model. Nor does letting ν_t depend on the number of events or μ_{it} depend directly on time. The constant power parameter still has the same estimate given above, quite far from a doubly stochastic Poisson process.

It is not clear why such dependencies on time and on the number of previous events should be detected because these accidents are continuing processes with no fixed time origin. The observations were taken over 245 days so that the model may be detecting a seasonal effect involving a simultaneous decrease in the location and dispersion parameters.

8.5 Change points

As we already have seen several times, a common problem in the study of stochastic processes is to determine whether or not the functioning of a process is being modified over time. If such a change is relatively abrupt, occurring at some point in time, determination of that time point is called the change-point problem.

8.5.1 Theory

One important class of stochastic processes involves modelling the occurrence of abrupt modifications at one or more points in time. Thus, when individuals are followed over time, they may change state at some unknown point in time. This means that the distribution of the response is altered at this time point. The response variable Y_t at time t may, then, be modelled as

$$\begin{array}{ll} f(y_t;\phi_1) & t<\tau \\ f(y_t;\phi_2) & t\geq\tau \end{array} \tag{8.28}$$

where $f(\cdot)$ is some given distribution and τ an unknown parameter giving the change-point time. Some or all of the elements of the vector of parameters ϕ in the distribution may change. In a more complex situation, the distribution function itself may be different before and after the change point. Notice that the derivative of ϕ does not exist at $t=\tau$ so that care must be taken in the technical procedure for estimating the parameters.

Consider the simple case of a Poisson process; this will be nonhomogeneous because of the change point. However, the process on each side of the change point can be homogeneous: the rate of occurrence of events can stay the same up to the change point and then change to some other constant. Equivalently, in each period, the interarrival times will have an exponential distribution with constant intensity or the counts within fixed intervals of time will follow a Poisson distribution (Section 4.1.2). Thus, at the change point, the intensity and the mean change. Thus, we shall have a special case of Equation (8.28):

$$\begin{array}{ll} f(y_t;\mu_1) & t=1,\ldots,\tau \\ f(y_t;\mu_2) & t=\tau+1,\ldots,R \end{array} \tag{8.29}$$

where $f(\cdot)$ is either an exponential or a Poisson distribution, depending on the approach taken, and τ is again the unknown parameter giving the change-point time. Hence, for a single series, there will be three parameters to estimate: the two intensities or means, before and after, and the change point.

Unfortunately, for such models, the form of the likelihood function can be very irregular, with local maxima; standard optimisation algorithms, especially those using derivatives, will have trouble finding the maximum. A simple approach is to fix a grid of values of the change-point parameter τ and, for each fixed value, to optimise over the other parameters. The resulting set of likelihood values then can be plotted.

Mine disasters The numbers of days between successive coal-mining disasters (explosions), each with at least ten deaths in Great Britain between 15 March 1851 and 22 March 1962 are given in Table 8.8. In the middle of this period, near the end of the nineteenth century, the legislation for mine safety was changed. I shall see whether or not I can detect any effects of this change.

The cumulative number of recurrent events is plotted in the left graph of Figure 8.6. The rate of occurrence of disasters seems to slow down after about 15 000 days or 40 years, that is, about 1890. The counts of disasters per year are plotted in the right graph of Figure 8.6. Again, we see that there are fewer disasters after about 40 years.

Let us look, first, at the intervals between disasters using the exponential distribution. The normed likelihood function for the change point, based on Equation (8.29) with exponential distributions, is plotted in the left graph of Figure 8.7,

Table 8.8. *Days between successive coal-mining disasters, each with at least 10 deaths, in Great Britain, from 15 March 1851 to 22 March 1962. (Jarrett, 1979; read across rows)*

157	123	2	124	12	4	10	216	80	12
33	66	232	826	40	12	29	190	97	65
186	23	92	197	431	16	154	95	25	19
78	202	36	110	276	16	88	225	53	17
538	187	34	101	41	139	42	1	250	80
3	324	56	31	96	70	41	93	24	91
143	16	27	144	45	6	208	29	112	43
193	134	420	95	125	34	127	218	2	0
378	36	15	31	215	11	137	4	15	72
96	124	50	120	203	176	55	93	59	315
59	61	1	13	189	345	20	81	286	114
108	188	233	28	22	61	78	99	326	275
54	217	113	32	388	151	361	312	354	307
275	78	17	1205	644	467	871	48	123	456
498	49	131	182	255	194	224	566	462	228
806	517	1643	54	326	1312	348	745	217	120
275	20	66	292	4	368	307	336	19	329
330	312	536	145	75	364	37	19	156	47
129	1630	29	217	7	18	1358	2366	952	632

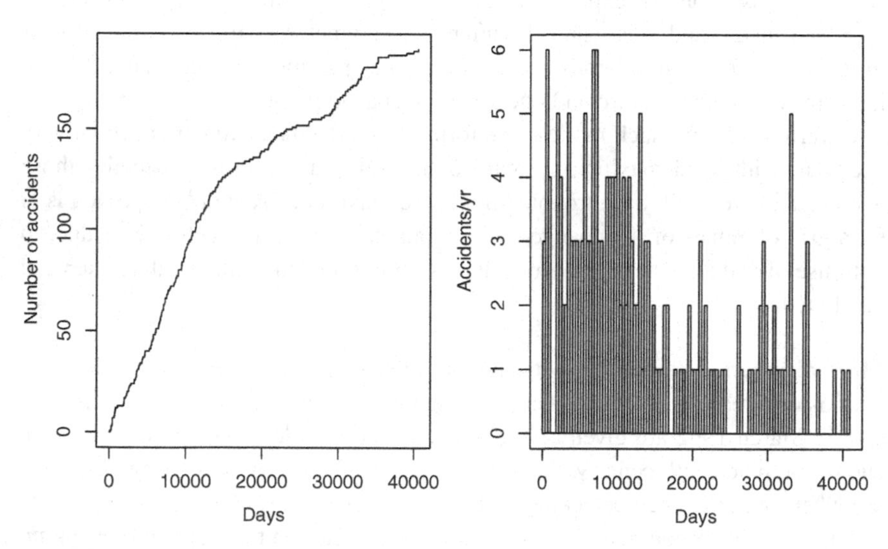

Fig. 8.6. The cumulative numbers of accidents and the numbers of accidents/yr for the mine disaster data of Table 8.8.

clearly pointing to interval 124. This means that the change occurred after 14 240 days or about 39 years, that is, 1890. The intensities are estimated to be 0.0087 disasters per day before the change and 0.0025 after, or 3.18/yr before and 0.92 after. The AIC for this model is 1176.4.

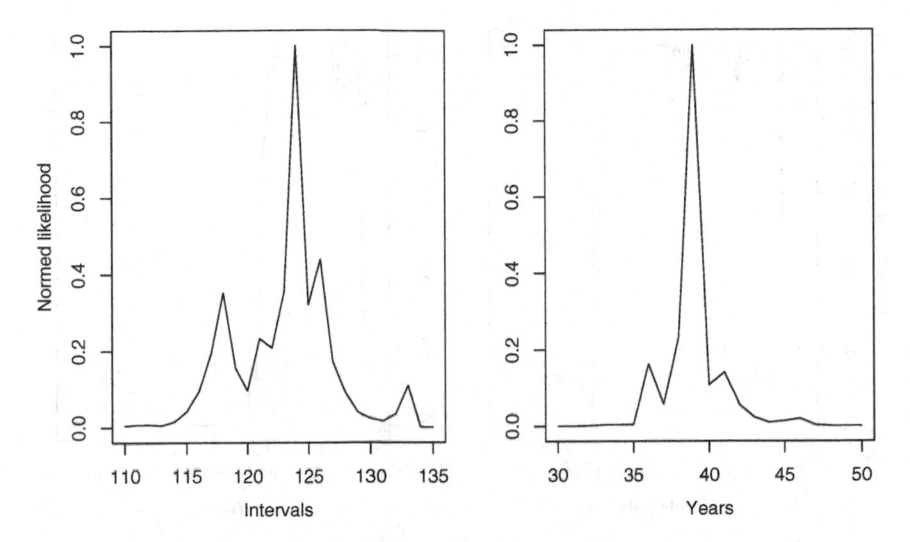

Fig. 8.7. The normed likelihood functions of the change point for the exponential (left) and Poisson (right) models for the mine disaster data of Table 8.8.

Now, let us, instead, look at the counts of disasters per year using the Poisson distribution. The normed likelihood function for the change point, again based on Equation (8.29), but now with Poisson distributions, is plotted in the right graph of Figure 8.7, again clearly pointing to 39 years. The mean numbers of recurrent events per year are estimated to be 3.18 before the change and 0.90 after. The AIC is 168.8; because the data are used differently, this is not comparable to the previous one.

8.5.2 Hidden Markov model

The hidden Markov models of Section 7.2 also can be used, with minor modifications, to detect a change point. Here, we need assume only two hidden states, with the process necessarily starting in the first. Thus, the initial distribution is $\pi_0 = (1,0)^\top$ instead of the stationary distribution of the hidden transition matrix. Only one change point is assumed, so that the second hidden state is absorbing: this matrix will have zero probability of return to the first state:

$$\mathbf{T} = \begin{pmatrix} \pi_{1|1} & \pi_{2|1} \\ 0 & 1 \end{pmatrix} \tag{8.30}$$

This means that it contains only one parameter to estimate: $\pi_{1|1} = 1 - \pi_{2|1}$.

Mine disasters For this model applied to the mine disaster data, the probability of being in the first hidden state is plotted in the left graph of Figure 8.8 for the exponential distribution of intervals between disasters and in the right graph for the Poisson distribution of counts of disasters.

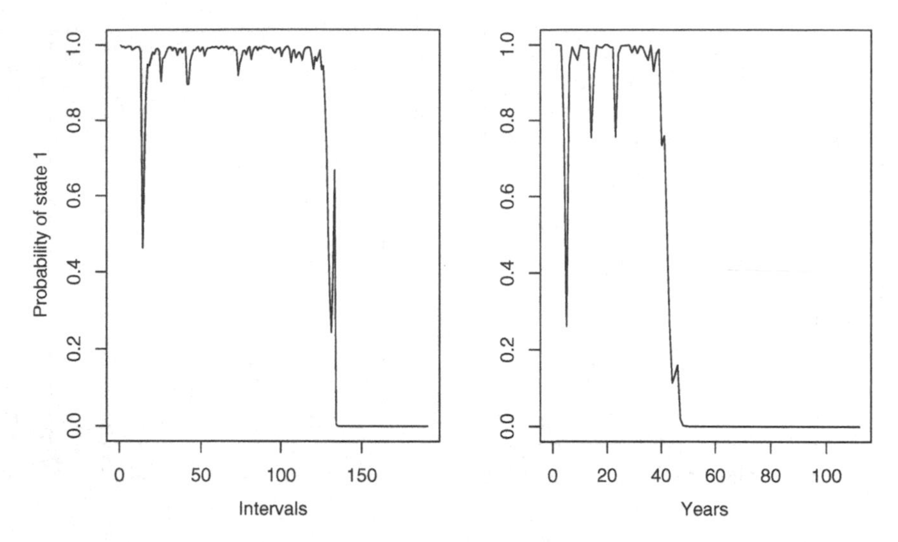

Fig. 8.8. The probability of being in the first hidden state of the hidden Markov model for the exponential (left) and Poisson (right) models for the mine disaster data of Table 8.8.

The first model points to a change after 133 intervals or about 45 years; the second indicates 46 years. These are both later than those found by the method in the previous section. The AICs are 1180.9 and 172.9, respectively, both worse than those for the corresponding change-point models above, although all have the same number of parameters. Here, the intensities are estimated to be 3.16 disasters per year before by both models and 0.93 and 0.92 by the two models after the change. The matrices of hidden transition probabilities are estimated, respectively, to be

$$\begin{pmatrix} 0.992 & 0.008 \\ 0.000 & 1.000 \end{pmatrix} \quad \text{and} \quad \begin{pmatrix} 0.974 & 0.026 \\ 0.000 & 1.000 \end{pmatrix}$$

showing a very high probability of staying in the first state.

A disadvantage of this approach is that hidden Markov models do not have an explicit parameter providing an estimate of when the change took place. Thus, it would be difficult to obtain a measure of precision of the estimate such as that given by the likelihood functions in Figure 8.7.

Further reading

For more details on marked point processes and doubly stochastic processes, see Cox and Isham (1980), Karr (1991), Kingman (1992), and Snyder and Miller (1991).

Exercises

8.1 In Section 7.1.2, I modelled animal learning data using autoregression.

Table 8.9. *Effect of treatment on disease course of ten EAE-induced mice over 40 days, with disease graded on an ordinal scale. 0: no abnormality; 1: floppy tail and mild hind leg weakness; 2: floppy tail with moderate hind leg weakness; 3: hind leg paresis without complete paralysis; 4: total paralysis of hind legs. (Albert, 1994)*

Placebo
0 2 3 4 4 4 4 4 3 3 3 2 2 2 1 1 2 2 3 4 4 4 4 4 4 4 4 4 4 4 3 3 3 3 3 2 2 2 2 3
0 0 1 2 3 3 3 3 3 2 2 1 1 1 1 1 1 1 1 1 0 0 0 0 1 0 1 2 4 4 3 4 3 3 3 2 2 2 3 4
0 0 1 2 2 2 2 2 1 1 1 1 1 1 1 1 1 1 2 3 2 2 2 1 0 0 2 2 3 3 3 3 3 3 3 3 3 3 3 3
0 1 2 3 4 4 3 3 3 2 2 2 2 2 1 1 2 2 2 3 2 2 2 2 2 1 0 0 0 0 0 1 1 1 1 1 2 3 1 3
0 0 1 2 3 3 3 3 2 2 2 1 1 1 1 1 1 1 2 2 2 2 2 1 1 1 1 2 2 3 3 3 2 2 2 2 3 3 2 2
Treated
0 1 1 2 2 2 2 2 1 1 0 0 0 0 0 1 3 4 4 3 3 2 2 1 1 0 0 0 0 0 0 0 0 0 1 1 1 1 2 3
0 0 0 0 1 2 1 1 1 0 0 0 0 0 0 0 0 0 0 0 0 0 0 0 0 0 0 0 1 1 1 1 1 1 1 1 1 1 2 3
0 0 0 0 0 0 0 1 1 2 2 1 1 0 0 0 1 1 1 1 2 1 1 0 0 0 0 1 1 1 2 2 2 2 1 1 0 0 0 3
0 0 0 0 0 0 0 0 0 0 0 0 0 0 0 0 0 0 0 0 2 1 1 0 0 0 0 0 0 1 1 1 1 1 1 1 1 1 2 1
0 0 0 0 0 0 0 0 0 0 0 0 0 0 0 0 0 0 0 0 0 0 0 0 0 0 1 1 1 1 1 1 1 1 1 2 0 0 0 0

(a) Can you fit an appropriate birth (learning) model to these data?

(b) Compare the results and the interpretation for the two approaches.

8.2 Let us look more closely at the data on precipitation at Snoqualmie Falls. Records are available for all days of the year between 1948 and 1983.

(a) First, extend the model in the text above for the January precipitation to a semi-Markov process.

(b) Now, fit appropriate models for the month of July and compare them to those for January.

(c) Finally, develop a model of the entire series over the 36 years. It is likely that you will have to incorporate a seasonality effect.

8.3 Experimental allergic encephalomyelitis (EAE) can be induced in mice; this is of interest because it closely mimics multiple sclerosis in human beings. Both are relapsing–remitting diseases characterised by inflammation and demyelination of the central nervous system. The diseases exhibit clinically nondeterministic fluctuations between worsening and improving symptoms.

EAE was induced in ten mice that were, then, randomly assigned either to placebo or to a therapeutic agent. Each was followed for 40 days for signs of disease, with possible states: no abnormality (0), floppy tail and mild hind leg weakness (1), floppy tail with moderate hind leg weakness (2), hind leg paresis without complete paralysis (3), and total paralysis of hind legs (4). The results are shown in Table 8.9.

(a) The goal of the study was to understand better the stochastic nature of the disease progression. Develop an appropriate model.

Table 8.10. *Times between four types of events in a baboon troop (Andersen* et al.*, 1992, p. 41, from Altmann and Altmann)*

Days	Troop size	Number	Event
41	40	1	Birth
5	41	1	Birth
22	42	1	Birth
2	43	1	Death
17	42	1	Death
26	41	2	Immigration
55	43	1	Birth
35	44	1	Immigration
20	45	1	Emigration
5	44	1	Death
6	43	1	Emigration
32	42	1	Death
4	41	2	Death
22	39	1	Death
10	38	2	Birth
7	40	1	Death
4	39	1	Birth
17	40	1	Death
11	39	1	Emigration
3	38	1	Birth
4	39	1	Death
8	38	1	Death
2	37	1	Death
5	36	1	Birth
10	37	1	Birth

(b) What differences are there between the placebo and the therapeutic agent?

8.4 A baboon troop was studied in the Amboseli reserve in Kenya for a period of a year. In the dynamics of social behaviour, birth–immigration and death–emigration are of interest. Thus, times of births, deaths, emigration, and immigration were recorded, as in Table 8.10. Here, we have four different events that may depend on the size of the troop and on the time since a previous event of the same type or since any previous event. There are ten births, 12 deaths, three immigrations, and three emigrations, for a total of 28 events in 373 days of observation.

(a) Is some type of birth process appropriate?

(b) Do the same factors influence all four types of events?

8.5 Baskerville *et al.* (1984) present a novel form of clinical trial for evaluating therapeutic differences. The general principle is that the best available treatment should be provided to each patient so that, when there are alternatives, the appropriate treatment must be determined for each patient.

Any given trial will have a fixed length, determined by the v regularly scheduled visits to a clinic, but will have variable treatment lengths. Thus,

Table 8.11. *Data for preferences from a cross-over trial involving the treatment of chronic asthma. (Baskerville* et al., *1984)*

Sequence		Sequence	
1	AABBBBBBB	4	BBBBBBBCA
	AAAAAAAAA		BBBBBBBCC
	AAAAABBBC		BCCCCCCCC
	ABBBBCCCC		BBBBBBCCA
	AAAAAABBC		BBCCCCCCA
	AAAAAAAAA		BCCCAAA
2	AAACBBBBB	5	CCABBBBBB
	AACCCCCCC		CAB
	ACCBBBBBB		CCAAAAABB
	AAAAAACCC		CAABB
	AAAAAAACC		CCCCCCABB
	ACB		CCAABBBBB
3	BBBBBBBBB	6	CBBBBBBBB
	BACCCCCCC		CBBBBBBBB
	BBBBBBAAA		CCCBBBBBB
	BBBBAC		CCCCBBBAA
	BACCCCCCC		CCCCBBBAA
	BBAAAAACC		CBBBA

each patient is assigned the series of treatments under study in a sequence determined randomly in the standard way for cross-over trials. However, as the trial proceeds, double blinded, each patient remains on the assigned first treatment in the sequence between all visits until either clinical deterioration or adverse effects of the treatment indicate that a change is necessary. Then, at the following visit, the patient is switched to the next treatment in the assigned sequence, and so on. Thus, certain patients may remain on the first treatment for all visits in the trial if that treatment proves effective or may run out of new treatments before the v visits and stop the trial if no treatment proves effective.

Baskerville *et al.* (1984) provide data on clinical decisions in a trial of three bronchodilators (A, B, C) in the treatment of chronic asthma involving 36 patients, as shown in Table 8.11. Nine weekly visits were programmed. Only three patients remained on the same treatment throughout the trial, two on A and one on B; six others completed the treatment sequence (that is, ran out of new treatments) and stopped early.

(a) Develop a model to describe the processes occurring in this study.

(b) Which treatment appears to be preferable?

8.6 Hæmolytic uræmic syndrome is a severe illness, primarily of infants and young children, and can be life threatening. It is associated with diarrhœa but its ætiology is unknown. Various bacterial and viral agents have been suspected, particularly an association with the level of *Escherichia coli* in

Table 8.12. *Annual numbers of cases of hæmolytic uræmic syndrome in Birmingham and Newcastle-upon-Tyne, UK, from 1970 to 1989. (Henderson and Matthews, 1993; read across rows)*

Birmingham									
1	5	3	2	2	1	0	0	2	1
1	7	11	4	7	10	16	16	9	15
Newcastle-upon-Tyne									
6	1	0	0	2	0	1	8	4	1
4	0	4	3	3	13	14	8	9	19

the environment. Health authorities in the UK were concerned that this illness may have increased in the 1980s. Annual numbers of cases were available from two specialist centres, in Birmingham and Newcastle-upon-Tyne, from 1970 to 1989, as shown in Table 8.12.

(a) Can you detect a change point in these data?
(b) If so, did it occur at the same time in the two centres?
(c) What peculiarities have you noticed with these data?

8.7 For the data in Table 4.3 on the numbers of accidents in a section of a mine:

(a) Fit a marked point process with the number of accidents as the mark.
(b) Compare your results with those for a standard Poisson process in Exercise 4.2 and for Markov chains in Exercise 5.10.

8.8 Above, I fitted a doubly stochastic process to the data on mine accidents in Table 8.6.

(a) These responses might also be thought to follow a marked process. Compare the fits of the two models.
(b) Still another possibility is a hidden Markov model. Does this provide any further insights?
(c) Which approach is more informative about what is occurring?
(d) Is this the best fitting model?

Part III

Continuous state space

9

Time series

In this chapter, I shall begin the exploration of models for stochastic processes having a continuous state space rather than a small set of states. The best known case involves the classical time series models, used so widely in econometrics.

Traditionally, any series of numbers over time would be considered to be a time series and standard methods based on the normal distribution applied, perhaps after taking logarithms. As we have already seen in Sections 1.2 and 7.1, such an approach certainly is not generally recommended; modern methods are available based on more reasonable distributional assumptions. Exceptionally, I shall follow this classical approach in this chapter in order to illustrate how time series models are still often applied.

Much of time series analysis, especially in econometrics, involves tests of various hypotheses. Well known cases include the Durbin–Watson, Chow, and reset tests. Here instead, I shall concentrate on developing appropriate *models* for a series. Equivalent information to that from such tests can be obtained by comparing models, but much more can be learnt from modelling than from testing.

9.1 Descriptive graphical techniques

A variety of graphical methods is available for preliminary analysis of time series data.

9.1.1 Graphics

As with any stochastic process, the first thing to do with a time series is to produce appropriate informative plots. With a continuous response, a fundamental plot will show the series against time. Often, series are first standardised to have zero mean and unit variance, primarily to obtain better numerical stability. I shall not follow that practice here.

Respiratory diseases Let us begin by looking at the data on deaths from bronchitis, emphysema, and asthma in the United Kingdom over a period of six years, given in Table 7.7 and plotted in Figure 9.1. As can be seen from this graph, a *seasonal component* is present, with the number of deaths varying regularly over

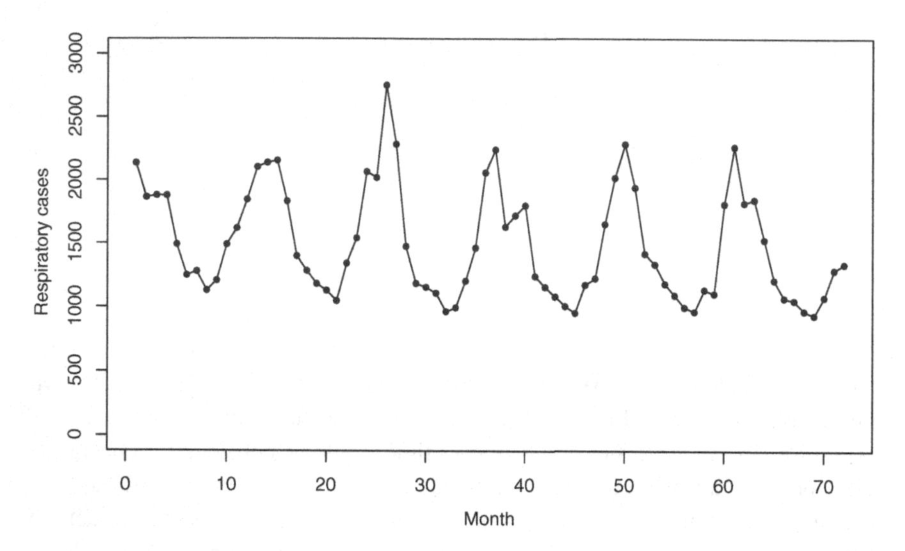

Fig. 9.1. Monthly male cases of bronchitis, emphysema, and asthma in the UK from 1974 to 1979, from Table 7.7.

each year. This is important short-term medical information. However, a more crucial question is to determine whether or not the number of deaths has changed systematically over the years, a *time trend.* The graph seems to indicate a slow but steady decrease with time. Thus, any model for this series should incorporate seasonality and a long term trend.

9.1.2 Correlograms

One important characteristic of time series data is that adjacent responses often may be related more closely than those more distantly separated in time. Several descriptive graphical methods are available for detecting such relationships.

Autocorrelation function

One useful tool for discovering dependencies in time series data is the empirical *autocorrelation function* (ACF). This function gives the correlation among responses in the series at various distances h in time, called *lags*. It is analogous to the autointensity function of Section 4.2.2, but will be used here to detect a quite different type of dependence. Thus, if the autocovariance between pairs of responses is

$$\gamma(h) = \text{cov}(Y_t, Y_{t-h}) \tag{9.1}$$

estimated by

$$\hat{\gamma}(h) = \sum_{t=h+1}^{R} (y_t - \hat{\mu})(y_{t-h} - \hat{\mu}) \tag{9.2}$$

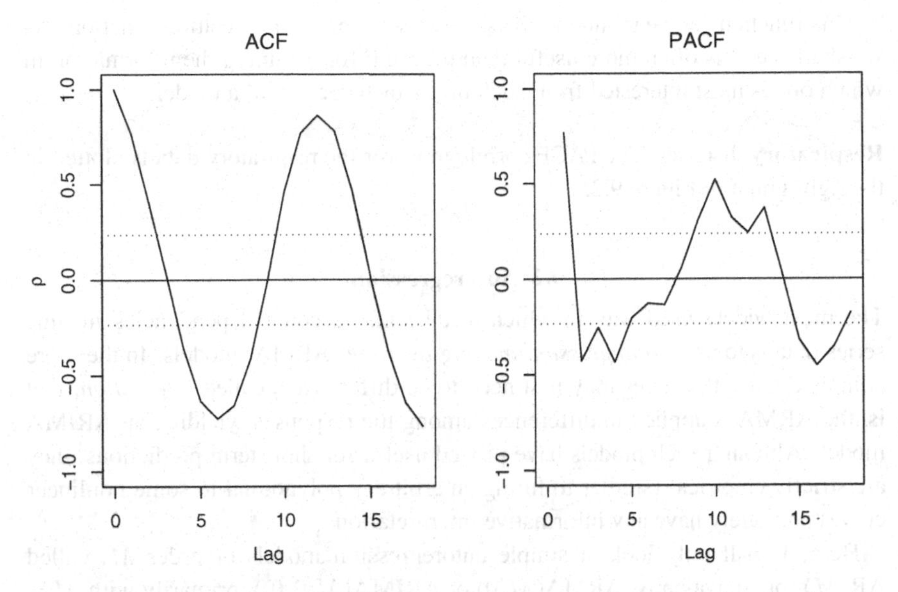

Fig. 9.2. The empirical autocorrelation and partial autocorrelation functions for the cases of bronchitis, emphysema, and asthma in Table 7.7.

then the autocorrelation is

$$\rho(h) = \frac{\gamma(h)}{\gamma(0)} \tag{9.3}$$

In practice, a reasonable value for the maximum of h to use in the calculations is usually about one-quarter of the length of the series.

A plot of the ACF is called a *correlogram*. Thus, a correlogram is a graph showing the correlations among responses in the series at a set of different distances apart in time.

Respiratory diseases For the respiratory data, the ACF correlogram is plotted in the left graph of Figure 9.2. We clearly can see the cyclical dependence induced by the seasonal changes in the series, with a period of about 12 months.

Partial autocorrelation function

The partial autocorrelation function (PACF) $\rho_p(h)$ is the correlation between Y_{t-h} and Y_t, with the latter conditional (autoregressed) on $y_{t-h+1}, \ldots, y_{t-1}$ (Sections 1.2 and 9.2). Such a procedure is computationally expensive, but a simple recursion can be used instead, based on the autocorrelation:

$$\begin{aligned} \widehat{\rho_p}(j,h+1) &= \widehat{\rho_p}(j,h) - \widehat{\rho_p}(h+1)\widehat{\rho_p}(h-j+1,h) \\ \widehat{\rho_p}(h+1) &= \frac{\hat{\rho}(h+1)-\sum_{j=1}^{h}\widehat{\rho_p}(j,h)\hat{\rho}(h+1-j)}{1-\sum_{j=1}^{h}\widehat{\rho_p}(j,h)\hat{\rho}(j)} \end{aligned} \tag{9.4}$$

where $\rho_p(j,h), j = 1, \ldots, h$ are the (auto)regression coefficients of y_t on $y_{t-1}, \ldots,$ y_{t-h}, with $\rho_p(h,h) = \rho_p(h)$ (Diggle, 1990, p. 168).

This function can be plotted in the same way as the autocorrelation function. As we shall see, it is often more useful than the ACF for obtaining the information in which one is most interested from such plots: the order M of a model.

Respiratory diseases The PACF correlogram for the respiratory data is plotted in the right graph of Figure 9.2.

9.2 Autoregression

The most widely used way in which to take into account dependencies in time series data is to use *autoregression-moving average* (ARMA) models. In the more complex cases, the series may first need to be differenced, called *integration*, that is, the ARMA is applied to differences among the responses, yielding an ARIMA model. Although such models have proved useful for short-term predictions, they are strictly empirical (similar to fitting an arbitrary polynomial to some nonlinear curve) and rarely have any informative interpretation.

Here, I shall only look at simple autoregression models of order M, called AR(M) (or equivalently, ARMA(M,0) or ARIMA(M,0,0)), primarily with $M = 1$, and at first differencing, ARIMA(M,1,0) models. In contrast to the more complex models, these can have an interpretation. I shall not cover models involving moving averages.

9.2.1 AR(1)

The simplest, and most easily interpretable, model is an autoregression of order one ($M = 1$). In its simplest form, without covariates, this is the state dependence model of Section 1.2.3; with covariates, it is the serial dependence model of Section 1.2.4. It has an ACF decreasing as a power of the autocorrelation parameter: $\rho^{|h|}$.

If one is not sure that a first order ($M = 1$) model is appropriate, a plot of the partial autocorrelation function (PACF) can be useful in attempting to determine the order. It should be zero for lags greater than M. In contrast, the ACF for such models will decrease slowly, not clearly indicating the order.

An AR(1) can be specified in a number of closely related ways.

(i) The most straightforward is conditionally, as a special case of Equation (1.7):

$$\mu_t^R = \rho[y_{t-1} - h(\boldsymbol{\beta}, \mathbf{x}_{t-1})] + h(\boldsymbol{\beta}, \mathbf{x}_t) \tag{9.5}$$

where μ_t^R is the conditional or recursive mean of a normal distribution. As we have seen (Sections 1.1.4 and 1.2.1), the advantage of this approach is that it involves a product of univariate conditional distributions.

(ii) A second possibility is to work directly with the multivariate normal distribution having an appropriately structured covariance matrix, as will be described shortly.

(iii) A third is to construct the model dynamically, as in Chapter 7, but here using the Kalman filter and smoother. I shall delay introduction of this approach until Section 11.1.

If there are no other covariates besides time, the marginal mean, $\mu_t = h(\boldsymbol{\beta}, \mathbf{x}_t)$, may just be a regression function describing a time trend and/or seasonal effects. Classically (and in this chapter), $h(\boldsymbol{\beta}, \mathbf{x}_t)$ has been taken to be a linear function of the parameters (which does not necessarily make Equation (9.5) linear); in Chapters 12 and 13, I shall concentrate on nonlinear models.

Conditional approach

Throughout previous chapters, I primarily have used a conditional approach to constructing models. This is the first specification of an AR(1) above. Thus, as we can see from Equation (9.5), a first-order autoregression model conditions on the previous residual $y_{t-1} - h(\boldsymbol{\beta}, \mathbf{x}_{t-1})$, called an *innovation* in this context. Except in simple cases, the regression function will be nonlinear in the unknown parameters, even if $h(\boldsymbol{\beta}, \mathbf{x}_t)$ is linear in them. This is because the model contains both $\rho h(\boldsymbol{\beta}, \mathbf{x}_{t-1})$ and $h(\boldsymbol{\beta}, \mathbf{x}_t)$.

Unfortunately, for the first response, the previous innovation will not be available; this observation is often ignored if the series is reasonably long. If this is unacceptable, the first response can be modelled by a separate normal distribution with a different variance (if several replications of the series are available, a different mean might also be estimated).

A third possibility is to assume that the time series is stationary. Then, the marginal distribution at the first time point, given the marginal mean, will be the stationary distribution. This is a normal distribution with variance $\sigma^2/(1-\rho^2)$, where σ^2 is the conditional variance and ρ the autocorrelation. This clearly implies that $|\rho| < 1$ for stationarity to hold. In contrast to the previous solution, here no new parameters are introduced.

Covariance matrix

When stationarity can be assumed, the second specification of an AR(1) above can also be used: direct construction of a multivariate normal distribution. Combining the univariate normal distributions, with means given by Equation (9.5), using Equation (1.1) yields

$$f(\mathbf{y}; \boldsymbol{\mu}, \boldsymbol{\Sigma}) = \frac{1}{\sqrt{2\pi|\boldsymbol{\Sigma}|}} \mathrm{e}^{\boldsymbol{\mu}^\top \boldsymbol{\Sigma}^{-1} \boldsymbol{\mu}/2} \tag{9.6}$$

where $\boldsymbol{\mu}$ is the vector of marginal means $\mu_t = h(\boldsymbol{\beta}, \mathbf{x}_t)$ and

$$\boldsymbol{\Sigma} = \frac{\sigma^2}{1-\rho^2} \begin{pmatrix} 1 & \rho & \cdots & \rho^{R-2} & \rho^{R-1} \\ \rho & 1 & \cdots & \rho^{R-3} & \rho^{R-2} \\ \vdots & \vdots & \ddots & \vdots & \\ \rho^{R-2} & \rho^{R-3} & \cdots & 1 & \rho \\ \rho^{R-1} & \rho^{R-2} & \cdots & \rho & 1 \end{pmatrix} \tag{9.7}$$

for a series of length R.

The inverse of this covariance matrix, called the *concentration matrix*, has a special form, with zeros everywhere except on the main and first minor diagonals:

$$\boldsymbol{\Sigma}^{-1} = \frac{1}{\sigma^2}\begin{pmatrix} 1 & -\rho & \cdots & 0 & 0 \\ -\rho & 1+\rho^2 & \cdots & 0 & 0 \\ \vdots & \vdots & \ddots & \vdots & \vdots \\ 0 & 0 & \cdots & 1+\rho^2 & -\rho \\ 0 & 0 & \cdots & -\rho & 1 \end{pmatrix} \tag{9.8}$$

In addition, it has a simple Cholesky decomposition $\boldsymbol{\Sigma}^{-1} = \mathbf{A}^\top \mathbf{A}$, where

$$\mathbf{A} = \frac{1}{\sigma}\begin{pmatrix} \sqrt{1-\rho^2} & 0 & \cdots & 0 & 0 \\ -\rho & 1 & \cdots & 0 & 0 \\ \vdots & \vdots & \ddots & \vdots & \vdots \\ 0 & 0 & \cdots & 1 & 0 \\ 0 & 0 & \cdots & -\rho & 1 \end{pmatrix} \tag{9.9}$$

More generally, for an AR(M), the number of non-zero minor diagonals in $\mathbf{A}$ corresponds to the order M of the autoregression.

Continuous time

Let us now consider the modifications required to work in continuous time. As we have seen, the autocorrelation function in Equation (9.7) can be rewritten as

$$\rho(h) = \rho^{|h|} \tag{9.10}$$

with $|\rho| < 1$, so that it decreases as the distance h in time increases. In discrete time, h only takes integer values, so that ρ can have negative values. However, negative values are not possible in continuous time because responses can be arbitrarily close together and, hence, must be positively correlated. Thus, in this latter case, a better formulation is obtained by setting

$$\rho(h) = \mathrm{e}^{-\kappa|h|}, \qquad \kappa > 0 \tag{9.11}$$

where $\rho = \mathrm{e}^{-\kappa}$. This is called an *exponential autocorrelation function*.

Such a continuous AR(1), or a CAR(1), process can be defined by the linear differential equation

$$\mathrm{E}(\mathrm{d}Y_t) = -\kappa y_t \mathrm{d}t \tag{9.12}$$

the mean of a normal distribution with variance $\sigma^2\mathrm{d}t$.

Autocorrelation functions

Other, more nonstandard, autocorrelation functions are also possible. One generalisation of the exponential ACF is

$$\rho(h) = \mathrm{e}^{-\kappa h^\nu}, \qquad \kappa, \nu > 0 \tag{9.13}$$

so that Equation (9.11) is a special case with $\nu = 1$.

Several other functions are also less common but may be useful in certain contexts.

Table 9.1. *Fits of various models to the monthly numbers of male deaths from bronchitis, emphysema, and asthma in Table 7.7.*

	Normal		Log normal	
	Independence	AR(1)	Independence	AR(1)
Null model	541.8	510.9	535.4	500.4
Linear trend	539.8	511.2	532.9	500.6
One harmonic	486.2	482.7	473.3	468.0
Two harmonics	481.5	479.3	471.6	467.1
Three harmonics	481.9	479.2	471.5	465.4
Trend + two harmonics	480.4	479.0	458.9	459.5

- The *Gaussian autocorrelation function* has

$$\begin{aligned} \rho(h) &= \rho^{h^2} \\ &= \mathrm{e}^{-\kappa h^2} \end{aligned} \tag{9.14}$$

another special case of Equation (9.13) with $\nu = 2$.

- The *Cauchy autocorrelation function* has

$$\rho(h) = \frac{1}{1+\rho h^2} \tag{9.15}$$

- The *spherical autocorrelation function* has

$$\rho(h) = \begin{cases} 0.5[(\rho h)^3 - 3\rho h + 2] & h \leq \frac{1}{\rho} \\ 0 & h > \frac{1}{\rho} \end{cases} \tag{9.16}$$

Still another function, which tends asymptotically to the exponential function of the standard CAR(1), will be introduced in Section 10.2.1: that for an integrated Ornstein–Uhlenbeck process.

Respiratory diseases Let us now attempt to model the series of male deaths from bronchitis, emphysema, and asthma in Table 7.7. These are count responses, unlikely to follow a normal distribution. Nevertheless, as is typical of time series applications, I shall apply such models here. I shall first determine the type of regression function required before adding the autoregression component. I shall use seasonal harmonics for variations of a 12-month period, fitting the models using the multivariate normal distribution of Equation (9.6). Thus, I shall be assuming stationarity, after accounting for seasonal and trend effects.

From the first column of Table 9.1, we see that a model with linear trend and two harmonics is required. This may be written

$$h(\boldsymbol{\beta}, \mathbf{x}_t) = \beta_0 + \beta_1 \cos(\beta_2 + \pi t/6) + \beta_3 \cos(\beta_4 + \pi t/3) + \beta_5 t \tag{9.17}$$

In the cosine function, $\pi/2$ is multiplied by 12 and 6 for the first two harmonics. The slope of the time trend is given by β_5.

When the AR(1) dependence is added, this greatly improves the simpler, poorly fitting, models, but makes only a small difference for the better models, as can be

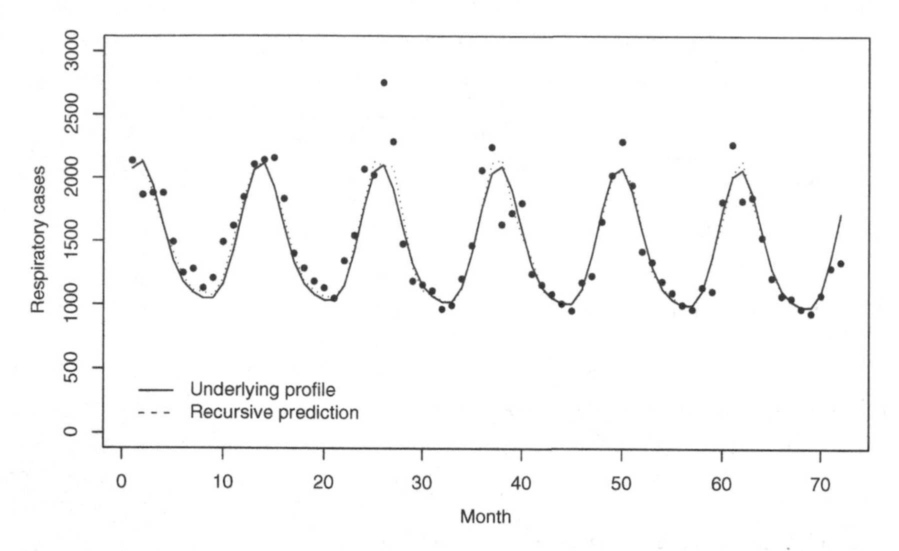

Fig. 9.3. The fitted normal AR(1) model for the cases of bronchitis, emphysema, and asthma in Table 7.7.

seen in the second column of Table 9.1. The autocorrelation for the best model is estimated to be $\hat{\rho} = 0.27$. The curve estimated for Equation (9.17) is plotted in Figure 9.3 as the underlying marginal profile. The *recursive predictions* in the same graph are $\hat{\mu}_t^R$ in Equation (9.5), that is, the marginal profile corrected by the previous marginal residual.

To check the model further, the empirical (P)ACF can be calculated and plotted for the *recursive residuals* obtained from it. These are the differences between the observed series and the recursive predictions. As might be expected, they are generally smaller than the marginal residuals. The results are shown in Figure 9.4. The fit appears to be reasonable: only two correlations out of 18 are outside the limits in the ACF and one in the PACF. Finally, the linear trend is estimated to be $\widehat{\beta}_5 = -0.0025$ indicating that the number of cases was decreasing over the observation period, as expected.

9.2.2 Transformations

Often in time series modelling, a log transformation is used, whether to 'stabilise' the variance or because the distribution is skewed. This yields a log normal distribution. Care must be taken to include the Jacobian of the transformation $(1/y)$ in the likelihood for the AICs to be comparable.

When the regression function of a log normal distribution is transformed back to the scale of the responses,

$$\mu_t^{gm} = \mathrm{e}^{h(\boldsymbol{\beta},\mathbf{x}_t)} \tag{9.18}$$

it refers to the *geometric* mean of the responses. For this distribution, the *arithmetic*

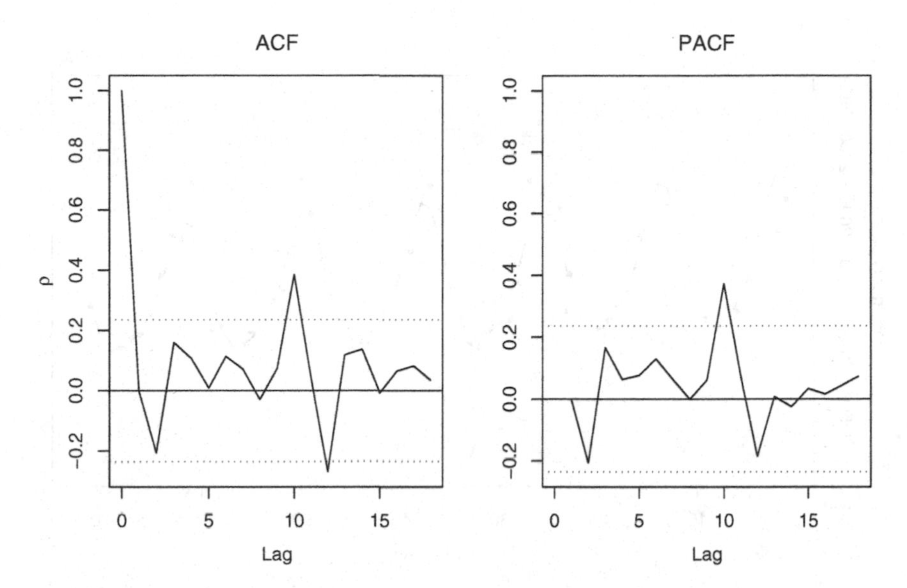

Fig. 9.4. The empirical autocorrelation and partial autocorrelation functions for the residuals from the fitted normal distribution AR(1) model for the cases of bronchitis, emphysema, and asthma in Table 7.7.

mean is obtained as

$$\begin{aligned} \mu_t^{am} &= \mathrm{E}(Y_t) \\ &= e^{h(\boldsymbol{\beta}, \mathbf{x}_t)+\sigma^2/2} \end{aligned} \tag{9.19}$$

so that it depends on the variance. For other distributions, it will be more complex.

Respiratory diseases Let us repeat the analysis of male deaths from bronchitis, emphysema, and asthma in Table 7.7 applying a log transformation to the responses. The results are shown in the last two columns of Table 9.1. We see that the fit is considerably improved. Moreover, here, when a suitable regression function in time is used, the AR(1) is no longer necessary. The marginal regression curve, transformed back to the original scale, is plotted in Figure 9.5. (Because there is no autocorrelation, there are no recursive predictions.) Inspection of the autocorrelation functions of the marginal residuals (there are no recursive ones) in Figure 9.6 reveal no anomalies. Here, the slope is $\widehat{\beta}_5 = -0.0030$, somewhat steeper than for the normal model, but the scale is not the same.

Because of the relationships between the arithmetic and geometric means given above, the graphs in Figures 9.3 and 9.5 are not strictly comparable. However, for these responses, the variance is estimated to be so small ($\hat{\sigma}^2 = 0.0080$) that the difference between the two is slight.

I shall look at another way to approach the periodicity of this series in Section 9.3.1 below.

In the above example, it was possible to eliminate the serial dependence after

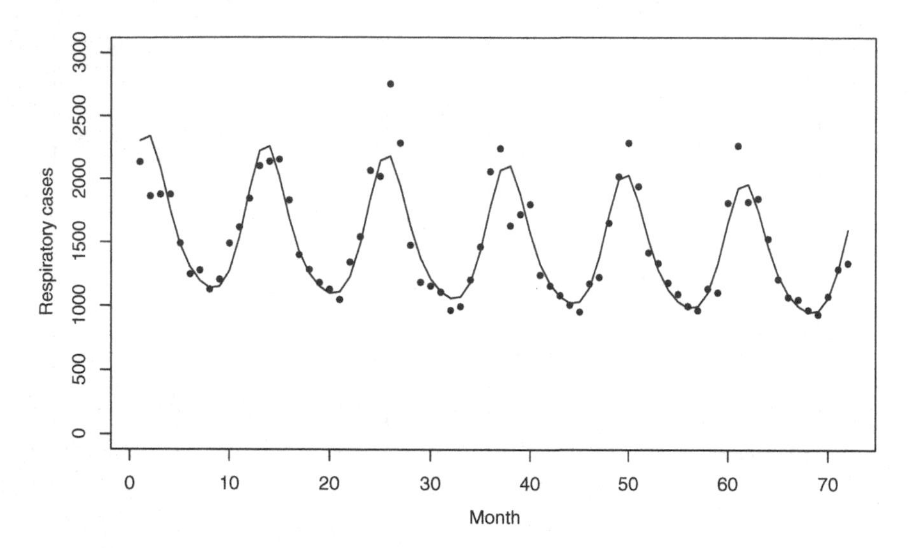

Fig. 9.5. The fitted independent log normal model for the cases of bronchitis, emphysema, and asthma in Table 7.7.

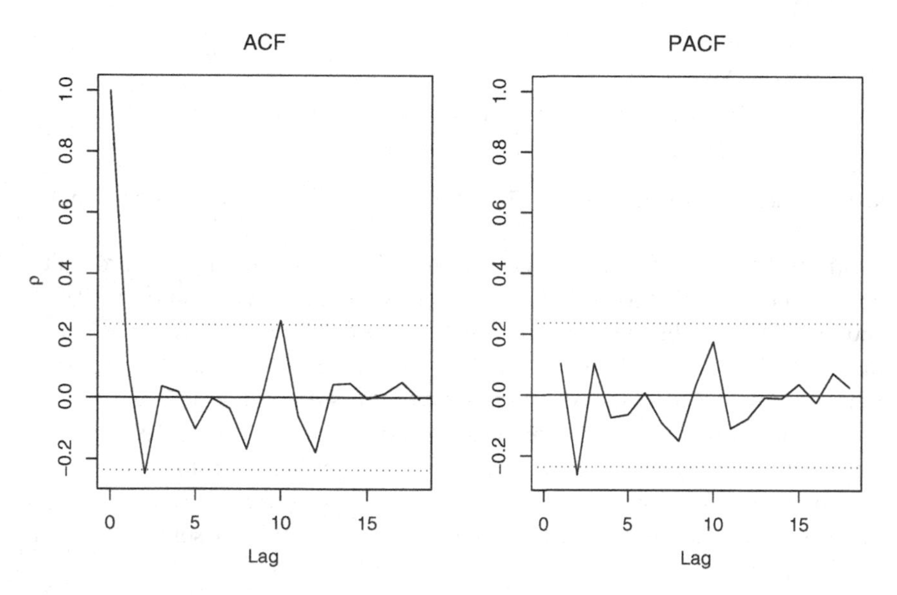

Fig. 9.6. The empirical autocorrelation and partial autocorrelation functions for the residuals from the fitted independent log normal model for the cases of bronchitis, emphysema, and asthma in Table 7.7.

finding an appropriate distribution (log normal) and regression function (seasonality and time trend). In principle, this often should be possible, but generally regression functions of time will not be sufficient. Usually, introduction of appropriate time-varying covariates will be necessary in order to eliminate serial dependencies.

9.2.3 Random walks

When a series is *nonstationary*, one possibility is to difference it to determine if the changes in response are stationary. If first differences are taken, this is equivalent to an AR(1) with $\rho = 1$, as can be seen from Equation (1.8)

$$\mu(t) - h(\boldsymbol{\beta}, \mathbf{x}_t) = \rho[y_{t-1} - h(\boldsymbol{\beta}, \mathbf{x}_{t-1})]$$

If there are no time-varying covariates so that $h(\boldsymbol{\beta}, \mathbf{x}_t) = h(\boldsymbol{\beta}, \mathbf{x}_{t-1})$, this simplifies to

$$\begin{aligned} E(Y_t) &= \mu(t) \\ &= y_{t-1} \end{aligned} \tag{9.20}$$

Such models provide another case of a *random walk* (Section 5.1.3). In this case, Y_t in Equation (5.12), which is here in fact the first difference $Y_t - Y_{t-1}$, can take any value on the real line.

If $h(\boldsymbol{\beta}, \mathbf{x}_t) - h(\boldsymbol{\beta}, \mathbf{x}_{t-1})$ is not equal to zero, the random walk has *drift*. If this difference is not a constant, the drift is changing over time. If an autoregression of order M is still required for the differences, this is an ARIMA(M,1,0) model. If not, the process has independent increments (Section 4.1.1).

Yale enrolment In Section 7.1, I looked at data on student enrolment at Yale University from 1796 to 1975. The series was given in Table 7.1 and plotted in Figure 7.1. As with the respiratory data previously modelled in this chapter, these are count responses for which normal distribution models really are not appropriate.

As it stands, this series is nonstationary in the mean. The empirical autocorrelation functions are plotted in Figure 9.7, showing very high correlations, a result of the nonstationarity. The series might well be modelled by some form of growth curve, to be presented in Chapter 12. However, here I shall look at another aspect of these data.

For the respiratory data above, I used the multivariate normal distribution of Equation (9.6), hence assuming that the series was stationary. Here instead, I shall use the conditional approach. Thus, I shall fit a standard univariate normal regression (Section 2.2) to the Yale series (except the first value), conditional on the lagged (previous) value, a state dependence model as in Equation (1.6). The estimate of the intercept is 52.9 and of the slope one (AIC 1353.3). Notice that this is a nonstationary model because $\hat{\rho} = 1$; it could not be fitted using the multivariate normal distribution directly. It is a random walk with drift, whereby enrolment is increasing on average by about 53 students per year. However, the drift is not necessary; eliminating it gives an AIC of 1352.9 with $\hat{\rho} = 1.01$, so that $\widehat{\mu_{1t}} \doteq y_{t-1}$.

This is essentially the same model as I fitted in Sections 7.1.1 and 7.1.3, except that the Poisson and negative binomial distributions there are here replaced by the normal distribution and the first value is not used. The estimated value of the slope

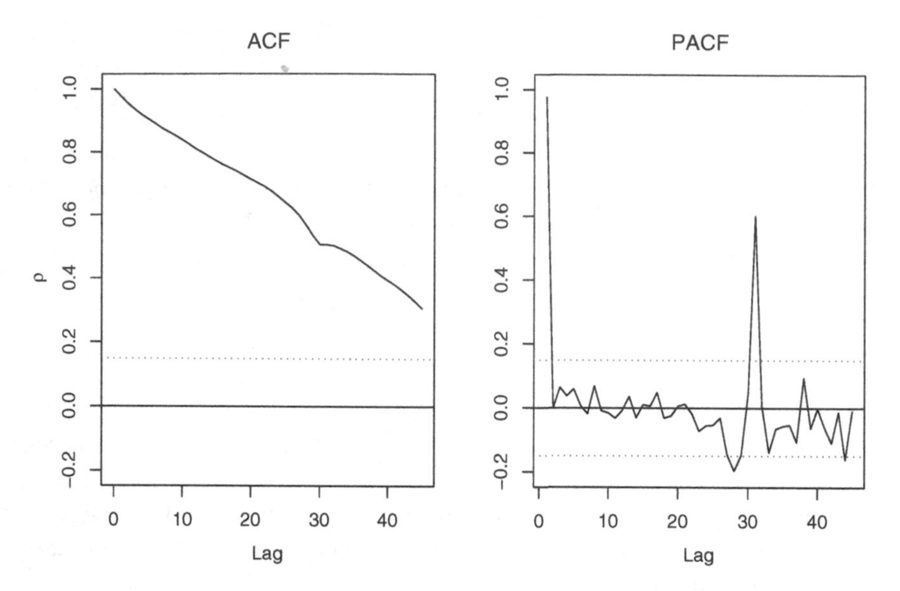

Fig. 9.7. The empirical autocorrelation and partial autocorrelation functions for enrolment at Yale University in Table 7.1.

is very close to that for the dependence parameter ρ in Chapter 7. However, the negative binomial serial dependence model fitted considerably better (AIC 1190.0, but using the first observation) than this normal model.

The changes (first differences) in enrolment were plotted in the middle graph of Figure 7.1. The curve is flat and the variability is low until about 1900, after which it increases greatly. The empirical autocorrelation functions for the first differences are plotted in Figure 9.8, showing that there is no longer an indication of autocorrelation (unless one is prepared to fit a very high order autoregression model): only a couple of values, of the 44 plotted, are outside the limits. If a first-order autocorrelation is added, the AIC increases. Thus, we have an ARIMA(0,1,0) model for the original series.

This random walk in discrete time is a special case of a Wiener diffusion process to be treated in more detail in Section 10.1.

9.2.4 Heteroscedasticity

One of the strong assumptions of standard normal distribution models is that the variance remains constant. If it does not, this is called *heteroscedasticity*. Much work has been done on this problem in time series analysis. Recall, first, that if a log transformation is applied, although the log responses will have constant variance, the original responses will not.

If a transformation is not sufficient, in simple cases, the variance may be allowed to depend deterministically in some way on time. Because the variance must be positive, it is usually most appropriate to construct regression functions in terms of

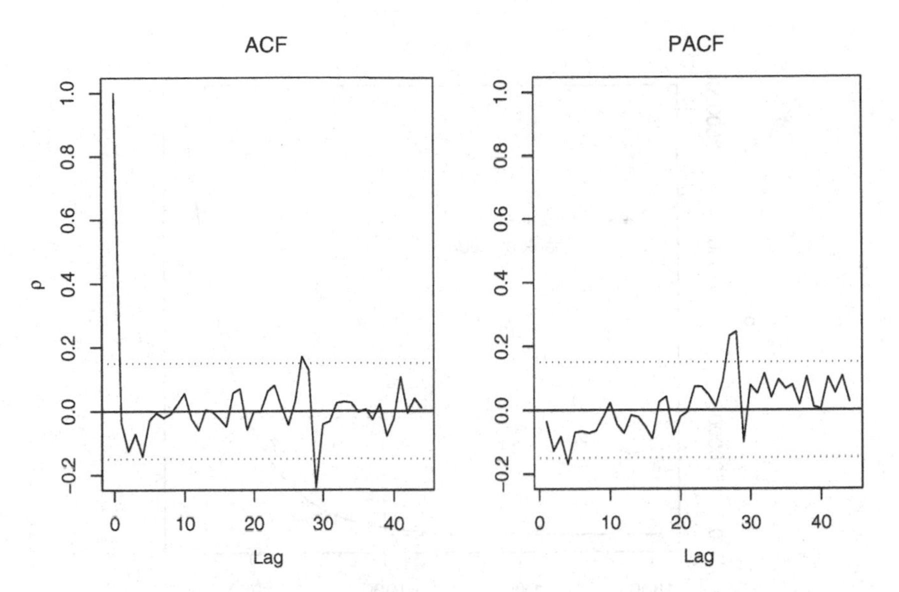

Fig. 9.8. The empirical autocorrelation and partial autocorrelation functions for the first differences of the enrolment at Yale University in Table 7.1.

the log variance. More complex methods allow the variance to depend on previous observed values or on the previous innovations (see Section 10.4). However, often heteroscedasticity indicates that the normal distribution assumption itself is inappropriate.

Yale enrolment For the Yale enrolment data, we saw in Figure 7.1 that the variability increases over time. For the negative binomial distribution, I already modelled this in Section 7.1.4. Here, let us look at two simple possibilities: allowing the log variance to depend linearly on time or to change abruptly at about 1900. The two estimated variance functions are plotted in Figure 9.9. For the random walk model (first differences), the first has an AIC of 1151.8 and the second 1164.1, both great improvements as compared to 1353.3 for the model with constant variance above. Thus, there does not appear to be a clear break, but rather a gradual change in variance.

A further possibility is that the variance (of the differences) depends on the previous level of enrolment. However, here this does not fit nearly as well: the AIC is 1307.4. A more sophisticated model might take the role of the two wars explicitly into account.

Notice that these models fit about as well as the corresponding negative binomial serial dependence models with nonconstant dispersion in Section 7.1 (the AICs here would be larger if the first response was included, as it was there). However, the negative binomial distribution is more acceptable for such data because it is designed for counts.

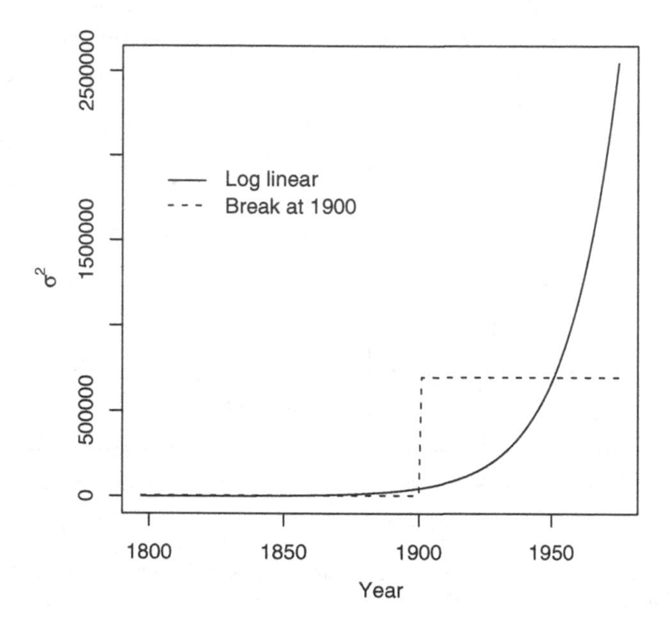

Fig. 9.9. Two variance functions for the changes in Yale University enrolment in Table 7.1.

9.3 Spectral analysis

Up until now, in this chapter, I have studied how a series of states evolves over time. This is called analysis in the *time domain*. Sometimes, the frequency with which the series varies is of special interest. One approach to examining this is by using harmonics in the regression function, as I did for the respiratory data in Section 9.2.1. Another way is to use Fourier or spectral analysis, called analysis in the *frequency domain*. This involves decomposing a function into a sum of sines and cosines. This is related to the examination of correlograms. Naturally, to model and to compare the periods of cyclic phenomena, fairly long series may be necessary, depending on the frequency of the cycles.

9.3.1 Periodograms

A complementary function to that expressed in a correlogram is the *spectral density function*. This can be written as

$$\begin{aligned} s(\omega) &= \sum_{h=-\infty}^{\infty} \gamma(h)\mathrm{e}^{-\mathrm{i}h\omega} \\ &= \gamma(0) + 2\sum_{h=1}^{\infty} \gamma(h)\cos(h\omega) \end{aligned} \tag{9.21}$$

where $\mathrm{i} = \sqrt{-1}$. The values ω are known as the *fundamental* or *Fourier frequencies*.

Estimation

The above definition involves an infinite series. However, in any empirical study, we only shall have a finite set of observations of length R. Thus, we shall require a representation of the spectral density in Equation (9.21) that is valid for a finite observed series. Let us, then, assume that the variation in the mean over time can be described by a sum of trigonometric functions

$$\begin{aligned} \mathrm{E}(Y_t) &= \frac{\beta_0}{2} + \sum_{m=1}^{[R/2]} [\beta_{1m}\cos(\omega_m t) + \beta_{2m}\sin(\omega_m t)] \\ &= \frac{\beta_0}{2} + \sum_{m=1}^{[R/2]} \upsilon_m \cos(\omega_m t + \nu_m) \end{aligned} \tag{9.22}$$

where $\upsilon_m = \sqrt{\beta_{1m}^2 + \beta_{2m}^2}$ is the *amplitude* of the mth harmonic of the series, $\nu_m = \tan^{-1}(-\beta_{2m}/\beta_{1m})$ is the *phase*, and $\omega_m = \frac{2\pi m}{R}$ is the *frequency* of the mth cycle with *period* $\frac{R}{m}$.

Using (least squares) estimates of β_{lm}, we can obtain

$$\begin{aligned} \frac{R}{4}(\widehat{\beta_{1m}}^2 + \widehat{\beta_{2m}}^2) &= \frac{1}{R}\left\{\left[\sum_{t=1}^{R} y_t \cos(\omega_m t)\right]^2 + \left[\sum_{t=1}^{R} y_t \sin(\omega_m t)\right]^2\right\} \\ &= \hat{\gamma}(0) + 2\sum_{h=1}^{R-1} \hat{\gamma}(h)\cos(h\omega_m) \end{aligned} \tag{9.23}$$

This is called the *periodogram*. It can be rewritten as

$$\mathrm{I}(\omega_m) = \frac{R\upsilon_m^2}{4\pi} \tag{9.24}$$

where an additional factor of $1/\pi$ has been included.

From these results, we see that the periodogram is the required representation. Thus, estimates of the autocovariances can be substituted into Equation (9.24) to give an estimate of the periodogram. If it is plotted against the frequency ω_m, the total area under the curve will equal the variance of the series. (This is why Equation (9.24) contains the extra factor.) Thus, the periodogram is a decomposition of the variance of the series.

The periodogram, as it stands, however, is often not a good estimate of the spectral density function. It can tend to fluctuate wildly. To be of use, it must, then, be smoothed, say with some filter or weighted average.

Respiratory diseases For the series of male deaths from bronchitis, emphysema, and asthma in Table 7.7, the raw and smoothed periodograms are plotted in Figure 9.10. The raw periodogram has a maximum at a frequency of about 0.52, corresponding to a period of $2\pi/0.52 = 12.1$ months, or one year, as would be expected. Smoothing was performed by taking a three-point moving average. In this case, it does not appear to help interpretability.

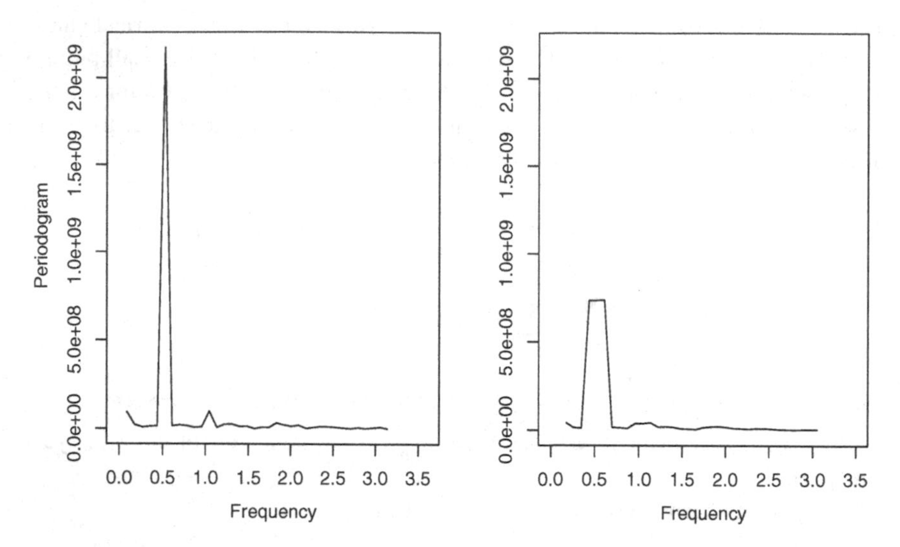

Fig. 9.10. The raw and smoothed periodograms for the cases of bronchitis, emphysema, and asthma in Table 7.7.

Cumulative periodogram

The cumulative periodogram may be used to check for departures from white noise, that is, from random sampling fluctuations. Under this assumption, it should increase approximately linearly. If it lies between a pair of parallel lines defining a likelihood interval, no such departures are indicated.

Respiratory diseases For the male deaths, the cumulative periodogram is plotted in Figure 9.11. Clearly, these data are not white noise.

9.3.2 Models for spectra

Because the periodogram involves variances, it has a χ^2 or gamma (in fact, exponential) distribution. Models can be fitted using this. Thus, periodograms from different series can be compared.

Autoregression

It is relatively simple to fit an autoregression of any order to a spectrum by using a nonlinear regression function with an exponential distribution (Cameron and Turner, 1987). For an AR(M), the theoretical spectral density function is

$$\begin{aligned}\mu_i &= \frac{\sigma^2}{2\pi|\sum_{j=0}^{M}\rho_j \mathrm{e}^{-\mathrm{i}j\omega_i}|^2} \\ &= \frac{\sigma^2}{2\pi\{[\sum_{j=0}^{M}\rho_j \sin(j\omega_i)]^2 + [\sum_{j=0}^{M}\rho_j \cos(j\omega_i)]^2\}} \qquad (9.25)\end{aligned}$$

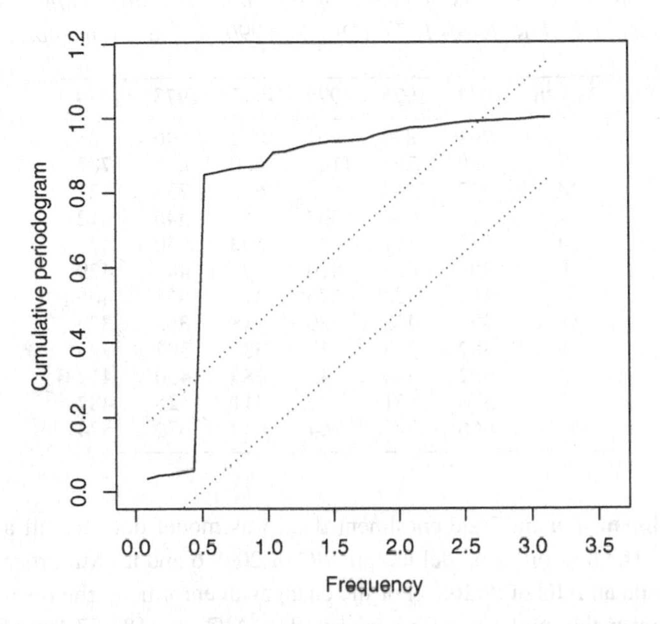

Fig. 9.11. The cumulative periodogram for the cases of bronchitis, emphysema, and asthma in Table 7.7, with likelihood intervals.

where the ρ_j are the negative of the coefficients of the autoregression with $\rho_0 = 1$. We can take σ^2 to be an additional unknown parameter.

Yale enrolment Consider the Yale enrolment data for which we found an AR(1) with $\rho = 1$ (here, this will be $-\rho_1$). The null (constant mean) model has an AIC of 2238.3. The above regression equation with $M = 1$ has 2007.3 whereas, with $M = 2$, it is 2008.3. The estimate of the autocorrelation is $-\widehat{\rho}_1 = 1.02$, confirming the previous results.

Applying the same method to the changes in enrolment yields AICs of 1878.8 for the null model, 1879.7 for $M = 1$, and 1879.4 for $M = 2$ (the data are different so that the AICs are not comparable with those above). Thus, as expected, there is no longer any indication of the need for an autoregression.

Exponential model

Bloomfield (1973) introduced an alternative to ARMA models that uses an exponential regression function for the spectral density:

$$\mu_i = \frac{\sigma^2}{2\pi} \exp\left[2 \sum_{j=1}^{M} \beta_j \cos(j\omega_i)\right] \tag{9.26}$$

This can more easily be fitted than the previous model because it is a generalised linear model with a gamma distribution having scale set to unity and a log link function.

Table 9.2. *Monthly numbers of female deaths from bronchitis, emphysema, and asthma in the UK, 1974–1979. (Diggle, 1990, p. 238, from Appleton)*

Month	1974	1975	1976	1977	1978	1979
J	901	830	767	862	796	821
F	689	752	1141	660	853	785
M	827	785	896	663	737	727
A	677	664	532	643	546	612
M	522	467	447	502	530	478
J	406	438	420	392	446	429
J	441	421	376	411	431	405
A	393	412	330	348	362	379
S	387	343	357	387	387	393
O	582	440	445	385	430	411
N	578	531	546	411	425	487
D	666	771	764	638	679	574

Yale enrolment For the Yale enrolment data, this model does not fit as well as the AR(1). The first-order model has an AIC of 2069.6 and a sixth order model is required, with an AIC of 2026.2. For the changes in enrolment, the results are the same as above: the model is not necessary. The AICs are 1879.7 for $M = 1$ and 1879.1 for $M = 2$ compared to 1878.8 for the null model.

Further reading

Chatfield (1989), Diggle (1990), and Janacek and Swift (1993) provide elementary introductions to time series. There are many texts containing more advanced material; see, among others, the standard texts by Brockwell and Davis (1996), Harvey (1989; 1993), and Kendall and Ord (1990).

For spectral analysis, see Bloomfield (1976) and Priestley (1981).

For a more general approach to continuous state spaces in discrete time than the classical time series literature, see Meyn and Tweedie (1993); their use of the term, Markov chain, is nonstandard.

Exercises

9.1 The data on monthly deaths from bronchitis, emphysema, and asthma are also available for women, as shown in Table 9.2.

(a) Does a similar model to that for the males provide an acceptable fit to these data?
(b) Is there a similar trend over time?
(c) Can you fit a model simultaneously to the data for the two sexes?
(d) Compare the periodograms of the two series using an exponential distribution model.

9.2 The counts of lynx trapped in the MacKenzie River District of Northwest

Table 9.3. *Canadian lynx trapped from 1821 to 1934. (Andrews and Herzberg, 1985, p. 14, from Elston and Nicholson; read across rows)*

269	321	585	871	1475	2821	3928	5943	4950	2577
523	98	184	279	409	2285	2685	3409	1824	409
151	45	68	213	546	1033	2129	2536	957	361
377	225	360	731	1638	2725	2871	2119	684	299
236	245	552	1623	3311	6721	4254	687	255	473
358	784	1594	1676	2251	1426	756	299	201	229
469	736	2042	2811	4431	2511	389	73	39	49
59	188	377	1292	4031	3495	587	105	153	387
758	1307	3465	6991	6313	3794	1836	345	382	808
1388	2713	3800	3091	2985	3790	374	81	80	108
229	399	1132	2432	3574	2935	1537	529	485	662
1000	1590	2657	3396						

Table 9.4. *Annual Wölfer sunspot numbers between 1770 and 1869. (Hand* et al. *1994, pp. 85–86; read across rows)*

101	82	66	35	31	7	20	92	154	125
85	68	38	23	10	24	83	132	131	118
90	67	60	47	41	21	16	6	4	7
14	34	45	43	48	42	28	10	8	2
0	1	5	12	14	35	46	41	30	24
16	7	4	2	8	17	36	50	62	67
71	48	28	8	13	57	122	138	103	86
63	37	24	11	15	40	62	98	124	96
66	64	54	39	21	7	4	23	55	94
96	77	59	44	47	30	16	7	37	74

Canada over 114 years, given in Table 9.3, are a classical time series that has been analysed many times. These data have traditionally been modelled by a log normal distribution.

(a) Is this an appropriate distribution for these data?
(b) What order of autoregression is necessary?
(c) Is the series stationary?
(d) Can you find any cycles in the data?

9.3 Some other classical data usually modelled by autoregression techniques involve the annual Wölfer sunspot numbers between 1770 and 1869 in Table 9.4. They measure the average number of sunspots on the sun each year.

(a) Can you find a Markov model to describe these data adequately? A number of different suggestions have been made in the literature.
(b) What order of model is required?

9.4 The yearly unemployment rates in the USA from 1890 to 1979 are given in Table 9.5. These are typical of time series modelled in econometrics.

Table 9.5. *Unemployment rates in the USA, 1890–1979. (Kramer and Sonnberger, 1986, pp. 166–167; read across rows)*

3.970	5.419	3.045	11.684	18.415
13.703	14.445	14.543	12.354	6.536
5.004	4.133	3.668	3.922	5.378
4.276	1.728	2.765	7.962	5.106
5.587	6.719	4.637	4.321	7.919
8.528	5.100	4.617	1.372	1.375
5.157	11.715	6.728	2.415	4.951
3.217	1.756	3.276	4.208	3.246
8.903	15.653	22.981	20.901	16.197
14.389	9.970	9.182	12.468	11.273
9.508	5.994	3.095	1.773	1.226
1.931	3.945	3.894	3.756	5.934
5.286	3.314	3.032	2.910	5.551
4.386	4.132	4.270	6.805	5.469
5.529	6.689	5.540	5.667	5.180
4.522	3.794	3.846	3.578	3.508
4.942	5.936	5.593	4.853	5.577
8.455	7.690	7.038	6.022	5.794

Table 9.6. *Annual snowfall (in) in Buffalo, New York, USA, 1910–1972 (Parzen, 1979; read across rows)*

126.4	82.4	78.1	51.1	90.9	76.2	104.5	87.4	110.5
25.0	69.3	53.5	39.8	63.6	46.7	72.9	79.7	83.6
80.7	60.3	79.0	74.4	49.6	54.7	71.8	49.1	103.9
51.6	82.4	83.6	77.8	79.3	89.6	85.5	58.0	120.7
110.5	65.4	39.9	40.1	88.7	71.4	83.0	55.9	89.9
84.8	105.2	113.7	124.7	114.5	115.6	102.4	101.4	89.8
71.5	70.9	98.3	55.5	66.1	78.4	120.5	97.0	110.0

(a) Develop a suitable model to describe these data.
(b) Can you detect any systematic periodicity?
(c) What predictions can you make for 1980?

9.5 Annual snowfall (in) in Buffalo, New York, USA, was recorded from 1910 to 1972 as shown in Table 9.6.

(a) Find an appropriate model to describe these data.
(b) Is there evidence of a trend or of a cyclical phenomenon?

10

Diffusion and volatility

Many observed stochastic processes may produce occasional very extreme values or periods with exceptional amounts of variability. Both of these phenomena are known as *volatility*. (In econometrics and finance, this term often has the more restricted meaning of models with changing variance over time.)

Two common approaches to modelling volatile phenomena are

(i) to use distributions with heavy tails and

(ii) to allow the dispersion to depend on some measure of the previous variability about the location regression function.

We already have met some volatility models related to the second approach in Sections 7.3 and 7.4. Here, I shall consider both approaches. However, I shall first look at an important class of processes in which the dispersion increases in a deterministic way with time: diffusion processes.

Because, in most non-normal models, the variance is a function of the mean, a location parameter changing over time implies that the variability, or volatility, is also changing. Thus, the distinctiveness of specifically modelling the volatility is most relevant for members of the location-scale family, such as the normal, Cauchy, Laplace, and Student t distributions, where the location and dispersion can vary independently of each other.

10.1 Wiener diffusion process

If the variance of a stochastic process increases systematically over time, a diffusion process may provide an appropriate model. One question will be whether the variance can increase without limit or slowly approaches some constant maximum. I shall consider the first possibility in this section.

10.1.1 Theory

Brownian motion may be one of the most famous stochastic processes, perhaps because of its connection with the movement of small particles floating on water. One model that commonly has been used to describe it is the Wiener diffusion process, defined by the Chapman–Kolmogorov differential equation

$$\frac{\partial f(y,t)}{\partial t} = -\delta\frac{\partial f(y,t)}{\partial y} + \frac{\nu}{2}\frac{\partial^2 f(y,t)}{\partial y^2} \tag{10.1}$$

The solution results in a multivariate normal distribution for Y with mean, variance, and covariance

$$\begin{aligned}\mu_t &= \alpha + \delta t \\ \sigma_t^2 &= \nu t \\ \gamma(Y_j, Y_k) &= \nu \min(t_j, t_k)\end{aligned} \tag{10.2}$$

where α gives the initial condition (location) of the process, assumed to be a fixed value without error, and δ is the drift. We see that, in this model, the variance can increase without limit.

Thus, this process is nonstationary. However, it has stationary, independent increments (Section 4.1.1) in time intervals Δt, each having an independent normal distribution with mean and variance

$$\begin{aligned}\mu_{\Delta t} &= \delta \Delta t \\ \sigma_{\Delta t}^2 &= \nu \Delta t\end{aligned} \tag{10.3}$$

(and, of course, zero covariance). Thus, $Y_t - Y_s$ is independent of Y_s for $t > s$. Hence, Y_t is a special case of a random walk (Section 5.1.3), one that can both increase and decrease. I mentioned, in Section 8.2, that this process is a member of the exponential dispersion family. As well, in Section 9.2.3, I fitted this type of model to the Yale enrolment data, as a random walk in discrete time with $\alpha = 0$.

An example of a Wiener process is plotted in Figure 10.1; we see that, even without drift, a Wiener process can stray far from its initial value. The path that such a process traces out has the particularity that it will be continuous but nowhere differentiable. This results because the correlations between nearby points are too small. Thus, if this process is used to describe the location of a particle, its velocity will not be defined. Although Wiener diffusion is sometimes called Brownian motion, it is, in fact, not a very good model for such movement because of this (Guttorp, 1995, p. 286).

When a Wiener diffusion process stops at a fixed boundary value $y_T = c$, the duration T of the process will have an inverse Gauss distribution, as given in Equation (2.14).

Measurement error

Suppose now that recording a Wiener process involves measurement error, also having a normal distribution, but with mean zero and constant variance σ^2. Then, the recorded responses y_t will have a multivariate normal distribution with the same mean and covariance as in Equation (10.2) but with variance now given by

$$\sigma_t^2 = \nu t + \sigma^2 \tag{10.4}$$

Notice that the first response, at $t = 0$, also is recorded with measurement error, so that it is now normal with mean α and variance σ^2.

Let us next examine what will happen to the first differences. These will no

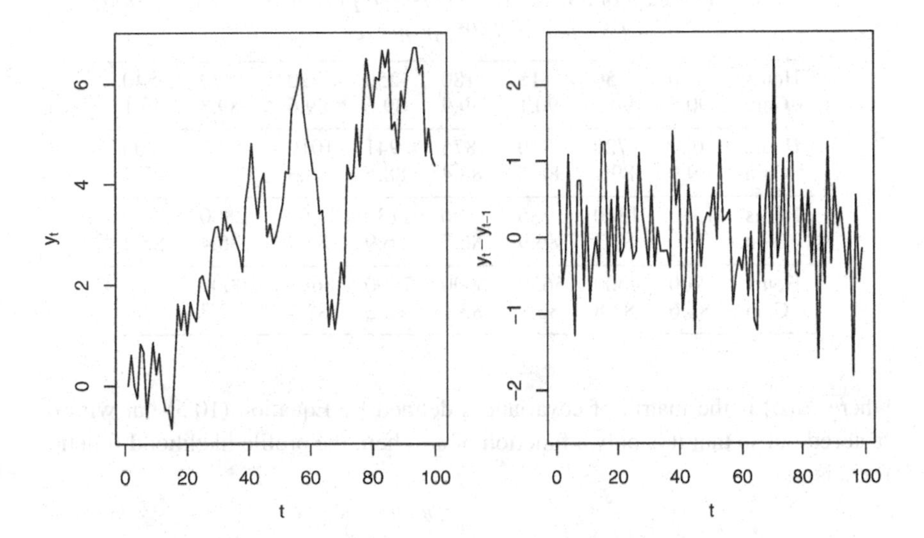

Fig. 10.1. A Wiener diffusion process with $\alpha = \delta = 0$, and $\nu = 0.5$ (left) and its increments (right).

longer be independent because the successive differences $y_{t+1} - y_t$ and $y_t - y_{t-1}$ both involve y_t and thus contain the same value of measurement error. One will be made larger and the other smaller. Instead of being independent, the first differences will have a multivariate normal distribution with same mean as in Equation (10.3), but with variance and covariance

$$\begin{aligned} \sigma^2_{\Delta t} &= \nu\Delta t + 2\sigma^2 \\ \gamma(\Delta Y_j, \Delta Y_k) &= \begin{cases} -\sigma^2 & j = k+1, k-1 \\ 0 & j > k+1 \text{ or } j < k-1 \end{cases} \end{aligned} \tag{10.5}$$

In order to be able to estimate α in a fitted model, the normal distribution of the undifferenced response at $t = 0$ must be included as the initial condition.

Likelihood function

As in many cases in which the normal distribution is involved, estimation of the parameters can be simplified by solving the score equations (first derivatives of the log likelihood function set to equal to zero) for certain parameters. Thus, if we set $\nu = \phi\sigma^2$,

$$\hat{\delta}(\phi) = \frac{\mathbf{\Delta t}^\top \mathbf{\Sigma}(\phi)^{-1}\mathbf{\Delta y}}{\mathbf{\Delta t}^\top \mathbf{\Sigma}(\phi)^{-1}\mathbf{\Delta t}} \tag{10.6}$$

and

$$\hat{\sigma}^2(\phi) = \frac{1}{R}(\mathbf{\Delta y} - \hat{\delta}(\phi)\mathbf{\Delta t})^\top \mathbf{\Sigma}(\phi)^{-1}(\mathbf{\Delta y} - \hat{\delta}(\phi)\mathbf{\Delta t}) \tag{10.7}$$

Table 10.1. *The gain of a transistor measured periodically over 10 000 h. (Whitmore, 1995, from Denton)*

Hours	0	50	115	180	250	320	420	540
Gain	90.9	90.3	90.1	89.9	89.6	89.6	89.3	89.1
Hours	630	720	810	875	941	1010	1100	1200
Gain	89.0	89.1	88.5	88.4	88.5	88.3	87.7	87.5
Hours	1350	1500	1735	1896	2130	2460	2800	3200
Gain	87.0	87.1	86.9	86.5	86.9	85.9	85.4	85.2
Hours	3900	4600	5650	6600	7800	8688	10000	
Gain	84.6	83.8	83.9	83.9	82.3	82.5	82.3	

where $\boldsymbol{\Sigma}(\phi)$ is the matrix of covariances defined by Equation (10.5) but with σ^2 factored out so that it is only a function of ϕ. Then, the profile likelihood function for ϕ is

$$L_p(\phi) = \frac{\mathrm{e}^{-\frac{R}{2}}}{[2\pi\hat{\sigma}^2(\phi)]^{\frac{R}{2}}\sqrt{|\boldsymbol{\Sigma}(\phi)|}} \tag{10.8}$$

from which it is easier to find the maximum likelihood estimate of ϕ than for all three parameters simultaneously (Whitmore, 1995).

10.1.2 Time transformations

In many situations, a process does not follow a linear drift. One way that has been suggested (Whitmore, 1995) to handle this is by transforming the times. A common type of nonlinear drift corresponds to the transform

$$g(t) = 1 - \mathrm{e}^{\lambda t^{\kappa}} \qquad \lambda, \kappa > 0 \tag{10.9}$$

This describes a slowing down of the process over real time.

The profile likelihood of Equation (10.8) will now contain the parameters λ and κ in addition to ϕ, complicating slightly the estimation. The equations for δ and σ^2 remain unchanged.

Transistor gains Engineers often are concerned with the degradation of materials or components. These may involve physical or chemical changes. Degradation tests attempt to measure such processes in the laboratory. The recordings made will incorporate the degradation process itself but also measurement errors due to the instruments, procedures, and environment.

The gain of a transistor is influenced by temperature and other stresses. It generally declines over time. Table 10.1 provides the results of one degradation test of a transistor. The plot of the responses given in Figure 10.2 shows that they roughly follow a curve such as that given in Equation (10.9). Whitmore (1995) suggests that $\kappa = 1$, obtained from a batch of similar transistors, where time is measured in thousands of hours.

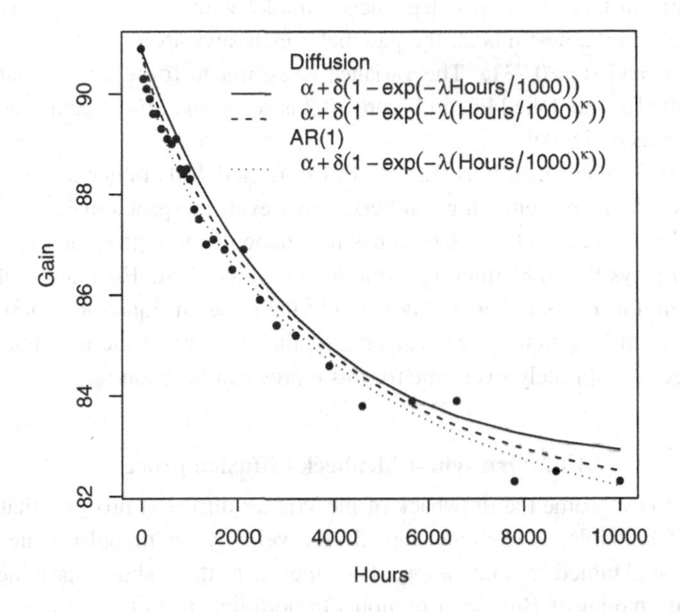

Fig. 10.2. The transistor gains from Table 10.1, with the curves for two Wiener diffusion models and the AR(1).

Table 10.2. *AICs for the models fitted to the transistor gains in Table 10.1.*

Model	Wiener	AR(1)	Ornstein–Uhlenbeck
$1-\mathrm{e}^{-\lambda t}$	11.4	8.3	8.2
$1-\mathrm{e}^{-\lambda t^{\gamma}}$	10.7	6.4	6.6

Fitting this model yields an AIC of 11.4. The parameter estimates are $\hat{\alpha}=90.9$, $\hat{\delta}=-8.31$, and $\hat{\lambda}=0.319$. The resulting regression curve is plotted as the solid line in Figure 10.2. The measurement error variance is estimated to be $\hat{\sigma}^2=0.0816$ and the diffusion variance to be $\hat{\nu}=0$. Estimating κ, instead of fixing it at unity, reduces the AIC slightly, to 10.7 with $\hat{\kappa}=0.855$. The measurement error variance is reduced to $\hat{\sigma}^2=0.0734$, whereas the estimates of the other parameters change very little. This curve also is plotted, as the dashed line, in Figure 10.2.

Because the diffusion variance is estimated to be zero, it will be informative to fit a standard nonlinear AR(1) model (Section 9.2.1) to these data. In order to have a model that varies over time in a way similar to the above time-transformed diffusion model, we can take the mean to change as

$$\mu_t=\alpha+\delta(1-\mathrm{e}^{-\lambda t^{\kappa}}) \tag{10.10}$$

The results are summarised in the first two columns of Table 10.2. The autoregression models fit considerably better than the Wiener diffusion models with measurement error. In fact, even an independence model with $\kappa = 1$ fits better, with an AIC of 8.6. In the best model, the parameter estimates are $\hat{\alpha} = 90.1$, $\hat{\delta} = -10.5$, $\hat{\lambda} = 0.338$, and $\hat{\kappa} = 0.731$. The variance is estimated to be $\hat{\sigma}^2 = 0.0865$. The curve plotted as the dotted line in Figure 10.2 is somewhat lower than those for the Wiener diffusion model.

Whitmore (1995) also finds that his diffusion model fits rather poorly, but tries to improve the fit by removing 'outliers'. However, the problem is that the time transform in Equation (10.9) determines the shape of the regression curve for the mean and plays the most important role in goodness of fit. Estimation of the parameters in it involves rather rigidly not only the mean in Equation (10.3) but also the variance in Equation (10.5). Apparently, the variance of these responses does not change appropriately over time for this regression function.

10.2 Ornstein–Uhlenbeck diffusion process

One way to overcome the drawback of the Wiener diffusion process, that its path is not differentiable, is to model directly the velocity. A model for the position can then be obtained by integration. This approach, then, should provide a more appropriate model of Brownian motion. In addition, it will yield a model with increasing variance over time that reaches a limiting constant value, instead of approaching infinity.

10.2.1 Theory

In order to describe a velocity, say $Z = \mathrm{d}Y/\mathrm{d}t$, the Chapman–Kolmogorov Equation (10.1) for the Wiener diffusion process must be modified. The drift will need to be linear in z and toward the origin (by changing the sign), rather than constant:

$$\frac{\partial f(z,t)}{\partial t} = \delta \frac{\partial z f(z,t)}{\partial z} + \frac{\nu}{2} \frac{\partial^2 f(z,t)}{\partial z^2} \tag{10.11}$$

In this way, the process is drawn back to the origin with a force proportional to its distance away. Then, the solution is a normal distribution for the velocity Z with mean and variance

$$\begin{aligned} \mu_t &= \alpha \mathrm{e}^{-\delta t} \\ \sigma_t^2 &= \tfrac{\nu}{2\delta}\left(1 - \mathrm{e}^{-2\delta t}\right) \end{aligned} \tag{10.12}$$

As $t \to \infty$, the model has an asymptotic stationary distribution which is normal with mean zero and variance $\nu/(2\delta)$. Thus, in contrast to the Wiener diffusion process, here the variance increases to a limiting maximum value (Gardiner, 1985, pp. 74–77).

The above model has the initial value α fixed, so that the variance is zero at $t = 0$. If instead the initial value is assumed to have a normal distribution with mean α and variance σ_0^2, the variance of the process becomes

$$\sigma_t^2 = \frac{\nu}{2\delta} + \left(\sigma_0^2 - \frac{\nu}{2\delta}\right) \mathrm{e}^{-2\delta t} \tag{10.13}$$

and the covariance function is

$$\operatorname{cov}(Z_t, Z_s) = \frac{\nu}{2\delta}\mathrm{e}^{-\delta|t-s|} + \left(\sigma_0^2 - \frac{\nu}{2\delta}\right)\mathrm{e}^{-\delta(t+s)} \tag{10.14}$$

Under these conditions, of course, the process also reaches the same stationary distribution (Gardiner, 1985, pp. 106–107).

In both cases, the stationary mean is zero and the covariance is

$$\operatorname{cov}(Z_t, Z_s) = \frac{\nu}{2\delta}\mathrm{e}^{-\delta|t-s|} \tag{10.15}$$

This is an AR(1) in continuous time, with $\rho = \mathrm{e}^{-\delta}$ in Equation (9.7). The Ornstein–Uhlenbeck process is the only stationary, normally-distributed, Markov process in continuous time.

Integrated Ornstein–Uhlenbeck process

The Ornstein–Uhlenbeck process is a model to describe velocity. The integrated Ornstein–Uhlenbeck process (Taylor *et al.*, 1994; Taylor and Law, 1998) will then describe the position Y_t. It has mean and variance

$$\begin{aligned} \mu_t &= 0 \\ \sigma_t^2 &= \tfrac{\nu}{\delta^3}\left(\mathrm{e}^{-\delta t} + \delta t - 1\right) \end{aligned} \tag{10.16}$$

with covariance given by

$$\operatorname{cov}(Y_t, Y_s) = \frac{\nu}{2\delta^3}\left[2\delta \min(t,s) + \mathrm{e}^{-\delta t} + \mathrm{e}^{-\delta s} + \mathrm{e}^{-\delta|t-s|} - 1\right] \tag{10.17}$$

This process is nonstationary.

Several special cases are noteworthy. With the ratio ν/δ^2 held constant, when $\delta \to \infty$ Brownian motion is obtained whereas, when $\delta \to 0$, a special case of a random effect model (Section 14.1) results.

10.2.2 Modelling velocity

Transistor gains The gain of a transistor might be considered to be a velocity. Hence, I shall now try fitting the Ornstein–Uhlenbeck process. (Note that the parameters will not have the same meaning as for the Wiener process.) For these responses, the asymptotic mean is not zero, so that the mean function has to be modified:

$$\mu_t = \alpha_0 + \alpha \mathrm{e}^{-\delta t} \tag{10.18}$$

Notice that this is just a reparametrisation of Equation (10.10) with $\kappa = 1$ so that this process has the same mean as that for the Wiener diffusion process with the time transform in Equation (10.9).

For these data, the AIC is 8.2, much better than the Wiener diffusion process and about the same as the AR(1) with $\kappa = 1$. If the power transformation of time is introduced into the mean function (but not the variance), the AIC is reduced to 6.6. These results are summarised in Table 10.2.

The parameter estimates are $\widehat{\alpha_0} = 81.0$, $\hat{\alpha} = 9.7$, $\hat{\delta} = 0.319$, $\hat{\sigma}_0^2 = 0.0158$, $\hat{\sigma}^2 =$

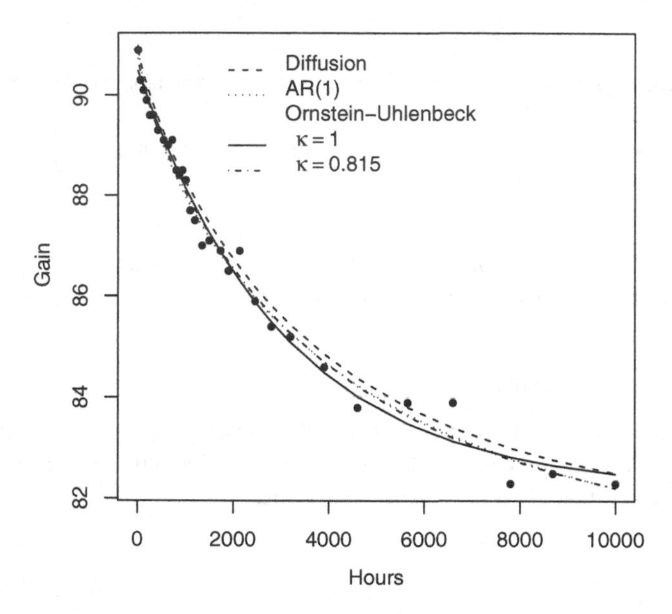

Fig. 10.3. The transistor gains from Table 10.1 with the Wiener diffusion model, the AR(1), and the two Ornstein–Uhlenbeck models.

0.0720, and $\hat{\kappa} = 0.815$. The two models are plotted in Figure 10.3, with the curves for the Wiener diffusion process and the AR(1) from Figure 10.2, the latter two both with κ estimated. The Ornstein–Uhlenbeck process and the AR(1), both with κ estimated, give very similar curves.

Thus, we can conclude that these data are closer to following an Ornstein–Uhlenbeck process than a Wiener process.

10.3 Heavy-tailed distributions

A distribution is called 'heavy-tailed' if the tails of its density are thicker than those of the normal distribution. Thus, there will be a higher probability of extreme values than with that distribution. A second approach to modelling volatility is to use such distributions, perhaps with changing dispersion or skewness to allow for the occasional or clustered occurrence of high variability or extreme values.

Processes in economics and finance often include considerable 'noise' as they evolve. For example, share prices in stock markets, even if they tend to show a global trend, are often quite variable. Such behaviour, resulting from a large number of external (usually uncontrollable) independent influences, might be modelled using distributions such as the normal or log normal, if it were not for rather common extreme responses. These 'outliers', although they may be of no direct interest in studying the overall evolution of such phenomena, cannot be ignored. They do not result from recording errors; they were actually produced by the stochastic process under study.

10.3.1 Stable distributions

The stable family forms one wide class of flexible distributions that can be useful for modelling stochastic processes when volatility is present.

Definition

A random variable Y is said to have a stable distribution if and only if it has a *domain of attraction*, that is, if a sequence of independent and identically distributed random variables $Z_1, \ldots, Z_n$, with accompanying constants a_n, b_n, exists, such that

$$\frac{Z_1 + Z_2 + \ldots + Z_n}{a_n} + b_n \tag{10.19}$$

converges in distribution to Y. In the special case where the Z_is have a finite variance, Y will be normally distributed.

This family provides the only possible limiting distributions for the sum of independent, identically distributed random variables. This property, which was originally used to derive the form of their characteristic function, generalises the central limit theorem to the case of infinite variance. Therefore, when an observed process results from a large number of external (usually uncontrollable), independent influences, as is often the case in economics and elsewhere, the family of stable distributions will provide the basis for one useful set of models.

The common way to specify a stable distribution is by its characteristic function $\phi(t)$, because its density function $f(y)$ is not available in an explicit form, except in three special cases discussed below. Until recently, this has been a major justification for not using stable distributions in practice, although their interesting theoretical properties indicate their potential importance in stochastic processes.

The characteristic function of the four-parameter family of stable distributions can be written

$$\begin{aligned} \phi(t) &= \int_{-\infty}^{\infty} \mathrm{e}^{ity} f(y) \mathrm{d}y \\ &= \mathrm{e}^{\mathrm{i}\gamma t - \delta |t|^{\alpha} [1 + \mathrm{i}\beta \ \mathrm{sign}(t) \omega(t,\alpha)]} \end{aligned} \tag{10.20}$$

where

$$\omega(t,\alpha) = \begin{cases} \tan(\pi\alpha/2), & \alpha \neq 1 \\ \frac{2}{\pi} \log|t|, & \alpha = 1 \end{cases}$$

Here, $-\infty < \gamma < \infty$ is a *location* parameter, $0 < \delta < \infty$ is a *scale* parameter, $-1 \leq \beta \leq 2$ is an index of *skewness*, and $0 < \alpha \leq 2$ is the *characteristic exponent* determining the type of stable distribution, especially the thickness of the tails. The distribution is respectively left- or right-skewed when $\beta > 0$ or $\beta < 0$, and symmetric when $\beta = 0$. Note that this has the opposite sign to the traditional coefficient of skewness based on third moments.

The distribution of Y can be standardised using $(Y - \gamma)/\delta^{1/\alpha}$. The properties of any stable distribution with given values of α and β can, then, be deduced from such a standardised distribution.

Special cases

In certain circumstances, the characteristic function in Equation (10.20) corresponds to a density function that can be written in an explicit form:

(i) for $\alpha = 2$, the normal distribution $N(\gamma, 2\delta)$;
(ii) for $\alpha = 1$ and $\beta = 0$, the (symmetric about γ) Cauchy distribution with density

$$f(y;\gamma,\delta) = \frac{1}{\pi\delta\left[1+\frac{(y-\gamma)^2}{\delta^2}\right]} \tag{10.21}$$

(iii) for $\alpha = \frac{1}{2}$ and $\beta = 1$, the (skewed) Lévy distribution with density

$$f(y;\gamma,\delta) = \frac{\delta}{\sqrt{2\pi(y-\gamma)^3}} \exp\left[-\frac{\delta^2}{2(y-\gamma)}\right], \quad y > \gamma \tag{10.22}$$

In all other situations, the density has to be generated numerically. However, as Hoffmann-Jørgensen (1994, I, pp. 406–411) points out, the characteristic function can be inverted and expressed in terms of incomplete hypergeometric functions. The latter generally are only defined as the sum of infinite series, but these can be approximated numerically.

Properties

Most stable random variables have $-\infty < Y < \infty$. Exceptions are the Lévy distribution and distributions with $\alpha < 1$ and $\beta = -1$, these having $\gamma < Y < \infty$. By definition, stable distributions are invariant under addition; that is, sums of independent variables with the same characteristic exponent α are stable with the same exponent.

Moments of order greater than or equal to α do not exist, except for $\alpha = 2$, in which case all moments are finite. Thus, all stable distributions, except the normal, have an infinite variance. Moreover, their mean is defined if and only if $1 < \alpha \leq 2$. Therefore, stable distributions are potential candidates to model heavy-tailed processes.

Estimation

The characteristic function can be written in a more convenient, alternative, but equivalent, form

$$\log[\phi(t)] \quad = \quad \mathrm{i}\gamma t - |t|^\alpha \delta^{*\alpha} \exp\left[-\mathrm{i}\beta^* \frac{\pi}{2}\eta_\alpha \operatorname{sign}(t)\right] \tag{10.23}$$

where $\eta_\alpha = \min(\alpha, 2-\alpha) = 1 - |1-\alpha|$. Here, the parameters play the same role as in Equation (10.20); they are related by

$$\beta^* \quad = \quad \frac{2}{\pi\eta_\alpha} \cos^{-1}\left[\frac{\cos(\pi\alpha/2)}{\Delta}\right]$$

Table 10.3. *Prices (pence) and relative returns of Abbey National shares between 31 July and 8 October, 1991. (Buckle, 1995)*

Day	Price	Return	Day	Price	Return
31/07	296	—	04/09	309	0.0000
01/08	296	0.0000	05/09	307	−0.0065
02/08	300	0.0135	06/09	306	−0.0033
05/08	302	0.0067	09/09	304	−0.0065
06/08	300	−0.0066	10/09	300	−0.0132
07/08	304	0.0133	11/09	296	−0.0133
08/08	303	−0.0033	12/09	301	0.0169
09/08	299	−0.0132	13/09	298	−0.0100
12/08	293	−0.0201	16/09	295	−0.0101
13/08	294	−0.0034	17/09	295	0.0000
14/08	294	0.0000	18/09	293	−0.0068
15/08	293	−0.0034	19/09	292	−0.0034
16/08	295	0.0068	20/09	297	0.0171
19/08	287	−0.0271	23/09	294	−0.0101
20/08	288	0.0035	24/09	293	−0.0034
21/08	297	0.0312	25/09	306	0.0424
22/08	305	0.0269	26/09	303	−0.0098
23/08	307	0.0066	27/09	301	−0.0066
26/08	—	—	30/09	303	0.0066
27/08	304	−0.0098	01/10	308	0.0165
28/08	303	−0.0033	02/10	305	−0.0097
29/08	304	0.0033	03/10	302	−0.0098
30/08	304	0.0000	04/10	301	−0.0033
02/09	309	0.0164	07/10	297	−0.0133
03/09	309	0.0000	08/10	299	0.0067

$$\delta^* = \left[\frac{\Delta\delta}{\cos(\pi\alpha/2)}\right]^{\frac{1}{\alpha}}$$

where $\Delta^2 = \cos^2(\pi\alpha/2) + \beta^2 \sin^2(\pi\alpha/2)$, $\text{sign}(\Delta) = \text{sign}(1-\alpha)$, and $\text{sign}(\beta^*) = \text{sign}(\beta)$. The location and tail parameters, γ and α, are unchanged.

The density corresponding to this characteristic function can be computed from it by using Fourier inversion,

$$f_\alpha(y;\gamma,\delta^*,\beta^*) = \frac{1}{\pi}\int_0^\infty \cos\left[(\gamma-y)\frac{s}{\delta^*} + s^\alpha \sin(\eta^*_{\alpha,\beta^*})\right] \mathrm{e}^{-s^\alpha \cos(\eta^*_{\alpha,\beta^*})} \frac{ds}{\delta^*} \qquad (10.24)$$

where $\eta^*_{\alpha,\beta^*} = \beta^* \pi \eta_\alpha / 2$. This integral can be evaluated numerically for fixed values of the parameters, thus approximating the density (and hence the likelihood) to any desired precision.

Abbey National returns The daily closing price (pence) of the shares of the British-based bank, Abbey National, between 31 July and 8 October 1991, are reproduced in Table 10.3. The relative share returns, defined as $(y_t - y_{t-1})/y_{t-1}$, are also given in the table. At first sight, these appear to display quite variable

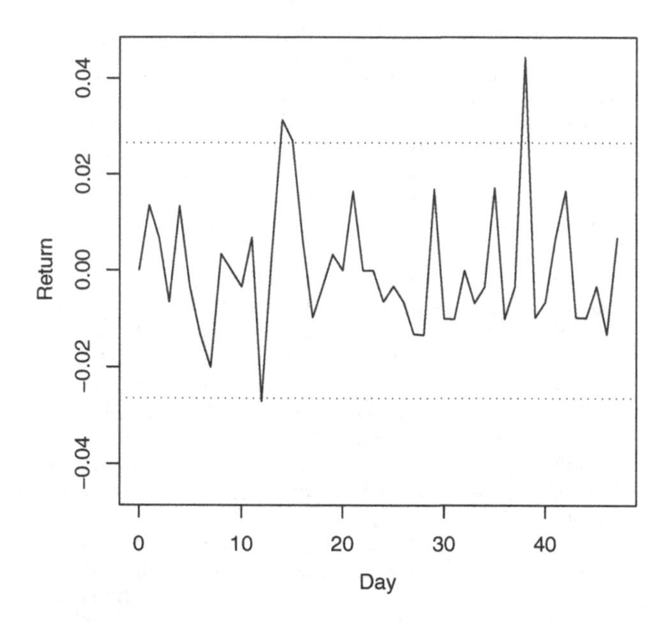

Fig. 10.4. The Abbey National returns, 31 July to 8 October, 1991, from Table 10.3, with an interval of two standard deviations (dotted).

behaviour, as can be seen in Figure 10.4 showing the relative share returns against time. However, the limits of two standard deviations only exclude three values out of 48, not exceptional for normally-distributed responses.

Various problems might be studied involving the evolution of such share returns:

(i) How can the occurrence of extreme values be understood and forecast in order to develop short-term investment strategies?
(ii) How does the shape of the region where most returns are observed change, occasional extreme responses being of no direct interest?
(iii) How is the most probable relative return changing with time, providing information about trends in likely future returns?

Here, I shall look at the last two questions.

Econometricians often analyse log returns, defined as $\log(y_t/y_{t-1})$, instead of relative share returns, with the argument that these constitute a stationary process. Using a Taylor's series approximation

$$\begin{aligned}\log\left(\frac{y_t}{y_{t-1}}\right) &= \log\left(1+\frac{y_t-y_{t-1}}{y_{t-1}}\right)\\ &\doteq \frac{y_t-y_{t-1}}{y_{t-1}}\end{aligned}$$

we can see that these reasonably approximate relative returns.

Consider first a model making the assumptions that the distribution of relative

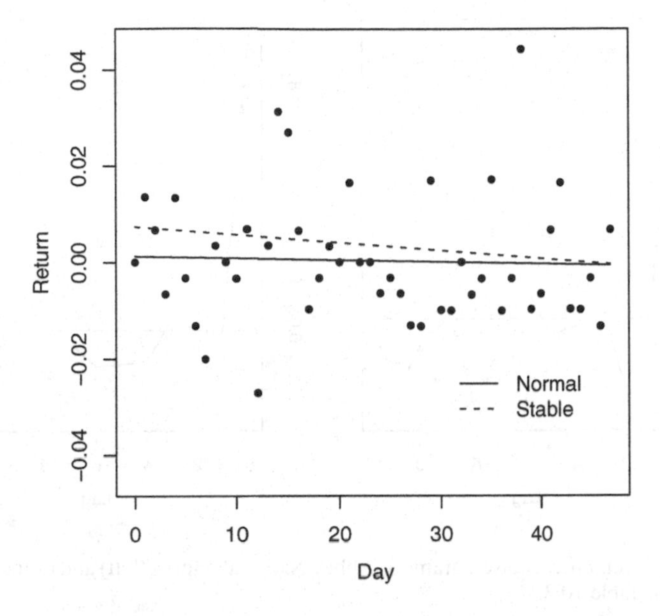

Fig. 10.5. The Abbey National returns from Table 10.3, with the normal and stable regressions of time on the location parameter.

returns is independent and not evolving with time. The AIC for the normal distribution is 135.7, whereas that for the general stable family is 134.3, although the latter has two more parameters. The maximum likelihood estimates for the stable distribution are $\hat{\gamma} = 0.0017$, $\widehat{\delta^*} = 0.0079$, $\widehat{\beta^*} = -0.82$ and $\hat{\alpha} = 1.53$. The latter indicates some distance from the normal distribution with $\alpha = 2$. There is also indication of skewness, with $\widehat{\beta^*} < 0$.

Next, I shall introduce time dependence for the location parameter. The AICs are respectively 136.7 and 134.4. Thus, there is no indication of a change in the location parameter over time. However, it is informative to plot the two curves; these are shown in Figure 10.5. In comparison to the stable regression curve, the normal curve is pulled down at the beginning by the more extreme negative responses.

An AR(1), with dependence of the location parameter on the previous return, is not necessary; the AIC is 135.2 for the stable model with linear dependence on time. This also is true of the other models to follow. Such a conclusion may not be surprising: relative returns are a form of first differences of the share values. This conclusion is confirmed by the empirical correlogram plotted in the left graph of Figure 10.6. (As well, the empirical correlogram of the squared returns, in the right graph of Figure 10.6, does not indicate autocorrelation, so that the ARCH models of Section 10.4 below are not indicated.)

Let us now investigate how the other parameters of the stable distribution may depend on time. The results are summarised in Table 10.4. We see that the

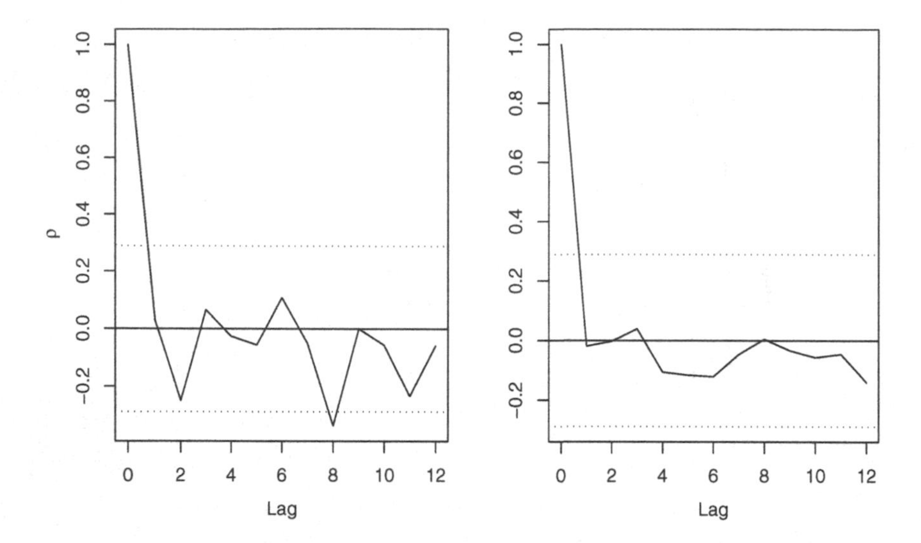

Fig. 10.6. The empirical correlograms of Abbey National returns (left) and of their squares (right), from Table 10.3.

Table 10.4. *AICs for dependence of various parameters of the stable distribution on time for the Abbey National returns of Table 10.3.*

Parameter	Time
Null	134.3
Location	134.4
Dispersion	135.0
Skew	131.3
Tail	132.4

strongest indication is for the skewness to change over time. This model is plotted in Figure 10.7. We see that, although the location parameter is constant, the mode is decreasing over this period, with skewness toward negative returns at the beginning and toward positive returns at the end. Because the location parameter γ is constant over time, the mode must be moving to allow the skewness to change. The estimate $\hat{\alpha} = 1.37$ indicates that the distribution has heavier tails than the previous models.

10.3.2 Other heavy-tailed distributions

Well known heavy-tailed distributions include the Cauchy, Laplace, and logistic, each with two parameters. The first of these is a member of the stable family.

As well, a number of three-parameter distributions also have relatively heavy tails. The Burr distribution, already given in Equation (8.25), is one of them; it is

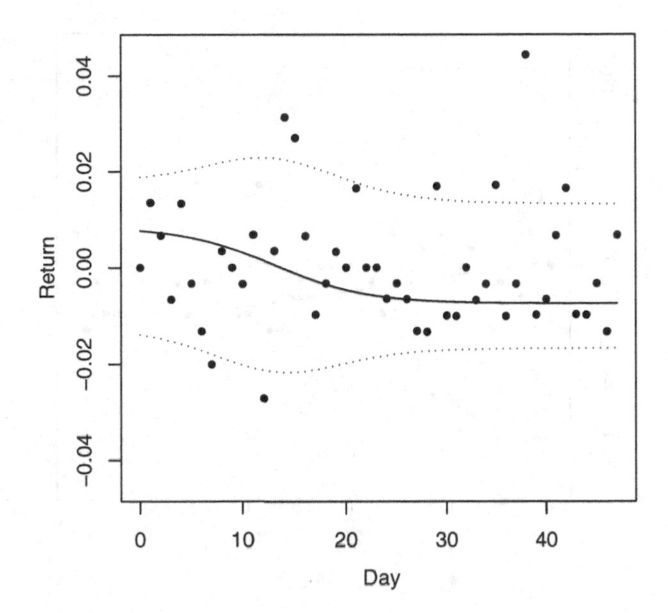

Fig. 10.7. The Abbey National returns from Table 10.3, with the stable regression of time on the skew parameter and the limits of 10 per cent of the height of the mode.

skewed, but has no explicit parameter for the skewness. The Student t distribution has density

$$f(x;\mu,\phi,\kappa)=\frac{\Gamma\left(\frac{\kappa+1}{2}\right)}{\sqrt{\kappa\pi}\Gamma\left(\frac{\kappa}{2}\right)}\left[1+\frac{1}{\kappa}\left(\frac{y-\mu}{\phi}\right)^2\right]^{-\frac{\kappa+1}{2}} \tag{10.25}$$

and the power exponential

$$f(y;\mu,\phi,\kappa)=\frac{1}{\phi\Gamma\left(1+\frac{1}{2\kappa}\right)2^{1+\frac{1}{2\kappa}}}\exp\left[-\frac{1}{2}\left|\frac{y-\mu}{\phi}\right|^{2\kappa}\right] \tag{10.26}$$

Both of these are symmetric but, for skewed responses, a log transformation can be used.

On the other hand, the skew Laplace distribution (Kotz *et al.*, 2001)

$$\begin{aligned} f(y;\mu,\phi,\kappa) &= \frac{\kappa e^{-\frac{\kappa(y-\mu)}{\phi}}}{(1+\kappa^2)\phi} \qquad y\geq\mu \\ &= \frac{\kappa e^{\frac{y-\mu}{\kappa\phi}}}{(1+\kappa^2)\phi} \qquad y<\mu \end{aligned} \tag{10.27}$$

does have a skew parameter (κ) that can be modelled. This is an ordinary Laplace distribution for $\kappa=1$.

Here, I shall apply this latter model, leaving the others until the next section.

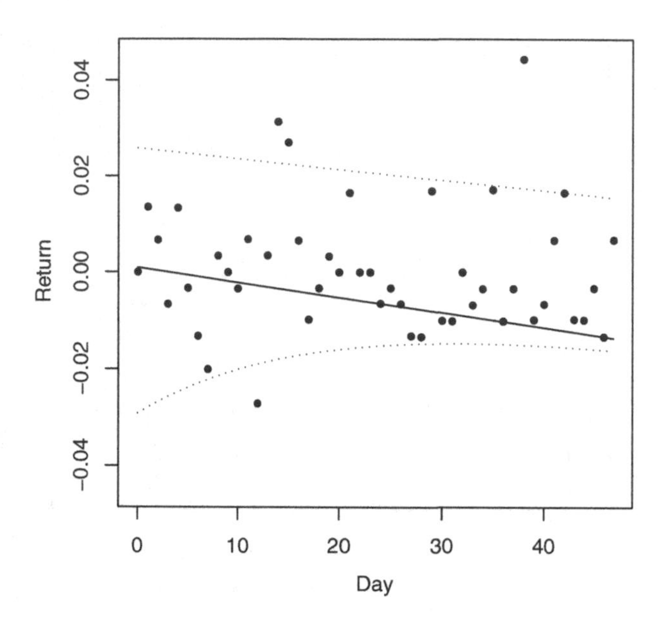

Fig. 10.8. The Abbey National returns from Table 10.3, with the skew Laplace regression of time on the location, shape, and skew parameters and limits of 10 per cent of the height of the mode.

Abbey National returns A similar analysis to that for stable distributions above can also be performed using the skew Laplace distribution of Equation (10.27). Here, we discover that all three parameters depend on time, with an AIC of 130.4, slightly better than the stable distribution with skew regression. The estimated regression equations are

$$\begin{aligned} \widehat{\mu}_t &= 0.000945 - 0.000309t \\ \widehat{\phi}_t &= e^{-4.43-0.0241t} \\ \widehat{\kappa}_t &= e^{0.0968-0.0275t} \end{aligned}$$

This model is plotted in Figure 10.8, showing some differences from Figure 10.7. As with the stable model, the skew switches from negative to positive returns over time.

These data are a very short time series. No substantive conclusions about relative share returns should be drawn from this analysis. With a longer series, it is quite likely that the skewness might, say, be oscillating back and forth. But, a clear conclusion is that, during the observed time period, the major mass of probability stayed in the same location, although fundamentally changing shape. In studying such data, among other things, we were interested in how the region in which most relative returns would lie was changing over time, and in how the most probable relative return varied over time. The stable and skew Laplace models have permitted us to answer both of these questions, as can be seen in Figures 10.7 and 10.8. On the other hand, a mean, with its standard deviation, could not have correctly answered either of them.

10.4 ARCH models

The diffusion and heavy-tailed models of the previous sections have allowed the variability to change in systematic ways over time. Now, I shall look at a family of models that allows it to vary stochastically.

10.4.1 Theory

Most classical regression models in statistics involve studying how a location parameter, usually the mean, changes with covariates. Much less attention has been paid to changes in shape of the distribution, except in restricted situations such as generalised linear models. In a similar way, autoregression models (Section 1.2 and Chapter 9) also concentrate on the location parameter, allowing it to depend on previous values of the response variable.

In Sections 7.1.4 and 9.2.4 and the previous sections of this chapter, I looked at some models that allow the dispersion to change in systematic ways over time, but this may not be sufficient to account for high volatility. Thus, parameters, other than the location, may display stochastic dependence on previous values of the response variable. Indeed, in Sections 7.3 and 7.4, I introduced some models where the distributional shape could change dynamically over time; in Section 11.3, I shall extend that particular family to continuous state spaces. However, here I shall look at still another possibility.

The basic idea underlying the models to be considered here is that extreme values may be clustered. This fact may, then, be used to predict their occurrence. Thus, such models cannot be expected to work well if extreme values are isolated rather than clustered.

Normal distribution

In a serial dependence model (Section 1.2.4), the location regression function is corrected by the previous residual or innovation. In a similar way, it often makes sense to correct the dispersion by the square of that value. The original *autoregressive conditional heteroscedasticity* (ARCH) model in discrete time, introduced by Engle (1982), allowed the variance of a normal distribution to depend on the squares of previous responses (assumed to have zero mean):

$$\sigma_t^2 = \theta_0 + \sum_{i=1}^{M} \theta_i y_{t-i}^2 \qquad (10.28)$$

Since then, many variations have been proposed.

In the ARCH(1), with $M = 1$, the variance is $\theta_0/(1-\theta_1)$ if $0 \leq \theta_1 < 1$. If $3\theta_1^2 < 1$, y_t^2 has autocorrelation function (Section 9.1.2)

$$\rho(h) = \theta_1^h \qquad (10.29)$$

This may be checked by plotting the empirical correlogram of y_t^2 (as I did above in Figure 10.6 for the Abbey National returns). In these models, the marginal distribution of y_t has heavier tails than a normal distribution, although its conditional distribution is normal. As well, although y_t^2 is second-order stationary (Section 1.1.4), y_t is not, except under even more strict conditions (Shephard, 1996).

Extensions

It is possible to apply this approach to the dispersion parameter of a variety of distributions, not just the normal distribution. Here, I only shall consider first-order dependence ($M = 1$). Useful possibilities include

$$\phi_t = \theta_0 + \theta_1^{\Delta t}(y_{t-1} - \mu_{t-1})^2 \tag{10.30}$$

$$\phi_t = \theta_0 + \theta_1^{\Delta t}|y_{t-1} - \mu_{t-1}| \tag{10.31}$$

and, because the dispersion parameter must be positive,

$$\phi_t = \theta_0 e^{\theta_1^{\Delta t}(y_{t-1} - \mu_{t-1})^2} \tag{10.32}$$

where ϕ_t is the dispersion parameter and μ_t is the location regression function for some distribution. In all cases, θ_0 and θ_1 must be positive (except θ_1 in Equation (10.32) if time is discrete). Of course, these functions can be used in combination with the more usual autoregression for the location parameter.

Consider again the simplest case of a conditional normal distribution in discrete time. A model based on Equation (10.30) only has finite variance if $\theta_1 < 1$. In contrast, for Equation (10.31), the variance is always finite and, for Equation (10.32), it never is. Similar results are not generally available for other distributions.

ARCH models can also be used with heavy-tailed distributions, as well as the normal distribution. This will permit us to have two means of accounting for extreme variability.

10.4.2 Biological variability

Although volatility models have primarily been applied in economics and finance, they have much wider applicability. For example, important applications can be found in human and veterinary medicine.

Cow hormones The level of luteinising hormone (ng/ml $\times$ 1000) in cows varies over time. Concentrations of this hormone are influenced by semi-periodic pulsing of the glands that produce it.

A study was performed to compare two groups of sixteen cows each, suckled and nonsuckled, by following them for ten days post-partum. The cows were ovariectomised on the fifth day post-partum. Over the observation period, their hormone levels were measured 15 times at unequally-spaced intervals. However, at the time when a sample was taken, a cow might, rather erratically, be in either the high or the low phase of the luteinising hormone cycle. The results were given by Raz (1989).

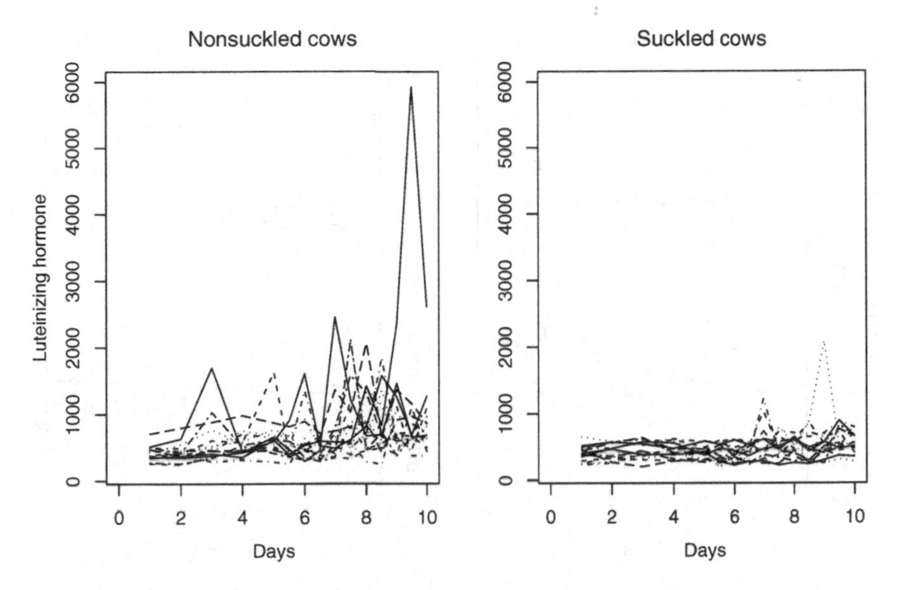

Fig. 10.9. The individual observed profiles for the luteinising hormone concentrations in cows.

Table 10.5. *AICs for several models fitted to the luteinising hormone data with an interaction between suckled and time.*

		ARCH(1)			
	None	None	Square	Absolute value	Exponential
		Dependence			
Distribution	None	AR(1)	AR(1)	AR(1)	AR(1)
Log normal	3239.4	3201.0	3184.9	3184.0	3188.0
Log Burr	3220.2	3172.4	3173.4	3174.5	3173.4
Log Student t	3228.1	3180.8	3166.0	3165.5	3168.4
Log skew Laplace	3229.8	3178.6	3164.1	3163.7	3166.5
Log power exponential	3229.2	3179.0	3165.2	3163.9	3169.0

To provide an overall idea of the variability, the individual observed profiles are plotted in Figure 10.9. We can see that variability increases as time goes by and that the nonsuckled cows show more variation than the suckled ones. Unfortunately, the series on individual cows are too short to be able to plot empirical correlograms of the squared residuals to check graphically for an ARCH model.

The fits of several models, mainly with heavy tails, are presented in Table 10.5. In all cases, the two additive ARCH models of Equations (10.30) and (10.31) fit better than the multiplicative one of Equation (10.32). The log skew Laplace distribution, with skew $\hat{\kappa} = 0.91$, fits best, better than the skew Laplace distribution

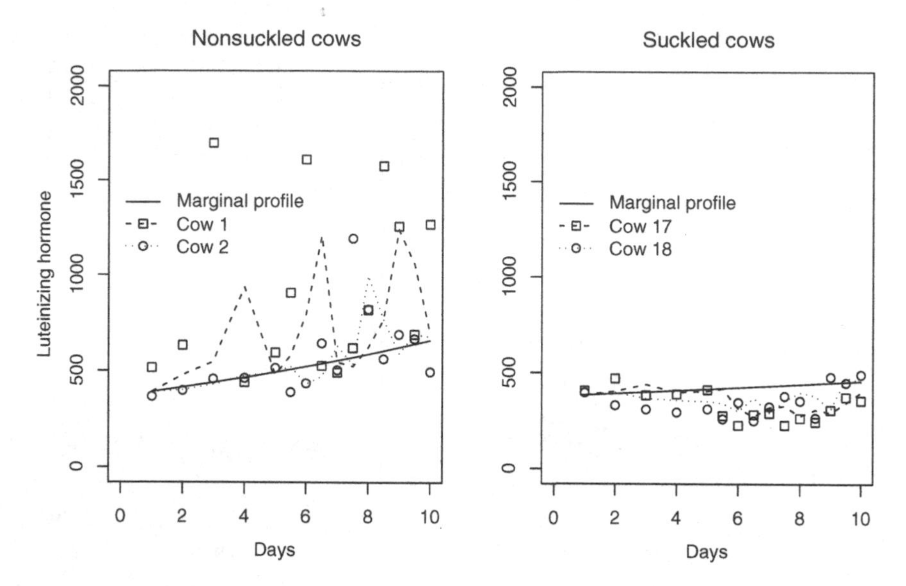

Fig. 10.10. The underlying profiles for the log skew Laplace ARCH model with AR(1) for the two groups of luteinising hormone concentrations in cows, with the individual recursive predictions for the first two cows in each group.

of Equation (10.27). The log power exponential of Equation (10.26) with $\hat{\kappa} = 0.56$ is a close second, followed by the log Student t distribution of Equation (10.25), with $\hat{\kappa} = 4.0$ degrees of freedom. The log Burr distribution, derived from Equation (8.25), fits better than the Burr distribution and is the only one that does not require the ARCH. All of the distributions have very heavy tails.

For the best model, the log skew Laplace with absolute value ARCH, the regression functions are estimated to be

$$\begin{aligned}\mu_{1t} &= 5.91 + 0.057t \\ \mu_{2t} &= 5.93 + 0.025t\end{aligned}$$

respectively for nonsuckled and suckled cows, where μ_{jt} refers to the log level of luteinising hormone. The slope over time is much smaller for the suckled cows than for the nonsuckled ones. The recursive predictions for four cows are given in Figure 10.10. These individual profiles show a clear attempt to follow the individual responses.

The dependence parameters are estimated to be $\hat{\rho} = 0.49$ for the AR(1) and $\widehat{\theta_1} = 0.078$ for the ARCH parameter in Equation (10.31). The volatility, defined as the square root of the dispersion parameter, is plotted for the same four cows in Figure 10.11.

I shall provide further analysis of these data in Chapters 11 and 14.

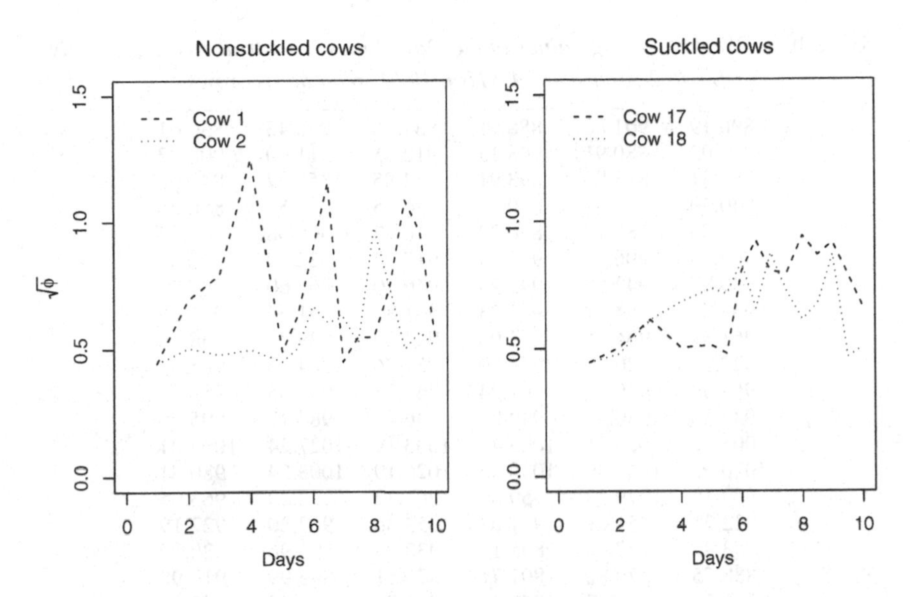

Fig. 10.11. The changing volatility for the first two cows in each group for the log skew Laplace ARCH model with AR(1).

Further reading

Bailey (1964, Ch. 14), Grimmett and Stirzaker (1992, Ch. 13), Guttorp (1995, Ch. 6), and Karlin and Taylor (1981, Ch. 15) provide introductions to diffusion processes. For an advanced coverage, see Rogers and Williams (1987, 1994). Berg (1993) gives applications in biology, Gardiner (1985) in chemistry, and Hoppensteadt and Peskin (2002) in medicine and biology.

For a systematic review of the use of stable distributions in stochastic processes, see Samorodnitsky and Taqqu (1994). Adler *et al.* (1998) provide a useful survey of applications of heavy-tailed distributions.

Shephard (1996) gives a good introduction to ARCH models and their many extensions.

Exercises

10.1 The data on student enrolment at Yale University from 1796 to 1975 in Table 7.1 show high variability during the two world wars.

(a) Can this be modelled by a heavier-tailed distribution than the normal?

(b) Can it be modelled by ARCH?

10.2 Do any of the following time series require an ARCH model:

(a) the Canadian lynx data of Table 9.3,

(b) the sunspot data of Table 9.4,

(c) the yearly unemployment in the USA of Table 9.5, or

Table 10.6. *Weekly closing values of the Dow-Jones Industrial Average, 1 July 1971 to 2 August 1974. (Hsu, 1979; read across rows)*

890.19	901.80	888.51	887.78	858.43	850.61
856.02	880.91	908.15	912.75	911.00	908.22
889.31	893.98	893.91	874.85	852.37	839.00
840.39	812.94	810.67	816.55	859.59	856.75
873.80	881.17	890.20	910.37	906.68	907.44
906.38	906.68	917.59	917.52	922.79	942.43
939.87	942.88	942.28	940.70	962.60	967.72
963.80	954.17	941.23	941.83	961.54	971.25
961.39	934.45	945.06	944.69	929.03	938.06
922.26	920.45	926.70	951.76	964.18	965.83
959.36	970.05	961.24	947.23	943.03	953.27
945.36	930.46	942.81	946.42	984.12	995.26
1005.57	1025.21	1023.43	1033.19	1027.24	1004.21
1020.02	1047.49	1039.36	1026.19	1003.54	980.81
979.46	979.23	959.89	961.32	972.23	963.05
922.71	951.01	931.07	959.36	963.20	922.19
953.87	927.89	895.17	930.84	893.96	920.00
888.55	879.82	891.71	870.11	885.99	910.90
936.71	908.87	852.38	871.84	863.49	887.57
898.63	886.36	927.90	947.10	971.25	978.63
963.73	987.06	935.28	908.42	891.33	854.00
822.25	838.05	815.65	818.73	848.02	880.23
841.48	855.47	859.39	843.94	820.40	820.32
855.99	851.92	878.05	887.83	878.13	846.68
847.54	844.81	858.90	834.64	845.90	850.44
818.84	816.65	802.17	853.72	843.09	815.39
802.41	791.77	787.23	787.94	784.57	752.58

(d) the annual snowfall data in Table 9.6?

10.3 Only the precipitation records at Snoqualmie Falls for the months of January were used in the example in Section 8.3.2. However, data are available for all days between 1948 and 1983 (see Exercise 8.2). The complete series shows high variability.

(a) Does an ARCH approach yield a reasonable model for the complete data?

(b) Compare this to a model using a marked point process similar to that used in Section 8.3.2 for the months of January.

(c) Which model provides a better explanation of the complete series?

10.4 The weekly Dow-Jones closing values for the period between July 1971 and August 1974 are given in Table 10.6. Various events involving the USA occurred in this period. These included Watergate, the Arab oil embargo, and climbing interest rates.

(a) Does a plot of the data reflect the timing of these events?

(b) Is a model of changing volatility required?

11

Dynamic models

As we saw in Chapter 7 for categorical state spaces, a useful approach to constructing models for stochastic processes is to have a distribution, or some of its parameters, change over time according to some (hidden) Markov process. I now shall look at the application of such models when the observed state space is continuous. If the assumption of normality can be made, development of dynamic models is especially simple.

11.1 Kalman filtering and smoothing

Suppose that we have a process (a time series, as in Chapter 9) for which the observed state space is continuous and that we can assume that the responses have a normal distribution, conditionally on some linear regression function. In a dynamic linear model, some of the parameters of this function are assumed to evolve randomly over time according to a stochastic process.

As often is the case, the assumptions of a normal distribution and a linear regression function can simplify greatly the models, allowing integration to be performed analytically instead of numerically. As usual, the accompanying cost is lack of realism. Models with these assumptions can be fitted by a procedure involving the *Kalman filter* and *smoother*. In this context, *filtering* refers to estimation of the current hidden state of a process, *smoothing* to estimation of previous hidden states, and *prediction* to estimation of future hidden states.

11.1.1 Theory

As for the dynamic models in Chapter 7, the standard linear regression function for the location parameter, here the mean of a normal distribution,

$$\mathrm{E}(Y_t) = \boldsymbol{\beta}^\top \mathbf{x}_t + \boldsymbol{\psi}_t^\top \mathbf{v}_t \tag{11.1}$$

will be called the *observation* or *measurement equation* because it is related directly to what actually can be observed. The two sets of observable covariates, $\mathbf{x}_t$ and $\mathbf{v}_t$, may vary over time and may contain elements in common. There are also two types of coefficients, $\boldsymbol{\beta}$ and $\boldsymbol{\psi}_t$, the former constant and the latter changing over time.

Recall that, in the hidden Markov models of Section 7.2 (see also Section 11.2 below), the entire distribution of responses differed in each of a (small) finite number of hidden states. On the other hand, in the overdispersed dynamic models of Sections 7.3 and 7.4, only parameters defining the shape of the distribution changed randomly over time. In contrast, here only parameters in the location regression function will be allowed to vary randomly over time.

Thus, in Equation (11.1), $\boldsymbol{\psi}_t$ will be a vector of random parameters defining the *hidden state* at time t, with a distribution conditional on the previous hidden state and on $\mathbf{v}_t$. This hidden state will be defined by the minimum past and present information necessary to predict a future state.

In an analogous way to the overdispersed dynamic models of Sections 7.3 and 7.4, the hidden states of the system will be assumed to evolve over time according to the *state transition equation*, already given in Equation (7.40):

$$\boldsymbol{\psi}_t^\top = \boldsymbol{\psi}_{t-1}^\top \mathbf{T}_t + \mathbf{b}_t^\top + \boldsymbol{\epsilon}_t^\top$$

where the residual or innovation $\boldsymbol{\epsilon}_t \sim \mathrm{N}(0, \sigma^2)$ is the random input at time t, $\mathbf{b}_t$ is the vector of deterministic input at that time, and $\mathbf{T}_t$ is a first-order Markovian *state transition matrix*. Here, this equation can also be written as

$$\mathrm{E}(\boldsymbol{\psi}_t^\top) = \boldsymbol{\psi}_{t-1}^\top \mathbf{T}_t + \mathbf{b}_t^\top \tag{11.2}$$

As in Sections 7.3 and 7.4, and in contrast to the hidden Markov chains in Section 7.2, here the state vector $\boldsymbol{\psi}_t$ does not contain probabilities and $\mathbf{T}$ is not a stochastic matrix. On the other hand, $\boldsymbol{\psi}_t$ may contain values before time t as well as present values. The distribution of $\boldsymbol{\psi}_t$ is assumed to be independent of the conditional distribution of Y_t.

Autoregression models

Often, specific models can be defined in several distinct, but equivalent, ways. As one simple case, in order to gain some familiarity with this type of model-building, let us look at one way to set up a discrete-time, dynamic linear model for an autoregression of order M. For simplicity, assume that there are no covariates $\mathbf{x}_t$. Then, one possible specification of the measurement and state equations for an AR(M) is

$$\mathrm{E}(Y_t) = [1,0,\ldots]\boldsymbol{\psi}_t$$

$$\mathrm{E}\left[\begin{pmatrix} \psi_t \\ \psi_{t-1} \\ \vdots \\ \psi_{t-M+1} \end{pmatrix}^\top\right] = \begin{pmatrix} \psi_{t-1} \\ \psi_{t-2} \\ \vdots \\ \psi_{t-M} \end{pmatrix}^\top \begin{pmatrix} \rho_1 & 1 & \cdots & 0 \\ \vdots & \vdots & \ddots & \vdots \\ \rho_{M-1} & 0 & \cdots & 1 \\ \rho_M & 0 & \cdots & 0 \end{pmatrix} \tag{11.3}$$

so that there is no deterministic input.

Consider next an AR(1) with covariates $\mathbf{x}_t$. Here, the corresponding measurement equation will be

$$\mathrm{E}(Y_t) = \boldsymbol{\beta}^\top \mathbf{x}_t + \psi_t \tag{11.4}$$

the mean of a normal distribution with conditional variance σ^2 and fixed initial conditions, and the state equation

$$\mathrm{E}(\psi_t) = \rho\psi_{t-1} \tag{11.5}$$

also the mean of a normal distribution but with variance ξ.

For a continuous time AR(1), that is, a CAR(1), the measurement equation remains the same, whereas the state equation for ψ_t becomes

$$\mathrm{E}(\mathrm{d}\psi_t) = -\kappa\psi_t\mathrm{d}t \tag{11.6}$$

This will be the mean of a normal distribution with variance $\xi\mathrm{d}t$, where $\rho = \mathrm{e}^{-\kappa}$ (Jones and Boadi-Boateng, 1991).

Model specification

Let us now look more closely at the general distributional assumptions for this type of model and how to fit it. As we have seen, filtering refers to estimating the current state given states up to the present. The Kalman filter is a sequential or recursive procedure, yielding new distributions at each time point. The procedure is similar to that described in Section 7.3.1.

From Bayes' formula,

$$p(\psi_t|\mathcal{F}_t) = \frac{f(y_t|\psi_t,\mathcal{F}_{t-1})p(\psi_t|\mathcal{F}_{t-1})}{f(y_t|\mathcal{F}_{t-1})} \tag{11.7}$$

where, as in Chapters 4 and 7, $\mathcal{F}_t$ denotes the history of the states up to and including time t, that is, the vector of observed states $(y_1,\ldots,y_t)$, with all pertinent relationships among them. Here, $f(y_t|\psi_t,\mathcal{F}_{t-1})$ is the usual distribution defined by the observation equation, whereas $p(\psi_t|\mathcal{F}_t)$ is called the *filtering* or *observation update*. The normalising constant

$$f(y_t|\mathcal{F}_{t-1}) = \int_{-\infty}^{\infty} f(y_t|\psi_t,\mathcal{F}_{t-1})p(\psi_t|\mathcal{F}_{t-1})\mathrm{d}\psi_t \tag{11.8}$$

is a mixture distribution. This conditional distribution will be used to calculate the likelihood function. The *one-step-ahead prediction* or *time update*,

$$p(\psi_t|\mathcal{F}_{t-1}) = \int_{-\infty}^{\infty} p(\psi_t|\psi_{t-1})p(\psi_{t-1}|\mathcal{F}_{t-1})\mathrm{d}\psi_{t-1} \tag{11.9}$$

is defined by the transition equation.

With the assumption of a normal distribution, these equations can be written in a simple closed form, only the mean and variance being required. Then, the procedure is relatively straightforward and efficient. On the other hand, both of these integrals usually are complicated when the distributions are not normal and/or the regression function is nonlinear.

The idea now is to move forward in time from the first response, estimating the expected value of each successive state before going on to the next. In this way, the total multivariate probability can be built up as a product of conditional probabilities using Equation (1.1), in this way creating the likelihood function.

For a discrete time series, the one-step-ahead prediction or time update for $\mathrm{E}[Y_t] - \boldsymbol{\beta}^\top \mathbf{x}_t$ has mean

$$\hat{\boldsymbol{\psi}}^\top_{t|t-1} = \hat{\boldsymbol{\psi}}^\top_{t-1}\mathbf{T}_t \tag{11.10}$$

and covariance matrix

$$\mathbf{A}_{t|t-1} = \mathbf{T}_t^\top \mathbf{A}_{t-1}\mathbf{T}_t + \hat{\boldsymbol{\Xi}} \tag{11.11}$$

where $\boldsymbol{\Xi}$ is a diagonal covariance matrix. In the AR(1) above, this would contain the covariance element ξ. $\mathbf{A}_{t-1}$ is the prior covariance of the estimation error

$$\mathbf{A}_t = \mathrm{E}[(\boldsymbol{\psi}_t - \widehat{\boldsymbol{\psi}_t})(\boldsymbol{\psi}_t - \widehat{\boldsymbol{\psi}_t})^\top] \tag{11.12}$$

The filtering or observation update, using the next observed state y_t, has posterior mean

$$\widehat{\boldsymbol{\psi}_t} = \widehat{\boldsymbol{\psi}_{t|t-1}} + \frac{1}{c_t}\mathbf{A}_{t|t-1}\mathbf{v}_t(y_t - \widehat{\boldsymbol{\psi}_{t|t-1}}^\top \mathbf{v}_t - \boldsymbol{\beta}^\top \mathbf{x}_t) \tag{11.13}$$

and posterior covariance matrix

$$\mathbf{A}_t = \mathbf{A}_{t|t-1} - \frac{1}{c_t}\mathbf{A}_{t|t-1}\mathbf{v}_t\mathbf{v}_t^\top \mathbf{A}_{t|t-1} \tag{11.14}$$

where

$$c_t = \mathbf{v}_t^\top \mathbf{A}_{t|t-1}\mathbf{v}_t + \sigma^2 \tag{11.15}$$

As is usual for time series, the initial conditions must be chosen for $\boldsymbol{\psi}_0$ and for $\mathbf{A}_0$.

To obtain the likelihood function, we can rewrite the observation equation (11.1) as

$$\mathrm{E}(Y_t) = \boldsymbol{\beta}^\top \mathbf{x}_t + \widehat{\boldsymbol{\psi}_{t-1}}^\top \mathbf{v}_t + (\boldsymbol{\psi}_t - \widehat{\boldsymbol{\psi}_{t-1}})^\top \mathbf{v}_t \tag{11.16}$$

from which the conditional distribution $f(y_t|\mathcal{F}_{t-1})$ has mean

$$\mathrm{E}(Y_t|\mathcal{F}_{t-1}) = \boldsymbol{\beta}^\top \mathbf{x}_t + \widehat{\boldsymbol{\psi}_{t-1}}^\top \mathbf{v}_t \tag{11.17}$$

with variance given by Equation (11.15). Then, the log likelihood function is

$$\log(\mathrm{L}) = -\frac{1}{2}\sum_{t=1}^{R}\left\{\log(2\pi c_t) + \frac{1}{c_t}[y_t - \mathrm{E}(Y_t|\mathcal{F}_{t-1})]^2\right\} \tag{11.18}$$

For a discrete-time AR(1) with the first response stationary, $\mathrm{E}(Y_1|\mathcal{F}_0) = \boldsymbol{\beta}^\top \mathbf{x}_1$ and $c_1 = \xi/(1-\rho^2)$, whereas $\mathrm{E}(Y_t|\mathcal{F}_{t-1}) = \boldsymbol{\beta}^\top \mathbf{x}_t + \rho y_{t-1}$ and $c_t = \xi$ for $t > 1$. These results are readily generalised to continuous time.

11.1.2 Continuous-time autoregression

For many time series, continuous-time models must be used because the responses are not recorded at equally-spaced time intervals. Then, dynamic linear models generally have an advantage, especially if the series is fairly long. For standard ARMA models, dynamic linear models will provide the same results as with the methods in Chapter 9. Thus, it is not necessary to present detailed examples.

Table 11.1. *AICs for several models fitted to the luteinising hormone data.*

	Normal		Log normal	
	Independence	AR(1)	Independence	AR(1)
Null	3568.8	3527.9	3300.3	3235.6
Suckled	3550.5	3519.5	3270.5	3225.3
Time trend	3553.7	3518.7	3277.1	3223.3
Both	3534.0	3508.9	3243.9	3210.0
Interaction	3529.1	3505.7	3239.4	3207.4

Luteinising hormone The data on luteinising hormone in cows analysed in Section 10.4.2 are highly variable. We have seen that, for these responses, with mainly relatively small positive values, but a few very large ones, a skewed distribution is necessary. When dynamic linear models are applied, the log normal distribution fits much better than the normal distribution, as can be seen in Table 11.1. However, as might be expected, these models fit more poorly than the ARCH models of Section 10.4.2. (The AR(1) log normal model fits more poorly than that in Table 10.5 because here the first response in the series is assumed to have a stationary distribution, whereas there it did not.)

A linear time trend is present and this is different for the two groups. The estimated regression functions are

$$\begin{aligned}\mu_{1t} &= 6.40 + 0.073t \\ \mu_{2t} &= 6.09 + 0.028t\end{aligned}$$

respectively for nonsuckled and suckled cows, where μ_{jt} is the marginal mean for the log level of luteinising hormone. Again, the slope over time is much smaller for the suckled cows than for the nonsuckled ones. Thus, the profile for the nonsuckled cows rises more steeply, as shown in Figure 11.1 for the same cows as in Figure 10.10.

The autocorrelation coefficient is estimated to be 0.16. The change at post-partum plays a minor role, lowering the AIC from 3207.4 to 3205.4. Hormone levels are lower after post-partum.

11.2 Hidden Markov models

As we saw in Section 7.2, a series may display spells at various periods in time. The dynamic hidden Markov models introduced there provide one way to handle such data. Their application to continuous state spaces is direct. The theory for such models was described in Section 7.2.1. The only difference here is that $f(y_t|\psi_t = m, \mathcal{F}_{t-1}; \boldsymbol{\kappa}_m)$ will be a density instead of a discrete probability distribution.

As noted in that section, these models are closely related to the classical dynamic models presented above in Section 11.1. Two important differences are:

(i) Here, the number of possible hidden states is finite, whereas there it was infinite.

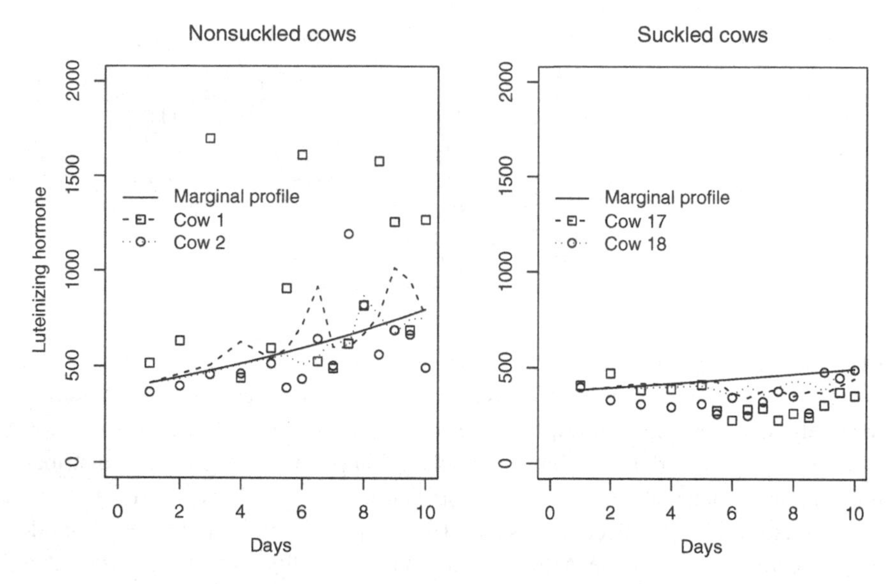

Fig. 11.1. The underlying marginal profiles from the AR(1) model for the two groups of luteinising hormone concentrations in cows, with the individual recursive predictions for the first two cows in each group.

(ii) Here, the transition matrix is stochastic, whereas there it was not.

A third, practical, difference is that calculations are not simplified for the normal distribution so that there is no advantage to choosing models based on such an often unrealistic distribution.

Luteinising hormone Let us look once again at the data on luteinising hormone in cows that I analysed above and in Section 10.4.2. Hidden Markov models with two hidden states may allow us to detect different phases in luteinising hormone, as well as differences between the two groups. From inspection of the data and our previous results, we know that a skewed distribution will be necessary. Here, I shall fit the heavy-tailed, three-parameter distributions introduced in Section 10.3.2.

The results for these distributions, and several others, are given in Table 11.2, without and with the hidden Markov chain. As in Section 10.4.2, the heavier tailed distributions fit better than the log normal, with the Burr distribution of Equation (8.25) best here. As well, the hidden Markov models fit better than the AR(1) used above and the ARCH models of Table 10.5. As we have seen (Section 7.3.1), the Burr distribution may be interpreted as a gamma mixture of Weibull distributions.

A different transition matrix clearly is necessary for suckled and for nonsuckled cows. A time trend (with the Burr distribution) is only necessary for the nonsuckled cows, lowering the AIC to 3130.8. For this group of cows, the regression equations in the two hidden states are estimated to be

$$\mu_{1t} = 314.0 + 23.97t$$

Table 11.2. *AICs for several models fitted to the luteinising hormone data with a difference between suckled or not but no time trend. The hidden Markov models have the same and different transition matrices for the two groups.*

		Hidden Markov	
Distribution	Independence	Same	Different
Log normal	3270.5	3198.3	3161.3
Gamma	3314.6	3225.9	3195.0
Weibull	3387.1	3262.3	3261.9
Inverse Gauss	3264.8	3187.0	3176.7
Burr	3250.6	3160.4	3150.3
Log Student t	3258.4	3176.7	3166.8
Log skew Laplace	3266.0	3177.1	3159.5
Log power exponential	3262.3	3184.6	3182.1

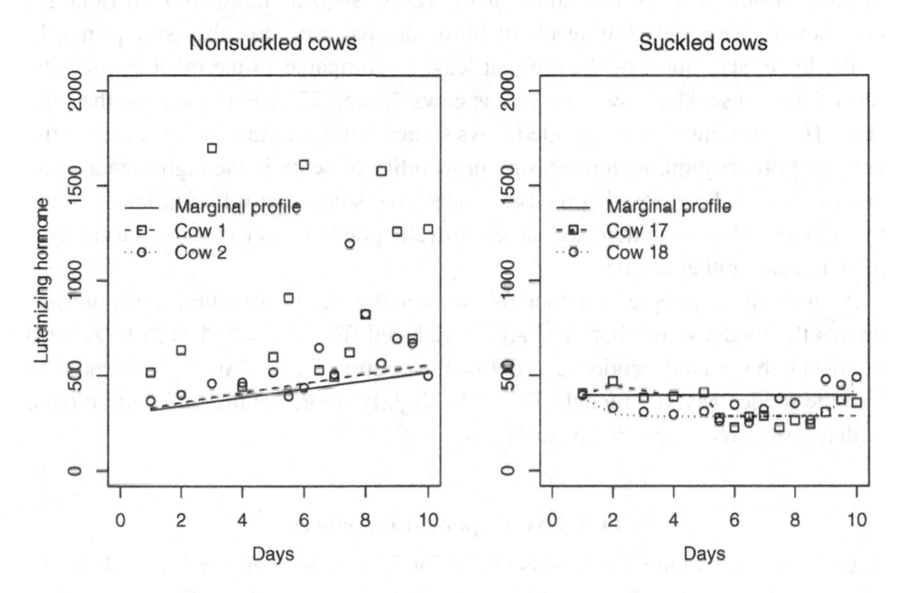

Fig. 11.2. The underlying profiles for the hidden Markov model for the two groups of luteinising hormone concentrations in cows, with the individual recursive predictions for the first two cows in each group.

$$\mu_{2t} \quad = \quad 225.7 + 13.04t$$

Here, μ_{mt} refers to the level of luteinising hormone, not to its logarithm. The other parameters are $\hat{\kappa} = 10.7$ and $\hat{\nu} = 0.24$ for both states. For the suckled cows (without time trend), the values in the two hidden states are $\hat{\mu} = 475.1$ and 285.0, with $\hat{\kappa} =$ 10.4 and $\hat{\nu} = 0.50$. Thus, according to this model, the hormone level is increasing over time for the nonsuckled cows but not for the suckled. The underlying profiles for the same cows as in Figures 10.10 and 11.1 are shown in Figure 11.2.

The intensity transition matrices, and the corresponding (one-day) probability transition matrices, the latter obtained by matrix exponentiation, are respectively

$$\mathbf{\Lambda} = \begin{pmatrix} -0.00021 & 0.00021 \\ 0.00084 & -0.00084 \end{pmatrix}, \quad \mathbf{T}^{(1)} = \begin{pmatrix} 0.99979 & 0.00021 \\ 0.00084 & 0.99916 \end{pmatrix}$$

for the nonsuckled cows and

$$\mathbf{\Lambda} = \begin{pmatrix} -0.039 & 0.039 \\ 0.056 & -0.056 \end{pmatrix}, \quad \mathbf{T}^{(1)} = \begin{pmatrix} 0.963 & 0.037 \\ 0.053 & 0.947 \end{pmatrix}$$

for the suckled ones. The stationary distributions are, respectively, (0.80, 0.20) and (0.59, 0.41). Thus, the nonsuckled cows have a somewhat higher probability of staying in the same state than the suckled ones. Both groups are considerably more often in the first state, with a higher level of hormone.

The (filtered) conditional probabilities of being in the first hidden state, with the larger location parameter, are plotted in Figure 11.3 for the first eight cows of each group. We see that some of the cows, such as numbers 6, 8 (and 10, not shown), have such low levels of hormone that they probably stay primarily in the lower state most of the time, at least as compared to the other cows. The rest of the nonsuckled cows, as well as cows 20 and 22, seem to stay in the high state. However, most of the suckled cows switch between the two states during the period of observation, with increasing probability of being in the high state as time passes. Recall, however, that the two states have somewhat different levels in the two groups. Thus, as anticipated, no clear cyclic phenomenon appears, the changes of state being rather erratic.

As with the autoregressive models, introducing the post-partum covariate improves the model somewhat. The AIC is reduced from 3130.8 to 3124.8. A form of AR(1), the serial dependence as defined in Section 1.2.4, also can be introduced into the hidden Markov model. This only slightly improves the fit for these data, with a further reduction in AIC to 3121.4.

11.3 Overdispersed responses

The overdispersed duration models of Section 7.3.1 can be adapted to apply to ordinary time series with little modification (see Lindsey, 2000). In Equation (7.19), we need to replace the time between recurrent events by the response variable:

$$f_m(y) = \frac{\alpha\beta^\alpha}{[\beta + \Lambda(y)]^{\alpha+1}}\lambda(y) \qquad (11.19)$$

Time only enters in the updating equations. This is another volatility model where the shape of the distribution is changing randomly over time.

As in Sections 7.3 and 7.4, the hidden states are defined by the random parameters α_i and β_i. Thus, the Markov update of Equation (7.45) becomes

$$\begin{aligned} \alpha_i &= \rho^{t_i - t_{i-1}}\alpha_{i-1} + (1 - \rho^{t_i - t_{i-1}})\delta + 1 \\ \beta_i &= \delta + \rho^{t_i - t_{i-1}}\Lambda(y_{i-1}) + \Lambda(y_i) \end{aligned} \qquad (11.20)$$

where $t_i - t_{i-1}$ is the time between the responses y_{i-1} and y_i.

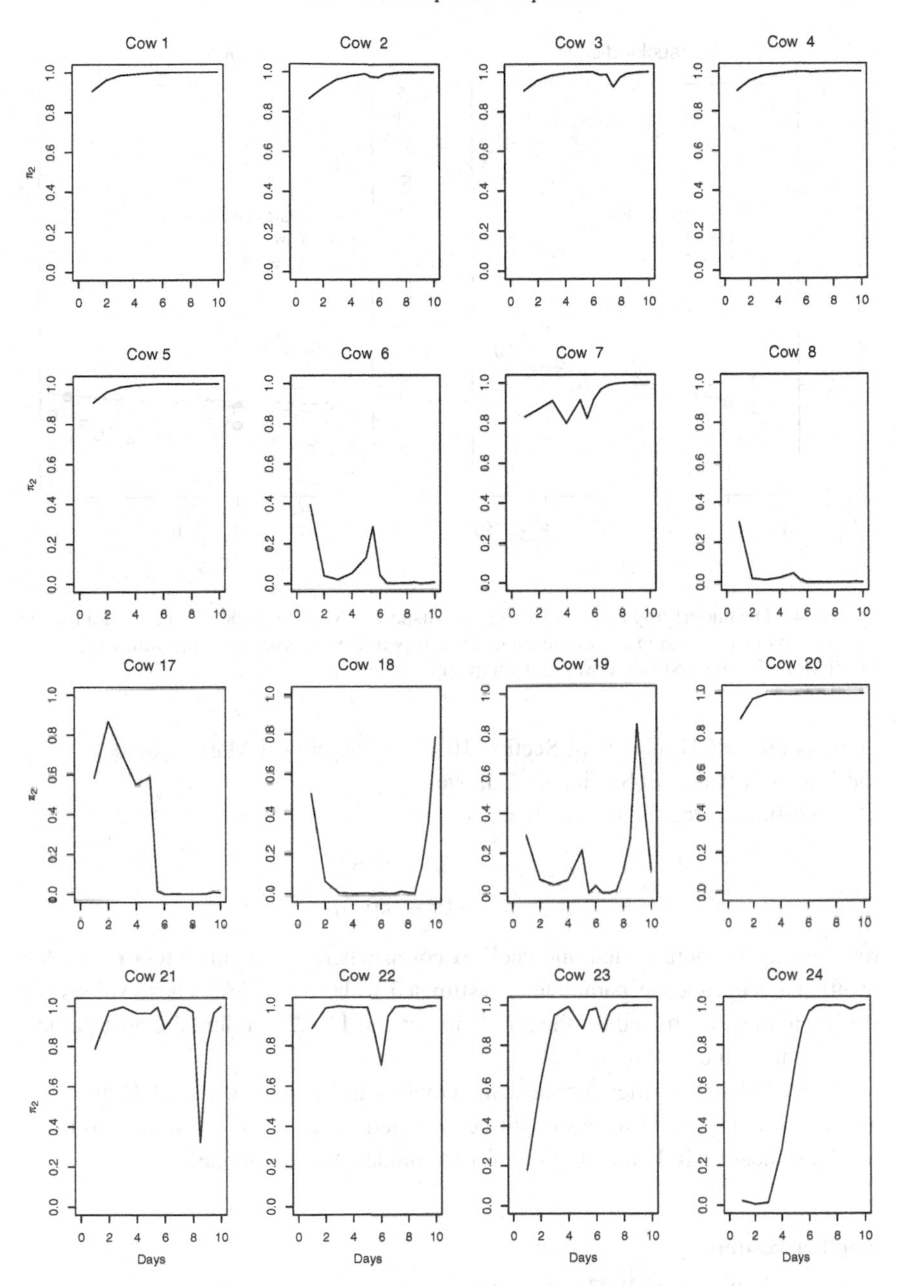

Fig. 11.3. The estimated probabilities of being in the hidden state with the higher level of luteinising hormone for the blood samples from the first eight cows of each group.

Luteinising hormone When this model is fitted to the luteinising hormone data with $\Lambda(y)$ being from the log normal distribution and with the regression function containing the interaction between suckled and time, the AIC is 3204.4. This is better than the standard AR(1) model fitted above in Section 11.1.2, but not as

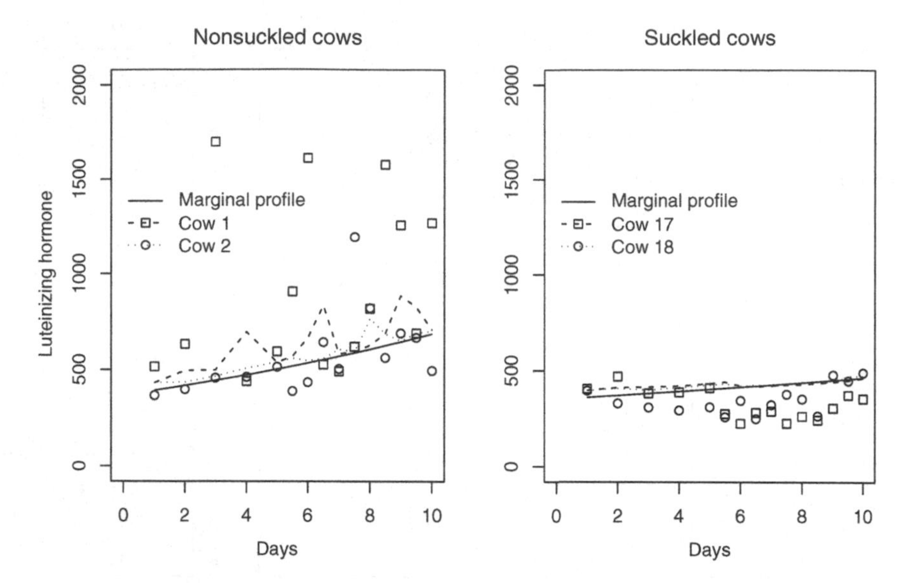

Fig. 11.4. The underlying profiles for the overdispersed Markov dependence model for the two groups of luteinising hormone concentrations in cows, with the individual recursive predictions for the first two cows in each group.

good as the ARCH models of Section 10.4.2 or the hidden Markov models using the Burr distribution in Section 11.2 above.

The estimated regression functions are

$$\begin{aligned}\mu_{1t} &= 6.30+0.062t\\ \mu_{2t} &= 6.04+0.027t\end{aligned}$$

respectively for nonsuckled and suckled cows, where μ_{jt} again refers to the log level. The dependence parameter is estimated to be $\hat{\rho}=0.24$. These values are similar to those estimated for the AR(1) in Section 11.1.2. Adding the post-partum effect reduces the AIC to 3197.9.

The underlying profiles for the same cows as in Figures 10.10, 11.1, and 11.2 are shown in Figure 11.4. As might be expected, these are more similar to those for the standard AR(1) than to those for the hidden Markov model.

Further reading

Harvey (1989), Jones (1993), and Durbin and Koopman (2001) provide good introductions to dynamic linear models. Although proposed originally for normally-distributed data, the dynamic linear model has been extended to other distributions, especially in the generalised linear model family, notably by West *et al.* (1985), Kitagawa (1987), Harvey (1989), and Fahrmeir and Tutz (1994).

For hidden Markov models, see the references in Chapter 7.

Table 11.3. *Reaction times (s) of rats at several times (min) after administration of four levels of an analgesic drug. Times are censored at ten seconds. (Crowder, 1998)*

Time after administration							
0	15	30	60	0	15	30	60
Level 1				Level 3			
1.58	1.78	2.02	2.19	2.74	5.10	7.00	3.54
1.55	1.40	2.20	1.73	1.89	8.00	9.80	6.63
1.47	2.66	3.05	3.76	1.47	9.77	9.98	10.00
2.16	2.28	2.04	2.12	3.13	9.04	9.92	6.25
2.19	2.24	1.99	1.58	1.53	4.72	5.56	3.25
2.07	1.97	1.50	1.24	2.12	4.73	7.90	6.42
1.28	1.67	2.25	1.54	1.28	4.32	6.24	4.73
1.53	2.68	1.79	2.03	1.50	8.91	9.22	4.30
2.62	2.15	1.60	2.91	2.05	4.21	10.00	3.03
1.67	2.52	1.53	1.98	1.53	7.10	9.88	3.72
Level 2				Level 4			
2.02	3.86	2.73	2.88	1.68	5.33	10.00	10.00
1.75	3.38	3.74	2.57	2.80	5.93	8.77	4.62
1.93	3.62	6.05	2.91	2.19	8.73	10.00	8.20
2.04	4.70	3.70	3.05	1.52	10.00	10.00	10.00
2.00	3.34	4.14	2.78	1.85	7.71	8.35	7.57
1.63	3.37	4.84	2.36	2.78	7.41	10.00	10.00
1.97	3.72	7.83	3.14	1.81	7.56	10.00	10.00
2.42	4.81	4.90	1.69	1.80	10.00	9.41	10.00
1.38	3.88	4.20	2.05	1.16	9.95	10.00	10.00
2.32	3.75	3.06	3.81	1.67	7.82	10.00	7.97

Exercises

11.1 Let us reconsider the monthly deaths from bronchitis, emphysema, and asthma in Tables 7.7 and 9.2.

(a) Which of the various dynamic models presented in this chapter fits best?

(b) Give reasons why the one you have chosen may be appropriate for these data.

11.2 Do any of the dynamic models fit well to the Yale enrolment data of Table 7.1?

11.3 In the context of studies of the effect of body lead concentration on response times to sensory stimuli, four groups of ten rats each were subjected to different dose levels of an analgesic drug. All rats had the same age and the same level of lead concentration. The experiment consisted in immersing the tail of the rat in hot water at times zero, 15, 30, and 60 min after drug administration. Then, the reaction time in seconds for the rat to flick its tail out again was recorded, as shown in Table 11.3. All reaction times are right-censored at ten seconds.

Table 11.4. *Recordings of DNA binding (DB) and serum complement (SC) over nine years on a patient with systemic lupus erythematosis. (Jones, 1984)*

Day	DB	SC	Day	DB	SC	Day	DB	SC
0	8.40	16.0	541	7.20	21.0	1717	4.66	10.5
8	8.50	NA	576	7.30	13.3	1738	5.41	10.5
23	8.70	9.0	611	8.00	13.2	1808	5.60	9.2
35	6.60	13.9	674	6.00	11.7	1850	5.26	13.4
44	4.40	13.8	709	7.60	13.4	1878	5.55	10.1
58	4.20	13.7	730	7.10	12.5	1906	5.61	12.0
100	7.90	17.2	807	6.20	12.5	1990	4.30	13.5
121	7.10	13.5	842	7.50	11.7	2025	3.79	11.8
149	9.10	8.6	884	5.30	12.9	2081	4.93	11.0
156	9.00	8.1	921	5.90	14.0	2102	5.23	10.8
184	10.00	7.9	1011	5.50	12.0	2144	4.52	11.4
198	10.00	6.9	1038	4.20	10.0	2186	3.55	11.0
254	9.84	12.3	1046	4.90	9.0	2218	2.87	2.9
280	10.00	10.0	1100	2.50	9.6	2263	3.20	12.8
281	9.30	10.4	1121	3.20	12.5	2284	3.09	12.8
288	8.80	8.2	1133	3.30	15.0	2333	3.33	10.0
295	7.60	8.6	1137	3.60	11.8	2361	2.80	12.7
302	4.80	9.6	1157	2.50	13.2	2424	3.06	12.7
309	5.50	8.2	1213	2.70	20.0	2452	2.03	11.2
316	5.70	10.7	1255	4.60	10.8	2458	2.35	10.6
323	4.30	9.6	1297	3.60	12.0	2501	1.88	12.2
330	6.00	10.7	1325	3.40	13.0	2564	1.76	10.4
337	5.90	10.7	1353	5.80	14.5	2599	1.17	12.8
343	6.90	11.7	1381	4.40	11.5	2641	2.01	13.6
351	6.90	10.2	1437	3.82	13.0	2676	1.16	13.8
358	7.90	9.1	1507	5.17	10.0	2711	0.39	11.6
364	8.40	9.1	1535	5.57	12.9	2739	1.25	10.7
370	6.00	10.7	1556	5.49	7.1	2767	0.90	12.7
380	9.20	11.0	1584	6.96	16.0	2823	1.49	11.7
401	7.20	10.6	1604	6.09	14.5	2886	2.05	10.3
440	7.30	14.4	1618	5.69	10.0	2928	1.97	5.4
478	7.90	16.3	1653	5.61	13.0	2991	1.11	15.0
506	6.10	16.8	1689	6.64	15.0	3061	0.84	10.0

(a) How do reaction times vary with drug concentration?

(b) The experimenters believed that 15 min between trials was sufficient so that responses on a rat would be independent. Check this assumption.

11.4 Levels of DNA binding and serum complement were observed at unequally spaced time intervals over nine years on a patient with systemic lupus erythematosis, as shown in Table 11.4.

(a) Fit an appropriate model to each series separately.

(b) Does either series show evidence of a trend or of a cyclical phenomenon?

(c) Is there any evidence of volatility?

(d) If so, is it the result of individual extreme values or clustering?
(e) Can you find any dependence relationship between the two series?

12

Growth curves

In certain stochastic systems, we cannot observe changes for individual elements but only in aggregation. For example, in a chemical reaction, we cannot observe the changes of state of the participating atoms but only the concentration of each reactant and product; in the growth of a biological organism, we cannot observe the addition of individual proteins, or even of cells, but only the increase in weight, length, or volume. In other words, records of change in such a system are averages of the stochastic changes of all the components involved.

Such systems can, nevertheless, generally be modelled mechanistically by rates of change of the individual elements, in ways similar to the intensity functions in Part 2. This procedure will yield some form of mean function for the aggregate change. However, a second level of stochastic variability usually also is present, resulting from random external influences on the system: changes in pressure or temperature of a chemical reaction, changes in food supply, stress, and so on, to a biological organism. Thus, modifications at the level of the individual components only can be modelled as a mean function, with variation about it arising from the second level of stochastic variability.

Here, and in Chapter 13, I shall look at some ways to model two of the most important types of such phenomena:

(i) in this chapter, processes where material is accumulating in a system;
(ii) in the next, when the material is entering, passing through, and leaving a system.

In both cases, model construction will involve solving differential equations.

12.1 Characteristics

One special type of series recorded over time involves measurements related to growth of some kind. Observations may be made repeatedly in one or more series. One goal of such studies is often to predict future growth. A series of responses on a given individual usually will not be independent so that some of the procedures of the previous chapters, such as autoregression, will have to be used in the model construction.

Because, in a growth curve process, material is accumulating, any chosen regression function generally will need to increase at first. However, growth usually has an upper limit, so that the curve eventually should flatten off. Thus, appropriate regression functions generally will have some S shape, similar in form to a cumulative distribution function. Such a function may be interpreted in terms of the individual elements accumulating in the growth process: the process of addition of material over time follows, on average, the density function corresponding to that cumulative distribution function.

Naturally, the shape of the curve, and the type of model, will depend heavily on the way in which growth is expressed. For example, in biology, one cannot expect growth of an organism, measured as the weight, or volume, to take the same form as its growth in length. The study of the relationships among these various possibilities is known as *allometry*. Most allometric relationships are expressed in terms of powers: thus, volume is roughly related to the cube of length.

As compared to other types of stochastic processes, growth curves have several special characteristics, only some of which are shared with other series of responses over time:

- By definition, growth is not stationary; occasionally, the increments, or innovations, may be.
- Random variability generally will increase with size, so that the dispersion is not constant.
- The growth profile generally will be a nonlinear function of time, most often eventually reaching an asymptote.
- If several series are observed, they may have different growth rates, maximum levels, or even overall profiles, either inherently or due to environmental effects.

A wide class of different functions is available for modelling such phenomena. I only shall look at a few of the most common ones here.

12.2 Exponential forms

The exponential growth curve is well known, for example in attempts over the last several centuries to describe increases in the world's population. However, as explained above, such a curve will be of limited use because growth phenomena have upper limits.

12.2.1 Exponential growth

Let us first make the simple, reasonable assumption that the rate of addition of new material, that is of growth, is proportional to the size already attained y_t:

$$\frac{\mathrm{d}y_t}{\mathrm{d}t} = \gamma y_t \tag{12.1}$$

with the initial condition that growth starts with size y_0 at time $t = 0$. This assumption implies that external factors are not influencing or limiting the process.

Solving this equation yields one of the simplest functions for describing growth, the *exponential* curve:

$$y_t = y_0 e^{\gamma t} \tag{12.2}$$

where γ is called the *intrinsic growth coefficient*. With $\gamma > 0$, such a function, usually, will be realistic only in the early stages of growth; nothing can continue growing exponentially forever. On the other hand, with $\gamma < 0$, it may be a reasonable model of exponential decline from some reasonably large initial value y_0.

Distributional assumptions

As it stands, Equation (12.2) describes a deterministic relationship between the state of the process (size) and time. It can be used to describe average growth. However, a stochastic element, to allow for external influences (the second level), usually will be required. This can be introduced in at least two ways.

(i) The size at each given time may have some specific distribution, say a normal or gamma, with mean given by Equation (12.1). Then, by using a log link, we can obtain the linear regression function

$$\log(\mu_t^y) = \log(\mu_0) + \gamma t \tag{12.3}$$

or

$$\mu_t^y = \mu_0 e^{\gamma t} \tag{12.4}$$

where the superscript on μ indicates that the mean refers to the observed measurements y.

(ii) The logarithm of the size may be assumed to have some distribution, such as the normal or logistic, yielding a log normal or log logistic distribution for the sizes themselves. With the identity link and a linear regression function, this yields

$$\mu_t^{\log(y)} = \log(\mu_0) + \gamma t \tag{12.5}$$

where μ here refers to the logarithms of the responses, as indicated by the superscript. Back transformation

$$e^{\mu_t^{\log(y)}} = \mu_0 e^{\gamma t} \tag{12.6}$$

now yields a function describing the geometric mean of size.

The resulting curves from these two approaches can differ rather significantly, both in their fit to observed responses and in the predictions they yield. This will depend primarily on the skew of the distribution of the responses, but also on what we mean by a prediction: the mean, the mode, the median, and so on? Thus, if μ is the mean, in the first model, the curve goes through the arithmetic mean of the responses (and the mode, if the chosen distribution is symmetric), whereas, in the second, it goes through the geometric mean.

The stochastic variability modelled by the two equations also is quite different. Thus, for the first model, with a normal distribution, the variance of the size is

Table 12.1. *Gross domestic fixed capital formation in the UK, 1948–1967, in millions of pounds. (Oliver, 1970)*

1948	1949	1950	1951	1952	1953	1954	1955	1956	1957
1422	1577	1700	1889	2106	2359	2552	2829	3103	3381
1958	1959	1960	1961	1962	1963	1964	1965	1966	1967
3492	3736	4120	4619	4731	4906	5860	6331	6686	7145

Table 12.2. *Comparison of models fitted to the gross domestic fixed capital formation data in Table 12.1.*

	Log normal		Normal	
	Independence	AR(1)	Independence	AR(1)
AIC	127.2	125.0	128.4	129.0
$\widehat{\mu_0}$	3327.39	3311.89	3349.76	3348.55
$\hat{\gamma}$	0.0841	0.0844	0.0814	0.00813
$\hat{\rho}$	—	0.55	—	0.21
1968 Prediction				
Marginal	8052.6	8036.6	7877.6	7866.7
Recursive	—	7894.5	—	7844.3

constant. In the second, again with a normal distribution, the variance of the log size is constant, implying that the variance of the size is increasing with size. If Equation (12.3) is used with a gamma distribution, the coefficient of variation, not the variance, is assumed constant. Other distributions will carry still other assumptions about how the variability is changing with mean size, that is, over time.

Capital accumulation Let us look at the gross domestic fixed capital formation in the UK from 1948 to 1967, as shown in Table 12.1. I shall study the performance of these models in the prediction of the value for 1968 using the exponential growth function

$$\mu_t^y = \mu_0 e^{\gamma(t-1957.5)} \tag{12.7}$$

with a normal distribution and the corresponding one for the geometric mean with a log normal distribution. Here, μ_0 will be assumed to be an unknown parameter. Note that, in this formulation, it is not the initial condition, but rather the value at $t = 1957.5$.

The results are given in Table 12.2. We see that the log normal distribution with an AR(1) fits best to the data. The autocorrelation is estimated to be $\hat{\rho} = 0.55$ whereas it is only 0.21 for the normal distribution. The curves from the two models are plotted in Figure 12.1. As would be expected from the smaller autocorrelation, the recursive predictions vary less around the marginal curve for the normal than for the log normal model.

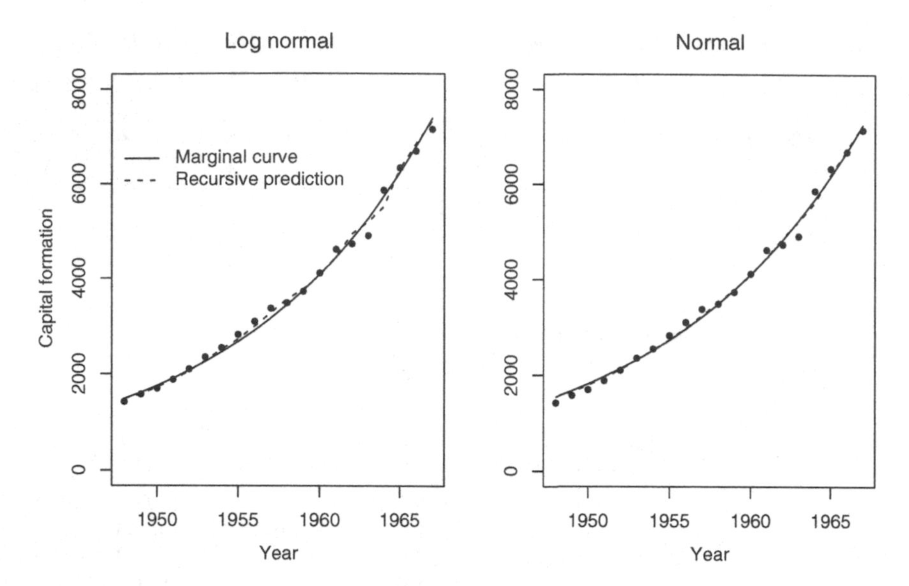

Fig. 12.1. The log normal and normal AR(1) models for the capital formation data of Table 12.1.

Recall from Equation (9.19) that the (arithmetic) mean response for the log normal distribution is given by $\exp(\mu+\sigma^2/2)$. I used this in Table 12.2 to make the predictions from the normal and log normal distributions comparable. In both models, the marginal prediction, that is the projected regression curve, is larger than the recursive prediction, that is the marginal prediction corrected by the product, $\hat{\rho}$ times the previous residual, because the 1967 residual is negative. The actual observed value for 1968 was 7734 so that, as can be seen from Table 12.2, all models overpredict, but the recursive predictions do better than the regression curve alone. Interestingly, the normal models predict this particular value better, even though they fit more poorly to the previously observed values.

12.2.2 Monomolecular growth

With the exponential growth curve, the size continues to increase indefinitely. For many phenomena, this is not reasonable, except perhaps over short periods of time. A second, alternative, simple and reasonable assumption is, then, that there is some physical maximum size limiting the system. Then, the growth rate can be made proportional to the remaining size. By definition, this necessarily leads to an asymptote, say μ_∞, with

$$\frac{d\mu_t}{dt}=\gamma(\mu_\infty-\mu_t), \quad \gamma>0 \tag{12.8}$$

so that

$$\begin{aligned}\mu_t &= \mu_\infty-(\mu_\infty-\mu_0)e^{-\gamma t}\\ &= \mu_\infty-\zeta e^{-\gamma t}\end{aligned} \tag{12.9}$$

Table 12.3. *Area (unspecified units) of a bean plant leaf grown at constant temperature for 36 consecutive days with measurements starting at day 17. (Scallon, 1985; read across rows)*

0.745	1.047	1.695	2.428	3.664	4.022	5.447	6.993
8.221	8.829	10.080	12.971	14.808	17.341	19.705	22.597
24.537	25.869	27.816	29.595	30.451	30.817	32.472	32.999
33.555	34.682	34.682	35.041	35.356	35.919	36.058	36.454
36.849	37.200	37.200	37.200				

where $\mu_0 = \mu_\infty - \zeta$ is the mean initial size. Growth occurs if $\mu_\infty > \mu_0 > 0$.

In this model, growth is not influenced by the present size. Thus, this curve shows rapid initial increase, only levelling off as the constraint of maximum size begins to take effect. Such a process is not too common.

In kinetics, this function is called the *monomolecular* growth; it has also been called the confined exponential and the Mitscherlich growth curve. When $\mu_\infty = \zeta$ so that initial size is zero, it is called the von Bertalanffy growth curve, used in ecology to describe animal growth. In population ecology, for this and the following growth curves, μ_∞ is called the *carrying capacity*.

Notice that, when the initial condition and asymptote are standardised so that $\mu_0 = 0$ and $\mu_\infty = 1$, this regression function has the form of an exponential cumulative distribution function.

Plant growth Let us look at the growth in area of a bean plant leaf, with values as recorded in Table 12.3. It is clear from these values that the growth has levelled off by the end of the study so that this growth function might be appropriate. However, the plots in Figure 12.2 indicate a slightly S-shaped curve, so that the initial rapid growth is not present.

The normal model with independence has an AIC of 85.7, whereas that with an AR(1) has 47.2, a great improvement. The underlying profile curve for the latter is estimated to be

$$\mu_t = 67.7 - 99.9\mathrm{e}^{-0.023t}$$

implying negative size at time zero. This function is plotted in the left graph of Figure 12.2 as the solid curve, along with the recursive predictions which follow the responses fairly well. The autocorrelation is estimated to be $\hat{\rho} = 0.95$.

We also can try a log normal model. With the log area following Equation (12.9), this implies that the geometric mean of the area follows the curve

$$\mu_t = \exp\left(\mu_\infty - \zeta\mathrm{e}^{-\gamma t}\right) \tag{12.10}$$

which is no longer monomolecular growth. Thus, the parameters do not have the same physical interpretation. This log normal model has an AIC of 52.8 with independence and 45.3 with an AR(1). The latter is only a small improvement on the normal distribution monomolecular growth with AR(1). The large improvement

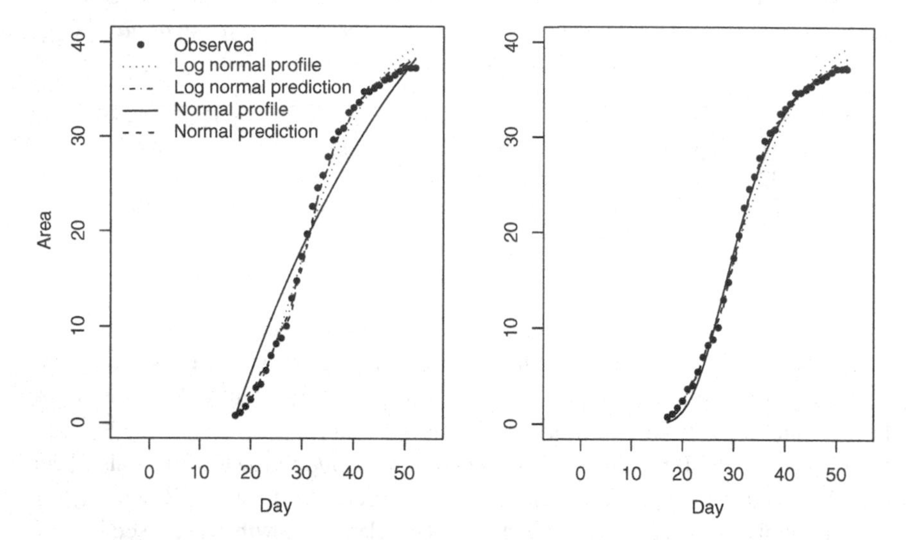

Fig. 12.2. The log normal and normal AR(1) models for the bean growth data of Table 12.3. The former is the same exponentiated monomolecular function in both graphs. For the normal distribution, the left function is monomolecular and the right one is exponentiated monomolecular.

in the independence model indicates that Equation (12.10) may be preferable to Equation (12.9). I shall investigate this further below.

Here, the underlying marginal curve for the log normal AR(1) model is estimated to be

$$\mu_t = \exp\left(3.7 - 28.9\mathrm{e}^{-0.116t}\right)$$

This is also plotted in the left graph of Figure 12.2 with the recursive predictions. (The variance is very small, so that the arithmetic and geometric mean curves are virtually identical.) The autocorrelation is estimated to be $\hat{\rho} = 0.62$. The marginal curve has a sigmoidal shape, more appropriate for these responses. The recursive predictions are very similar to those from the previous model, as might be expected from the similarity of the AICs.

The next step is, then, to try Equation (12.10) with the normal distribution. Indeed, the AIC for the AR(1) model with this regression function is 28.0. The underlying marginal curve is estimated to be

$$\mu_t = \exp\left(3.7 - 66.8\mathrm{e}^{-0.149t}\right)$$

This curve and the recursive predictions are plotted in the right graph of Figure 12.2 along with those for the log normal distribution. The autocorrelation is estimated to be $\hat{\rho} = 0.79$. Thus, we can conclude that Equation (12.10) is preferable to monomolecular growth for these responses and that the normal distribution fits better than the log normal.

As can be seen in the graphs, exponentiating the monomolecular function results in a sigmoidal-shaped growth curve. Let us now investigate some other possible functions yielding such a shape.

12.3 Sigmoidal curves

We may expect that combining the assumptions of the two previous models, given in Equations (12.1) and (12.8), could yield a more flexible and appropriate model. Thus, let us now assume that the growth rate depends both on the current size and on the remaining size. In other words, the growth *rate* should increase to a maximum before steadily declining to zero. To allow this model to be as general as possible, let us first only assume that the rate is proportional to some arbitrary functions of the current and maximum or remaining sizes:

$$\frac{\mathrm{d}\mu_t}{\mathrm{d}t} = \gamma g_1(\mu_t) g_2(\mu_t, \mu_\infty) \tag{12.11}$$

The resulting growth curves will have sigmoidal behaviour. Several specific functions of size most often have been used. These produce different sigmoidal forms. (I have already used some growth curves falling into this large family in Sections 2.4.3 and 7.1.2, the latter for binary counts.)

12.3.1 Logistic growth

The simplest possibility is that the rate is directly proportional to the two sizes:

$$\frac{\mathrm{d}\mu_t}{\mathrm{d}t} = \frac{\gamma}{\mu_\infty} \mu_t (\mu_\infty - \mu_t) \tag{12.12}$$

yielding the well known symmetric S-shaped *logistic* function

$$\mu_t = \frac{\mu_\infty}{1 + \zeta \mathrm{e}^{-\gamma t}} \tag{12.13}$$

where again μ_∞ is the asymptotic maximum value of the size. We already have met this in another form in Equation (2.19).

We can transform this function to a linear form by using a logit link:

$$\log\left(\frac{\mu_t}{\mu_\infty - \mu_t}\right) = -\log(\zeta) + \gamma t \tag{12.14}$$

If μ_∞ were known, this could be fitted as a generalised linear model. Here, when the asymptote is standardised so that $\mu_\infty = 1$, this regression function has the form of a logistic cumulative distribution function.

One handicap of this function, as a growth curve, is that it is symmetric, reflecting the symmetry of the logistic distribution. Addition of material is assumed to be symmetric over time, with the maximum, corresponding to the most rapid increase in size, occurring exactly at the middle of the growth period. This is reflected in the shape of the curve, with the form of increasing early growth being identical to that of decreasing late growth. Often, there is no biological reason for these assumptions to hold.

Table 12.4. *Numbers of diagnosed AIDS cases reported in the UK from January 1982 to April 1987. (Healy and Tillett, 1988)*

	Month											
Year	J	F	M	A	M	J	J	A	S	O	N	D
1982	1	0	2	1	0	0	1	1	1	1	1	2
1983	1	0	2	0	1	1	5	4	3	0	5	7
1984	6	5	8	4	5	6	10	14	7	14	8	16
1985	16	15	18	16	14	12	17	24	23	25	22	21
1986	25	30	26	35	38	27	24	29	39	34	38	47
1987	33	44	35	35								

Table 12.5. *Comparison of the Poisson growth models fitted to the data on AIDS cases in Table 12.4.*

	Exponential	Logistic	Gompertz	Complementary Gompertz
AIC	167.5	148.2	149.8	148.9
Asymptote	—	44.1	105.6	37.5

AIDS cases The acquired immune deficiency syndrome (AIDS) has had a substantial impact on the costs of health care for some time now. Thus, early in the epidemic, it was very important to be able to project its growing size accurately into the future. The early numbers of cases for the UK are given in Table 12.4. At the time, people wanted to determine whether or not there was any indication that the epidemic was levelling off and, if so, at what level.

Because the numbers of cases diagnosed are counts, it may be reasonable to start with a nonhomogeneous Poisson process (Section 4.3). The exponential growth curve does not have an asymptote, but I shall fit it as a point of comparison. The logistic curve fits considerably better, as can be seen in Table 12.5. It predicts that the asymptotic mean number of cases will be about 44. However, this and all the following predictions are based primarily on the shape of the lower part of the curve which certainly is not reasonable. The exponential and logistic curves are plotted in Figure 12.3. Although the fit appears reasonable for the series observed at that time, it is clear that prediction will not be reliable.

There is no evidence that these responses are either overdispersed (Section 3.6.2) or dependent over time (Section 7.1): adding a negative binomial or AR(1) assumption does not improve the fit.

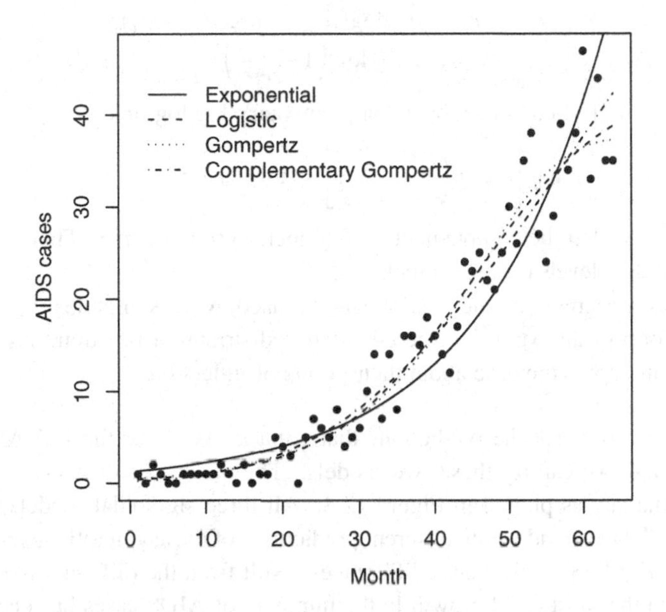

Fig. 12.3. Various Poisson-based models for the AIDS cases in the UK each month from January 1982 to April 1987, from Table 12.4.

12.3.2 Gompertz growth

Another commonly used S-shaped growth curve with an asymptote has growth rate defined by

$$\begin{aligned} \frac{d\mu_t}{dt} &= \gamma\mu_t[\log(\mu_\infty) - \log(\mu_t)] \\ &= -\gamma\mu_t \log\left(\frac{\mu_t}{\mu_\infty}\right) \end{aligned} \tag{12.15}$$

The solution to this is called the *Gompertz* growth curve,

$$\mu_t = \mu_\infty e^{-\zeta e^{-\gamma t}} \tag{12.16}$$

In contrast to the logistic curve, this one is not symmetric about its point of inflection. As with the logistic curve, we can obtain a linear form, this time by means of a log log link:

$$\log\left[-\log\left(\frac{\mu_t}{\mu_\infty}\right)\right] = \log(\zeta) - \gamma t \tag{12.17}$$

This curve rises quickly and then levels off more slowly.

If we invert the asymmetry, we have

$$\mu_t = \mu_\infty\left(1 - e^{-\zeta e^{\gamma t}}\right) \tag{12.18}$$

obtained from

$$\begin{aligned}\frac{d\mu_t}{dt} &= \gamma(\mu_\infty - \mu_t)[\log(\mu_\infty) - \log(\mu_\infty - \mu_t)] \\ &= -\gamma(\mu_\infty - \mu_t)\log\left(1 - \frac{\mu_t}{\mu_\infty}\right)\end{aligned} \tag{12.19}$$

The linear form is, here, given by a complementary log log link:

$$\log\left[-\log\left(1 - \frac{\mu_t}{\mu_\infty}\right)\right] = \log(\zeta) + \gamma t \tag{12.20}$$

This may be called the complementary Gompertz growth curve. This curve rises slowly and then levels off more quickly.

When these regression functions are standardised, with asymptotes $\mu_\infty = 1$, they have the form of an extreme value cumulative distribution function. As we have seen, they are not symmetric about their points of inflection.

AIDS cases To check the predictions that I made above for the UK AIDS data in Table 12.4, we can try these two models. The results also are given in Table 12.5 and the curves plotted in Figure 12.3. All three sigmoidal models fit about equally well, but provide quite different predictions of the asymptotic mean level of monthly UK AIDS cases. These differences result from the different assumptions about when the most rapid growth in the numbers of AIDS cases has occurred or will occur.

12.4 Richards growth curve

Let us now attempt to construct a simple, but more realistic, biological model for growth. In terms of weight or volume, the growth rate of a living organism is the difference between the metabolic forces of anabolism and catabolism, the synthesis of new body matter and its loss. Catabolism can be taken proportional to weight whereas anabolism can be assumed to have an allometric relationship to weight. This can be written

$$\frac{d\mu_t}{dt} = \frac{\gamma}{\kappa}\mu_t\left[1 - \left(\frac{\mu_t}{\mu_\infty}\right)^\kappa\right], \quad \kappa \neq 0 \tag{12.21}$$

with solution

$$\mu_t = \mu_\infty\left\{1 - \left[1 - \left(\frac{\mu_\infty}{\mu_0}\right)^\kappa\right]e^{-\gamma t}\right\}^{-\frac{1}{\kappa}} \tag{12.22}$$

called the *Richards* curve. The parameters γ and κ both control the rate of growth. This curve describes initial exponential growth that is increasingly damped as the size increases until eventually it stops.

When $\kappa = -1$, this is the monomolecular function, when $\kappa = 1$, the logistic function, and, when $\kappa \to 0$, the Gompertz function. If $\gamma < 0$ and $\mu_\infty > \mu_0$ or $\gamma > 0$ and $\mu_\infty < \mu_0$, there will be negative growth or decay.

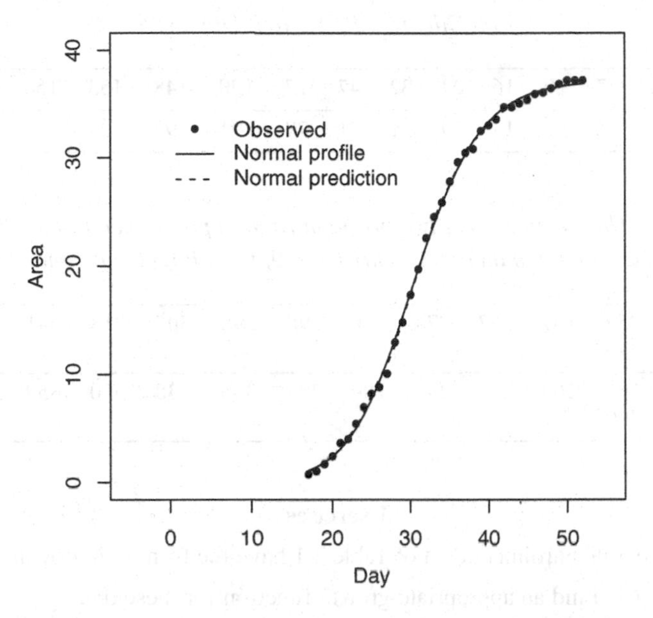

Fig. 12.4. The Richards growth curve with a normal AR(1) model for the bean growth data of Table 12.3.

Plant growth When applied to the bean growth data of Table 12.3, the Richards model shows a large improvement over those fitted above. For a normal distribution with AR(1), the AIC is reduced to 22.1 (26.6 without the autoregression). The autocorrelation is estimated to be $\hat{\rho} = 0.51$. The underlying profile curve is estimated to be

$$\mu_t = 37.2\left\{1 - \left[1 - \left(\frac{37.2}{0.0077}\right)^{0.73}\right] e^{-0.22t}\right\}^{-\frac{1}{0.73}}$$

This curve and the recursive predictions are plotted in Figure 12.4.

Although the Richards curve is meant to describe growth in weight, it does quite well here with area. For these data, with $\hat{\kappa} = 0.73$, the curve is closer to that from a logistic function than from a monomolecular, but is still some distance from either.

Further reading

Banks (1994) presents a wider variety of growth curves and their applications, but with no discussion of appropriate statistical methods for fitting them. Lindsey (2001) provides detailed discussion of fitting growth curves, with a number of complex examples of applications to medical data. For other examples in biology and epidemiology, see Brauer and Castillo-Chavez (2001).

Brown and West (2000) give a good introduction to recent issues in allometry.

Table 12.6. *The heights (cm) of two sunflowers measured weekly. (Sandland and McGilchrist, 1979, from Doyle)*

4	4.5	7	10	16	23	32	47	77	130	148	153	154	154
2	3	5	9	13	19	31	51	78	91	93	96	91	93

Table 12.7. *The weights (g) of two pregnant Afghan pikas over 14 equally spaced periods. (Sandland and McGilchrist, 1979, from Puget and Gouarderes)*

251	254	267	267	274	286	298	295	307	318	341	342	367	370
258	263	269	266	282	289	295	308	338	350	359	382	390	400

Exercises

12.1 The Yale enrolment data of Table 7.1 have the form of a growth curve.

(a) Find an appropriate growth function for these data.
(b) Is an autoregression model still required?
(c) Do you still have to allow for a changing variance?
(d) Can you now interpret in a different way the models fitted in Sections 7.1.1 and 7.1.3?
(e) How does the fit of your model compare to that of the random walk model in Section 9.2.3?

12.2 Table 12.6 gives the heights (cm) of two sunflower plants measured at weekly intervals.

(a) Find a growth curve to describe these data.
(b) Is the difference between the two plants due to the smaller initial size of the second or to the rate of growth?

12.3 Weights (g) of two pregnant Afghan pikas were recorded over 14 equally spaced periods from conception to parturition, as shown in Table 12.7.

(a) Find a growth curve to describe these data.
(b) Can you detect any difference between the two animals?

12.4 In the study of glucose turnover in human beings, 26 volunteers were injected with deuterium-labelled glucose and the deuterium enrichment (atom% excess) measured at various time points. The results for subject 15 are given in Table 12.8. It is expected that the resulting curve will be asymptotically zero after infinite time.

(a) Try to find an appropriate model.
(b) After how much time do you predict a zero measurement?

Table 12.8. *Measurements (atom% excess) of deuterium-labelled glucose over time in one individual. (Royston and Thompson, 1995, from Edwards)*

Minutes	2	3	4	5	6	8	10	12.5	15	20
Glucose	1.49	1.39	1.35	1.22	1.22	1.14	1.05	0.98	0.92	0.88
Minutes	30	40	50	60	75	90	120	150	180	
Glucose	0.79	0.65	0.65	0.59	0.48	0.38	0.40	0.32	0.31	

Table 12.9. *Dissolved oxygen concentration (mg/l) in a tank of water. (Banks, 1994, p. 66)*

Minutes	0	1	2	3	4	5	6	7
Concentration	0.36	1.54	2.55	3.37	3.98	4.56	5.03	5.32
Minutes	8	9	10	11	12	13	14	15
Concentration	5.67	5.88	6.13	6.22	6.42	6.49	6.59	6.68

12.5 In the study of water quality in lakes and rivers, the oxygen balance of the body of water is important. This involves the rate of transfer of oxygen from the atmosphere to the water and may depend on the velocity of the water and/or wind, on wave action, and on rainfall. Laboratory experiments to determine oxygen transfer rates start with a tank of water in which initial oxygen concentration is reduced to a very low value by addition of some chemical such as sodium sulphite. Then, the tank is submitted to the test variable, such as wind, and the dissolved oxygen measured at various time intervals. In this way, the oxygen transfer coefficient, called the reæration coefficient, can be determined. The results for one such experiment are shown in Table 12.9.

(a) What form of growth curve should be suitable for these data?
(b) What is the interpretation of μ_∞?
(c) Which parameter might be called the reæration coefficient?
(d) What is its estimated value?

12.6 A second important factor in measuring the quality of water is the biochemical oxygen demand (BOD). This is the amount of oxygen used by micro-organisms to stabilise organic matter in waste water. This can be determined by incubating a sample of water from the lake or river for several days at a constant temperature, usually 20° C, and taking samples to measure oxygen content. The results from one such study are given in Table 12.10.

(a) Is the same form of curve applicable here as in Exercise 12.5?
(b) What is the estimated maximum BOD for this water?
(c) Which parameter might be called the deoxygenation coefficient?
(d) What is its estimated value?

Table 12.10. *Biochemical oxygen demand (BOD, mg/l) of water from samples. (Banks, 1994, p. 68)*

Day	0.5	1	2	3	4	5	6	8	10	12
BOD	14	30	48	71	82	102	109	131	139	152

Table 12.11. *Reported AIDS cases, by quarter, as diagnosed in the UK, 1982–1992. (Healy and Tillett, 1988; de Angelis and Gilks, 1994)*

Year	Quarter			
1982	3	1	3	4
1983	3	2	12	12
1984	14	15	30	39
1985	47	40	63	65
1986	82	120	109	120
1987	134	141	153	173
1988	174	211	224	205
1989	224	219	253	233
1990	281	245	260	285
1991	271	263	306	258
1992	310	318	273	133

12.7 One major problem with reported cases of AIDS is the underreporting in the most recently available quarter years, because all cases will not yet have reached the central recording office. Any projections into the future must take this problem into account.

(a) Consider first the AIDS cases reported, by quarter, as diagnosed in the UK, 1982–1992; these are given in Table 12.11. Try various growth curves for these data.
(b) For each curve, state explicitly your assumptions about how growth is occurring.
(c) Now look at the numbers of AIDS cases diagnosed in the USA between 1982 and 1990; these are shown in Table 12.12. Does the growth curve that you choose resemble that for the UK?
(d) What projections can you make for each country?
(e) Find more recent data to check your predictions.

12.8 One application of growth curves is to the spread of innovations. In one such study, the number of ranchers in Uruguay having adopted the new technique of fertilised grass–legume pastures between 1961 and 1976 was recorded, as shown in Table 12.13.

(a) Find an appropriate growth curve for these data.
(b) In what year was the new procedure being adopted at the maximum rate?

Table 12.12. *Reported AIDS cases, by quarter, as diagnosed in the USA, 1982–1990. (Hay and Wolak, 1994)*

Year	Quarter			
1982	185	201	293	381
1983	536	705	769	851
1984	1148	1372	1573	1746
1985	2157	2578	2997	3107
1986	3775	4263	4692	4935
1987	5947	6409	6756	6920
1988	7560	7677	7674	7625
1989	8109	8224	7818	6935
1990	5922	329		

Table 12.13. *The numbers of ranchers in Uruguay having adopted improved pasture techniques. (Banks, 1994, p. 38, from Jarvis)*

Year	1961	1962	1963	1964	1965	1966	1967	1968
Number	141	261	397	697	944	1445	2060	3247
Year	1969	1970	1971	1972	1973	1974	1975	1976
Number	5284	7099	9554	11465	13767	14678	14998	15473

(c) How many ranchers do you predict will eventually adopt this approach?

12.9 Three closed colonies of *Paramecium aurelium* were raised in a nutritive medium consisting of a suspension of the bacterium *Bacillus pyocyaneous* in salt solution. At the beginning of each experiment, 20 *Paramecia* were placed in a tube containing 5 ml of the medium at a constant 26°C. Each day, starting on the second, the tube was stirred, a sample of 0.5 ml taken, and the number of individuals counted. The remaining suspension was centrifuged, the medium drawn off, and the residue washed with bacteria-free salt solution to remove waste products. After a second centrifuging to remove this solution, fresh medium was added to make up the original volume. The results are shown in Table 12.14.

(a) Are there any differences in the growth curves among the three colonies?

(b) What is the estimated carrying capacity?

(c) Is there any evidence of serial dependence in these data?

(d) Can you explain this?

Table 12.14. *The sizes of three closed colonies of* Paramecium aurelium *(Diggle, 1990, p. 239, from Gause).*

Day	0	2	3	4	5	6	7	8	9	10
	2	17	29	39	63	185	258	267	392	510
	2	15	36	62	84	156	234	348	370	480
	2	11	37	67	134	226	306	376	485	530
Day	**11**	**12**	**13**	**14**	**15**	**16**	**17**	**18**	**19**	
	570	650	560	575	650	550	480	520	500	
	520	575	400	545	560	480	510	650	500	
	650	605	580	660	460	650	575	525	550	

13

Compartment models

In Chapter 6, we studied how the event histories of individuals can be modelled as stochastic processes passing through a number of different states. A similar phenomenon occurs when a quantity of some material moves through the different parts (the states) of some system. If the system can be divided into a number of compartments (corresponding to the states in event histories), elements of the material can be assumed to move stochastically among them, possibly in both directions

In such a context, similar models to those in Chapter 6 should be applicable. However, in that chapter, we were able to observe the individual movements among compartments. Here, in contrast to event histories, I shall assume that we can only observe aggregate movement. Thus, the situation will be the same as for the growth curve models in Chapter 12: we have here a second set of phenomena where we cannot observe individual changes. However, in contrast to growth curves, the models will allow material, not to accumulate in the system, but rather to enter, flow through, and leave it.

Thus, once again, records of change in such a system will be averages of the unobservable stochastic changes of the elements involved. Generally, we shall either be interested in how the total amount of material in the system changes over time or in changes in the amount in one particular compartment of special interest. As in the case of growth curves, a second level of external stochastic disturbances will also be present.

13.1 Theory

Following this line of thought, we find that one simple way to begin construction of mechanistic models for a process of material moving through a system is to divide that system into *compartments* and to assume that the rate of flow of the substance between these obeys first-order kinetics. In other words, the rate of transfer to a receiving or *sink compartment* is proportional to the concentration in the supply or *source compartment*. Then, the differential equations will be linear. These are called the *mass balance equations*. There may be inputs into and outputs from one or more compartments.

For example, in a study of the body, a three-compartment model might correspond to blood, soft tissue, and muscle, plus the outside environment from which the substance comes and to which it is eliminated. Two important assumptions in such models are that the substance under study

(i) is well mixed in all compartments and
(ii) has identical kinetic behaviour within each one.

The idea is that the material is made up of a very large number of small particles moving independently through the compartments and that their individual behaviour cannot be observed. Thus, only their average behaviour is available, in contrast to the records of individuals moving through the states of a system in Chapter 6.

The growth curves of Chapter 12 modelled net accumulation of material in a system; here, instead, material entering the system generally will be eliminated later. If input to the system eventually stops, whereas elimination from the system continues, the resulting regression function for each compartment, after standardisation of the initial conditions, generally will have the form of a probability density function over time. Such functions will describe the mean of the stochastic process of movement of material into and out of each compartment.

13.1.1 First-order kinetics

As in Section 6.1, we shall need to set up an intensity matrix describing the rate of passage of the elements of material through the system. Let us first consider the simple case of first-order kinetics with no inputs to the system after the process begins at $t = 0$.

In such a situation, the system of linear differential equations will have the form

$$\frac{\mathrm{d}\boldsymbol{\mu}^\top(t)}{\mathrm{d}t} = \boldsymbol{\mu}^\top(t)\mathbf{A} \tag{13.1}$$

where $\boldsymbol{\mu}(t)$ is a column vector of length P giving the mean content of each compartment in the system at time t and $\mathbf{A}$ is a $P \times P$ transfer matrix containing rate constants of movement between the compartments.

In direct analogy to the solution of one such equation, the general solution to Equation (13.1) is

$$\boldsymbol{\mu}^\top(t) = \boldsymbol{\mu}^\top(0)\mathrm{e}^{\mathbf{A}t} \tag{13.2}$$

where matrix exponentiation is defined as in Section 6.1.3. By comparison with the results in that section, we see that $\boldsymbol{\mu}(t)$, giving the mean amount of material in each compartment at time t, corresponds to $\boldsymbol{\pi}_t$, the marginal probability of being in each state at time t. The transfer matrix $\mathbf{A}$ corresponds to the intensity matrix $\boldsymbol{\Lambda}$. Thus, elements of material are following a continuous-time Markov chain and the times that they spend in compartments will have exponential distributions.

If there are inputs to the system over time, the function describing these, say $\mathbf{b}(t)$, must be added to Equation (13.1). The solution, then, involves an integral

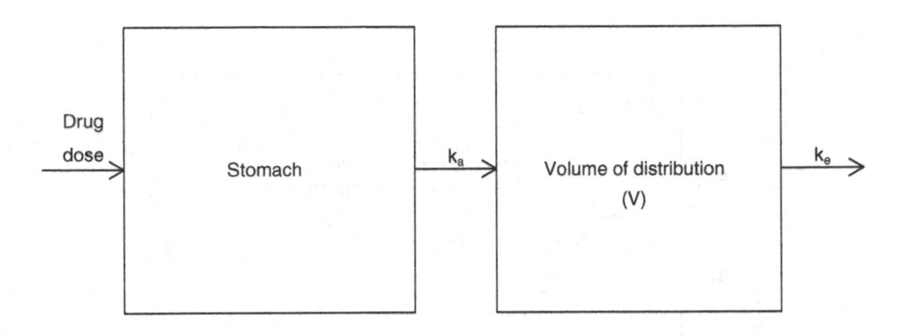

Fig. 13.1. A graphical presentation of an open, first-order, one-compartment model.

added to Equation (13.2):

$$\boldsymbol{\mu}^\top(t) = \boldsymbol{\mu}^\top(0)\mathrm{e}^{\mathbf{A}t} + \int_0^t \mathbf{b}(u)\mathrm{e}^{\mathbf{A}(t-u)}\mathrm{d}u \tag{13.3}$$

In simple cases, Equation (13.2) or (13.3) can be solved analytically, but often only a numerical solution may be available.

In this construction, we have used a stochastic process to describe the passage of material through a system. However, the difficulties in measuring concentrations in most situations implies that considerable measurement error will be present. In addition, especially in biological systems, further variation, including serial dependence, can arise through the changes over time in metabolic processes, environmental influences, and so on. Thus, the solution of these equations provides a mean over time about which random variation will be occurring. Because concentrations are positive-valued, usually varying from close to zero to fairly large values, a skewed distribution with nonconstant variance will usually be found to be most appropriate to model this.

13.1.2 Open, first-order, one-compartment model

As an example, let us look at a model often used in the pharmacokinetic study of the flow of a medication through the body. Suppose that a substance is ingested at one point in time (not continuously over the study period). The process to be modelled is illustrated in Figure 13.1. It corresponds to the differential equations

$$\begin{aligned} \frac{\mathrm{d}\mu_1(t)}{\mathrm{d}t} &= -k_a\mu_1(t) \\ \frac{\mathrm{d}\mu_2(t)}{\mathrm{d}t} &= k_a\mu_1(t) - k_e\mu_2(t) \end{aligned} \tag{13.4}$$

where μ_1 is the average amount at the absorption site (in Figure 13.1, the stomach) and μ_2 is the concentration that interests us, that of the substance in the body, usually measured in the blood. Two important parameters are the absorption rate k_a and the elimination rate k_e, both at this site. Generally, direct information is

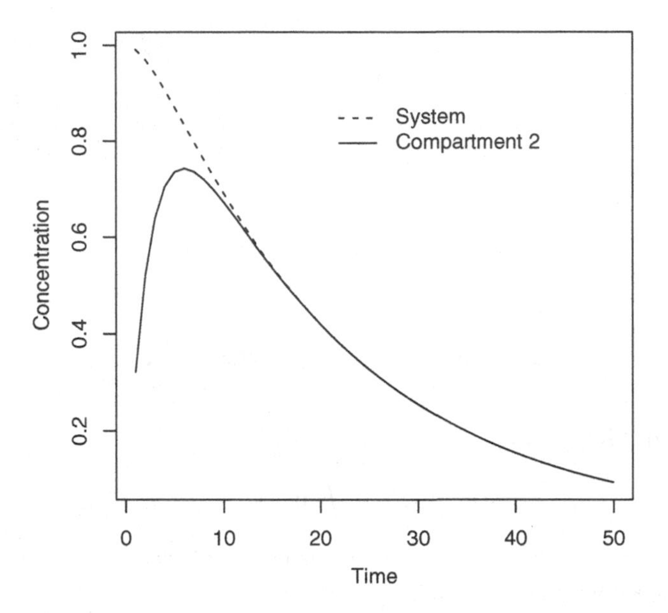

Fig. 13.2. Plots of total concentration in the system and concentration in the second compartment for the model of Figure 13.1 with $k_a = 0.4$, $k_e = 0.05$, and a dose of $x = 1$.

available only about $\mu_2(t)$. Indirect information about $\mu_1(t)$ comes from the dose administered.

Then, in Equation (13.1),

$$\mathbf{A} = \begin{pmatrix} -k_a & k_a \\ 0 & -k_e \end{pmatrix} \tag{13.5}$$

If a dose of size x is the input to the first compartment, the initial condition for Equation (13.2) will be $\boldsymbol{\mu}(0) = (x,0)^\top$. Then, at any time t, we shall be interested in the amount in the second compartment; this is the second element of $\boldsymbol{\mu}(t)$, which can be obtained from Equation (13.2). For given, fixed values of the parameters, this can be calculated numerically using matrix exponentiation, as in Equation (6.4).

Suppose, say, that $k_a = 0.4$, $k_e = 0.05$, and $x = 1$. Then, the curves of total concentration in the system and of concentration in the second compartment are plotted in Figure 13.2. We see that the concentration in the second compartment rises quickly and then drops off slowly. The absorption rate k_a controls the steepness of the increasing curve and the elimination rate k_e, usually smaller than k_a, controls the descent. After about time 12, almost all of the material in the system is in the second compartment, so that the two curves coincide, and after about time 50, almost all has left the system.

In fact, in this case, the differential equations can be solved analytically so that numerical exponentiation of the transfer matrix is not necessary; see Equation

(6.15). The resulting nonlinear function for the compartment of interest is

$$\mu_2(t) = \frac{xk_a}{k_a - k_e}\left(e^{-k_e t} - e^{-k_a t}\right) \tag{13.6}$$

Because not all of the total dose x administered may reach the site of measurement, generally a proportionality constant V, called the 'apparent volume of distribution', is included:

$$\mu_2(t) = \frac{xk_a}{V(k_a - k_e)}\left(e^{-k_e t} - e^{-k_a t}\right) \tag{13.7}$$

This commonly used nonlinear regression function is called the open, first-order, one-compartment model. (Only the second compartment appears in the final regression function.) The parameters to be estimated are k_a, k_e, and V. This function describes the average movement through the compartment if no outside influences are operating. In order to fit the model to data, appropriate assumptions also must be made about these external influences.

13.2 Modelling delays in elimination

The procedures described above provide the basis for constructing more elaborate and more realistic models. However, the compartments included in a model do not necessarily all need to correspond directly to physical parts of the system under study. Extra compartments may need to be added to modify the stochastic properties of the system. For example, in the passage of some substance through a biological organism, absorption often is relatively straightforward. Modelling elimination can be more difficult because the material can transit through several organs before leaving. Let us look at two ways to handle this.

In contrast to the open, first-order, one-compartment model of Section 13.1.2, for simplicity let us here assume that input and elimination occur in the same compartment (for example, a substance is injected directly into the blood, where it is measured over time). This means that the concentration curve will not rise at the beginning, as it did for the compartment of interest in Figure 13.2.

13.2.1 Random walk

Consider Figure 13.3, ignoring for the moment compartment four. Assume that concentration measurements are made in compartment three, where both input and elimination occur. However, there are two fictitious compartments, one and two, to the left of this. Thus, in this model, some of the material can move through the compartments to the left of the input, one compartment at a time, with rate k_l, and back with rate k_r. Elimination will not occur until it comes back through compartment three. What is the effect of adding these extra compartments?

In fact, this construction yields a random walk (Sections 5.1.3 and 9.2.3) with a reflecting barrier at compartment one and rates k_l to the left and k_r to the right. This random walk allows retention of the material longer in the system; for example, for a drug in the body, this might be by movement through sites neighbouring the organ

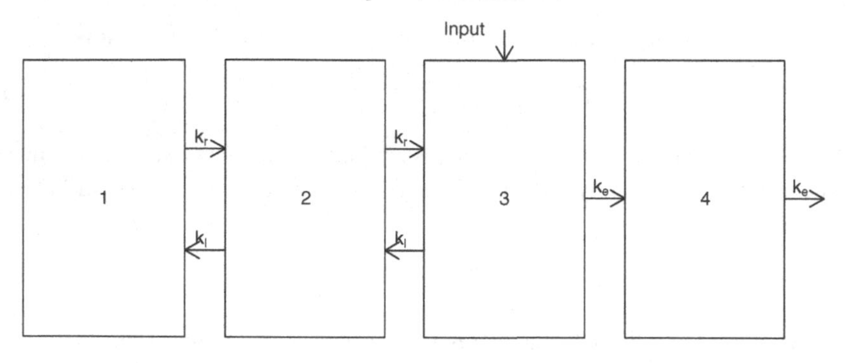

Fig. 13.3. A graphical presentation of a four-compartment, gamma distribution random walk model. Compartments one and two yield the random walk and three and four the gamma distribution.

Table 13.1. *Renal gentamicin concentrations (μg/g) in four sheep after administration of a bolus of 3 mg/kg body weight. (Faddy, 1993)*

Day	2	5	11	18	25	32	46	60
Concentration	71.9	38.6	10.1	10.7	7.1	6.7	4.1	3.3
Day	2	5	11	18	25	32	60	74
Concentration	102.2	49.8	24.8	14.1	10.0	7.1	3.1	4.8
Day	1	4	10	17	24	45	59	73
Concentration	183.0	90.6	25.0	7.9	8.4	5.8	7.1	7.1
Day	4	10	17	24	31	45	59	73
Concentration	108.7	69.2	20.4	16.4	13.0	9.1	7.5	7.4

in which measurements are made. A larger number of compartments to the left of input will indicate a longer delay in elimination (depending also on the two rates). With a sufficiently large number, this will approximate a diffusion process (Chapter 10).

Antibiotic residues Gentamicin, an antibiotic, is administered to animals to control infection. There is concern about residues staying in the kidneys. To study the process of elimination of this drug, four sheep were given gentamicin at 3 mg/kg body weight. Renal gentamicin concentrations were measured over a period of about ten weeks after this, as shown in Table 13.1. For all sheep except the last, the measured concentration *increases* at certain time points, indicating the presence of considerable measurement error.

Because the sheep were of different sizes, it is necessary to include a different proportionality constant, the apparent volume of distribution in Equation (13.7), for each sheep; for simplicity, I shall set $x = 1$ so that the actual dose is included in V. I shall be interested in the total amount remaining in the system at a given time: the sum of the elements of $\boldsymbol{\mu}(t)$.

Table 13.2. *Random walk models, with differing numbers of compartments, fitted to the renal gentamicin concentration data of Table 13.1.*

Compartments	Log normal	Gamma
1	122.0	122.7
2	92.8	92.6
3	92.4	92.1
4	92.2	92.0
5	92.2	92.0

Let us fit random walk models with input to the last compartment, as above, so that the final elimination rate is exponential. The results for a variety of lengths of random walk are shown in Table 13.2 when variability is handled by the log normal and gamma distributions. Notice that all models, except that with one compartment, have the same number of parameters. The fit is only little improved by increasing the number of random walk compartments. All are an improvement on direct exponential elimination (one compartment). The log normal distribution fits about as well as the gamma, whereas a normal distribution fits much more poorly. As well, serial correlation is not necessary.

The gamma-variability, four-compartment, random walk model (with no compartments to the right of input) is plotted in Figure 13.4, showing the separate curve for each sheep. The rate parameters are estimated to be $\widehat{k_l} = 0.017$, $\widehat{k_r} = 0.026$, and $\widehat{k_e} = 0.22$. The (reciprocal) volumes are $1/\widehat{\mathbf{V}} = (118.4, 163.8, 189.1, 290.2)^\top$; these are quite different from those for the log normal model (given by Faddy, 1993), in part because the latter refer to the geometric mean.

13.2.2 Gamma distribution

As we have seen, the assumption of first-order kinetics in standard compartment models implies that they are equivalent to the continuous-time Markov chains of Section 6.1.3. Thus, the time spent by a particle of the material in a compartment has an exponential distribution. This is why the duration time for material in the last compartment of the random walk model above has an exponential distribution.

If this is not reasonable, one simple correction, without dropping the first-order assumption, is to divide up the (physical) compartment having the nonexponential distribution into a series of subcompartments through which the material must pass before leaving the system, all with the same rate k_e. Here, in contrast to the random walk system, movement will be in one direction only. The amount of delay in elimination will depend on the number of extra compartments (as well as on the elimination rate k_e).

Thus, such a system will consist of a series of compartments whereby a given quantity of material enters through the first compartment in the series. Output only occurs by passing through all of the compartments to the right and out of the last compartment, all with the same rate k_e. With two compartments, this model would

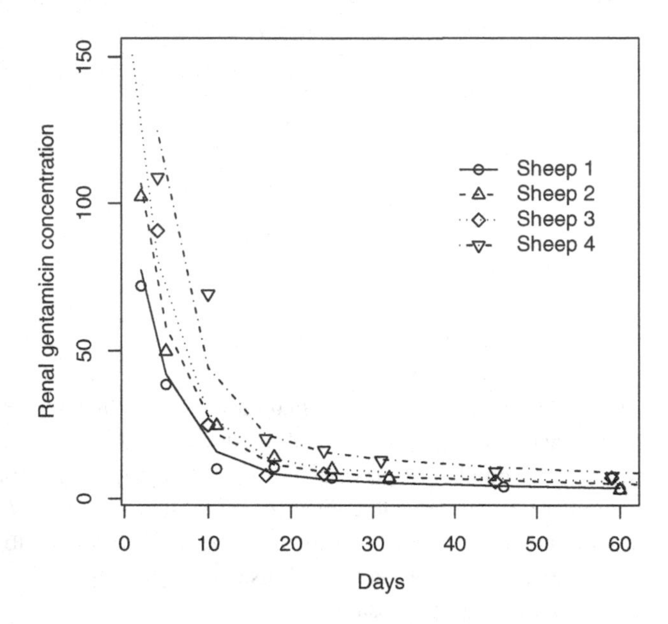

Fig. 13.4. The random walk compartment model fitted to the renal gentamicin concentration in sheep in Table 13.1.

be illustrated by compartments three and four in Figure 13.3, ignoring the random walk produced by compartments one and two. Input occurs to compartment three and the material must pass through compartment four before leaving the system.

Elimination through the compartments to the right of input corresponds to a gamma-distributed clearance time, instead of being exponentially distributed. The shape parameter ϕ in the gamma distribution of Equation (2.7) will equal the integer number of elimination compartments, including the input compartment (Matis, 1972). In Figure 13.3, these are compartments three and four, so that $\phi = 2$.

Thus, the complete model in Figure 13.3 has two components: diffusion within the site of input and gamma-distributed clearance from that site (Faddy, 1993). For this model, the transfer matrix will be

$$\mathbf{A} = \begin{pmatrix} -k_r & k_r & 0 & 0 \\ k_l & -k_l - k_r & k_r & 0 \\ 0 & k_l & -k_l - k_e & k_e \\ 0 & 0 & 0 & -k_e \end{pmatrix} \tag{13.8}$$

and $\boldsymbol{\mu}(0) = (0, 0, x, 0)^\top$ for an input dose of x. One important characteristic of this model is that the number of parameters to estimate does not change with the number of compartments, either of the random walk or of the gamma elimination.

Two special cases often are used in pharmacokinetics. If there is no random walk component and clearance has an exponential distribution, a zero-order, one-compartment model results. This is applicable when a bolus dose is injected directly into the compartment (usually the bloodstream) where concentrations will be measured directly. If this is too simple, a zero-order, two-compartment model can be produced by using a two-step random walk with exponential clearance.

Antibiotic residues For the sheep data above, if we add a compartment to the random walk model after (to the right of) input, elimination time will have a gamma distribution with $\phi = 2$. Although the number of parameters does not change, this yields a poorer fit. Thus, the four-compartment gamma-variability model (three of which involve the random walk) has an AIC of 93.5. Because compartments to the right of input are not required, clearance follows an exponential distribution ($\hat{\phi} = 1$).

13.3 Measurements in two compartments

In most pharmacokinetic studies, only concentration of the material of interest in the blood plasma is measured. However, it is possible to obtain valuable complementary information about the system by also recording concentration eliminated, for example in the urine. This can make it possible to follow the process in more than one compartment.

13.3.1 Models for proportions

Certain types of studies, such as those using radioactive tracers, do not measure concentrations, but only the proportion of the material introduced into the system that is remaining at each point in time. In such a situation, standard distributions may not be suitable. Two distributions specifically for proportions are the beta

$$f(y;\mu,\phi) = \frac{\Gamma(\phi)}{\Gamma(\mu\phi)\Gamma(\phi(1-\mu))} y^{\mu\phi-1}(1-y)^{\phi(1-\mu)-1} \tag{13.9}$$

and the simplex

$$f(y;\mu,\phi) = \frac{\exp\left(-\frac{(y-\mu)^2}{2\mu^2(1-\mu)^2 y(1-y)\phi}\right)}{\sqrt{2\pi\phi y^3(1-y)^3}} \tag{13.10}$$

distributions.

In studies that use radioactive tracers, it may be possible to measure the (relative) concentrations in two compartments. Thus, we need to look more closely at a general two-compartment model. Assume that input is to the first compartment, but that both may eliminate material. In addition, once in the system, material may pass in both directions between the two compartments before elimination.

For the case where a one-time injection of material originally enters the first compartment, this is illustrated in Figure 13.5. The corresponding transfer matrix is

$$\mathbf{A} = \begin{pmatrix} -k_{12}-k_{1e} & k_{12} & k_{1e} \\ k_{21} & -k_{21}-k_{2e} & k_{2e} \\ 0 & 0 & -k_{1e}-k_{2e} \end{pmatrix} \tag{13.11}$$

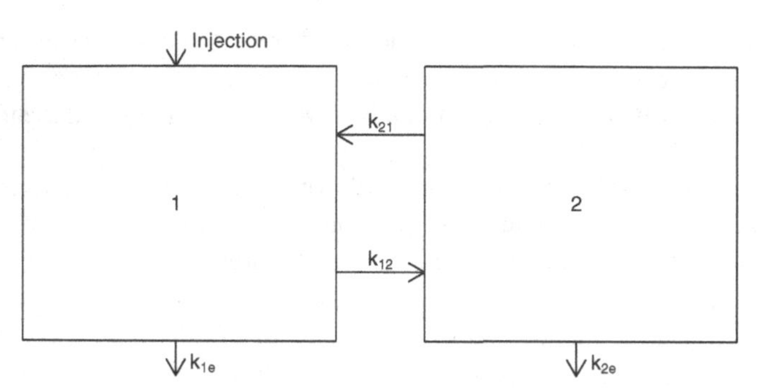

Fig. 13.5. A two-compartment model with injection to the first compartment and elimination from both.

Table 13.3. *Sulphate metabolism in a patient with initial injection of a dose of radioactive sulphate into compartment one. (Kodell and Matis, 1976, from Galambos and Cornell)*

Time	Compartment one	Compartment two
2	0.84	0.06
3	0.79	0.07
5	0.64	0.15
8	0.55	0.15
12	0.44	0.16
24	0.27	0.19
48	0.12	0.22
72	0.06	0.23

and the initial condition is $\boldsymbol{\mu}(0) = (x,0)^\top$ for an input dose of x. Here, input and elimination both occur in compartment one, but only elimination in compartment two. This means that the concentration curve for the first compartment will not rise at the beginning, whereas it will for the second.

Radioactive tracers A radioactive tracer was used in a study of the metabolism of sulphate in patients. The radioactivity was measured both in the serum and in the urine, the latter being exterior to the system. Serum corresponds to the first compartment, extracellular fluid, whereas the second is the metabolic pool. Biologically, sulphate should only be eliminated from compartment one so that we would expect to have $k_{2e} = 0$ in Figure 13.5.

In this study, measurements were made as fractions of the injected dose. The results for one patient are shown in Table 13.3. Investigation showed no indication of a serial correlation, so that only independence models will be considered here.

The fits of several models, both more traditional ones and those specifically for proportions, are shown in Table 13.4. As expected, in no case is there any indication that $k_{2e} \neq 0$. The two distributions for proportions are superior, with the

Table 13.4. *AICs for several models fitted to the radioactive sulphate data of Table 13.3.*

Distribution	With k_{2e}	Without k_{2e}
Normal	46.7	45.7
Log normal	47.0	46.0
Gamma	47.0	46.0
Beta	45.7	44.7
Simplex	44.4	43.4

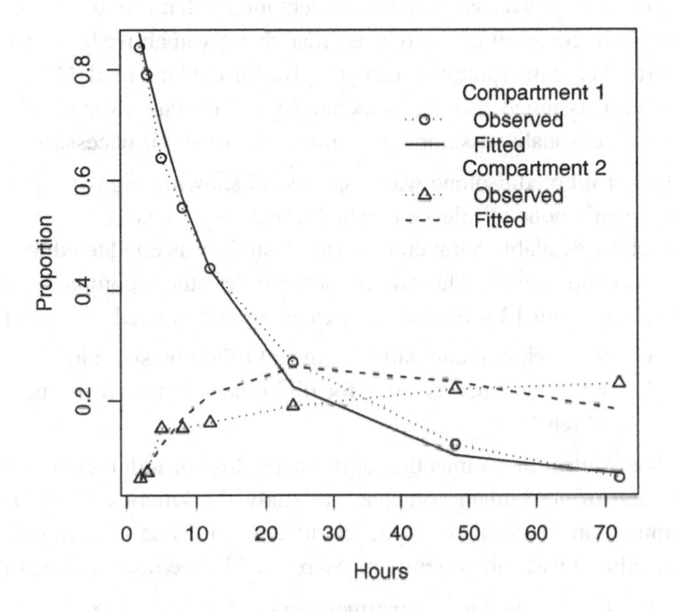

Fig. 13.6. The two-compartment model fitted to the radioactive sulphate data in Table 13.3.

simplex distribution best. The rate parameters are estimated to be $\widehat{k_{12}} = 0.030$, $\widehat{k_{21}} = 0.022$, and $\widehat{k_{1e}} = 0.043$. The curves are plotted in Figure 13.6. The proportions in compartment one appear to be followed more closely than those in compartment two. This may be because the former are measured directly whereas the latter are not.

Further reading

Gutfreund (1995) gives a useful introduction to the basic principles of kinetics. Standard texts on pharmacokinetics include Gibaldi and Perrier (1982) and Notari (1987). Lindsey (2001) provides detailed discussion of statistical fitting of compartment models to medical data.

Matis and Kiffe (2000) present applications of compartment models to stochastic processes of birth, death, and migration; they also discuss stochastic models that do not carry the assumption of uniform mixing (exponential passage times) in each compartment. Anderson and May (1991) show how compartment models are used in modelling the spread of infections among biological organisms, such as people, and Nowak and May (2000) how they can be used to model virus infections within biological organisms.

Exercises

13.1 Let us look more closely at the sulphate metabolism data of Table 13.3.

(a) Can you develop a compartment model that follows the proportions in compartment two better than that given above?
(b) What interpretation can you give for the best model?
(c) I assumed that the dispersion was constant over time. Is this a reasonable assumption? Correct the model if necessary.

13.2 The steroid prednisolone was suspected of showing high-risk potential for therapeutic nonequivalence due to differences in bioavailability in its commercially available 5 mg tablet form. A study was conducted to verify this *in vitro* and *in vivo*. The results for the latter study, comparing two types of tablets, with 12 subjects taking each, are reproduced in Table 13.5.

(a) What kind of compartment model might be suitable?
(b) What evidence is there for differences between the two types of tablets?

13.3 A bolus intravenous injection of the same dose of indomethicin was given to each of six human volunteers to study the kinetics of its absorption, elimination, and enterohepatic circulation. Plasma concentration of indomethicin was subsequently measured at 11 time points, as in Table 13.6.

(a) Find a suitable compartment model for these data.
(b) Develop appropriate stochastic elements for the model by considering different possible distributional assumptions as well as dependence among responses on each subject.

13.4 In a Phase I clinical trial, an oral dose of Sho-seiryu-to was administered to each of eight healthy male volunteers. Blood concentrations of ephedrine were measured over the following 24 h, with the results as shown in Table 13.7.

(a) Fit a suitable compartment model to these data.
(b) Develop appropriate stochastic elements for the model by considering different possible distributional assumptions as well as dependence among responses on each subject.

13.5 The rate of gastrointestinal absorption of Declomycin was studied under a

Table 13.5. *Plasma concentrations (ng/ml) in 24 subjects following ingestion of a 5 mg tablet of prednisolone in two different formulations. (Sullivan* et al., *1974)*

Hours									
0.25	0.5	1	2	3	4	6	8	12	24
Tablet A									
103.0	194.0	254.0	170.0	149.0	127.0	64.8	41.6	8.6	11.8
228.0	345.0	290.0	225.0	203.0	158.0	67.2	40.4	8.0	0.0
5.4	57.8	181.0	204.0	142.0	149.0	106.0	34.6	21.9	0.0
10.3	70.8	214.0	179.0	170.0	173.0	68.9	32.3	23.9	9.8
209.0	288.0	266.0	210.0	218.0	121.0	48.8	53.1	25.3	0.0
81.0	202.0	263.0	209.0	200.0	220.0	130.0	62.1	25.4	0.0
3.0	138.0	390.0	310.0	305.0	222.0	142.0	73.0	31.6	11.3
0.0	73.0	194.0	189.0	152.0	148.0	63.0	34.2	19.4	0.0
14.5	125.0	201.0	209.0	228.0	129.0	82.5	47.0	3.9	1.3
81.1	217.0	222.0	206.0	169.0	123.0	76.2	25.0	22.8	0.0
19.0	96.1	114.0	165.0	148.0	146.0	94.6	46.6	23.4	0.0
6.2	30.0	173.0	202.0	168.0	135.0	95.6	69.9	20.4	9.4
Tablet B									
62.3	268.0	249.0	192.0	190.0	129.0	53.2	23.5	11.4	11.2
116.0	178.0	205.0	219.0	161.0	116.0	32.2	17.2	7.3	0.0
29.8	138.0	224.0	196.0	192.0	158.0	105.0	70.2	22.5	0.9
28.5	79.2	173.0	286.0	305.0	229.0	126.0	79.5	25.2	15.7
21.6	154.0	244.0	246.0	357.0	222.0	119.0	58.5	26.8	1.7
76.6	253.0	267.0	242.0	211.0	164.0	98.4	57.2	22.5	2.3
0.0	50.8	168.0	289.0	255.0	170.0	109.0	67.1	26.2	8.3
6.7	160.0	184.0	197.0	246.0	165.0	103.0	69.0	2.0	0.0
22.4	125.0	222.0	166.0	134.0	148.0	52.9	19.0	26.3	12.2
0.0	77.5	163.0	174.0	199.0	170.0	102.0	84.4	27.4	2.3
8.3	56.0	149.0	174.0	130.0	153.0	83.1	39.1	22.0	0.0
25.0	63.4	145.0	148.0	143.0	140.0	95.6	45.9	11.2	24.0

Table 13.6. *Plasma concentrations (μg/ml) following intravenous injection of indomethicin for six human subjects. (Davidian and Giltinan, 1995, p. 18, from Kwan* et al.*)*

Hours										
0.25	0.5	0.75	1	1.25	2	3	4	5	6	8
1.50	0.94	0.78	0.48	0.37	0.19	0.12	0.11	0.08	0.07	0.05
2.03	1.63	0.71	0.70	0.64	0.36	0.32	0.20	0.25	0.12	0.08
2.72	1.49	1.16	0.80	0.80	0.39	0.22	0.12	0.11	0.08	0.08
1.85	1.39	1.02	0.89	0.59	0.40	0.16	0.11	0.10	0.07	0.07
2.05	1.04	0.81	0.39	0.30	0.23	0.13	0.11	0.08	0.10	0.06
2.31	1.44	1.03	0.84	0.64	0.42	0.24	0.17	0.13	0.10	0.09

number of experimental conditions to determine the decrease due to heavy-metal complexation. The four conditions were: A, after eight hours fasting; B, after a meal without dairy products; C, after a meal with 8 oz of

Table 13.7. *Blood concentrations of ephedrine (ng/ml) in eight healthy male volunteers after an oral dose of Sho-seiryu-to. (Yafune, 1999)*

Hours										
0.25	0.5	0.75	1	2	3	4	6	8	10	24
2.5	5.1	14.3	17.1	20.5	21.6	23.3	16.2	14.6	12.0	4.7
1.9	10.8	21.1	28.5	30.9	30.9	30.2	20.0	17.3	13.5	3.5
2.4	11.5	18.1	31.2	35.7	36.4	30.6	24.2	20.0	12.2	2.4
4.9	9.8	18.2	20.1	31.2	27.5	26.0	22.6	15.5	13.4	2.7
3.3	14.7	25.6	38.0	35.6	29.8	29.7	24.6	20.5	16.0	2.5
3.9	13.5	17.8	21.0	22.8	23.0	19.1	14.4	11.8	9.8	1.1
0.5	9.0	15.5	20.9	24.7	24.3	22.8	18.0	14.4	8.9	3.2
2.0	5.3	16.2	21.7	25.4	28.0	23.0	21.0	13.6	10.0	1.2

Table 13.8. *Average serum concentrations of Declomycin over time under four experimental conditions. (Notari, 1987, p. 186)*

Hours	Fasting	No dairy products	Whole milk	Aluminium hydroxide
0	0.0	0.0	0.0	0.0
1	0.7	1.0	0.1	0.2
2	1.1	1.2	0.3	0.3
3	1.4	1.7	0.4	0.4
4	2.1	2.0	0.4	0.5
5	2.0	—	0.4	0.5
6	1.8	2.1	0.4	0.5
12	1.4	1.8	0.3	0.4
18	—	—	0.2	0.3
24	0.8	1.1	0.1	0.2
48	0.4	0.7	0.0	0.1
72	0.2	0.3	—	0.0
96	0.1	0.2	—	—

whole milk; and D, coadministration with 20 ml of aluminium hydroxide gel. In each group, 300 mg of Declomycin were administered orally, with the results shown in Table 13.8.

(a) Fit appropriate models to determine which parameters vary under the various treatment conditions.
(b) What is the proportional reduction in total absorption when milk is taken with a meal as compared to fasting?
(c) What is the reduction with aluminium hydroxide gel?
(d) Compare these to the situation of a meal without dairy products.
(e) As stated in Table 13.8, these are averages. There were six volunteers in the fasting group and four in each of the other groups. What would be the advantages of having the individual data for all participants instead of just mean values?

13.6 The adult yellow lace birch bug is a flying insect that, as a nymph, goes through a number of instars. Data were collected in the summer of 1980,

Table 13.9. *Counts of yellow lace birch bugs at six stages in a small tree in northern Ontario. (Kalbfleisch* et al., *1983)*

Day	Instar					Adult
	1	2	3	4	5	
0	31	0	0	0	0	0
4	200	0	0	0	0	0
8	411	58	0	0	0	0
13	435	320	97	1	0	0
16	496	294	250	48	0	0
19	514	316	299	214	6	0
22	492	339	328	332	79	0
28	509	390	353	325	326	4
31	478	374	356	369	476	83
34	359	382	344	404	549	202
38	270	261	339	446	617	460
44	142	186	209	400	666	745
47	103	159	198	329	669	900
50	63	73	183	237	616	1095
54	28	40	66	196	451	1394
58	11	26	41	105	340	1581
68	0	1	6	26	97	1826

starting on July 2, on the numbers of this insect on one small tree in northern Ontario, Canada, as shown in Table 13.9. Mortality occurred at various stages; all adults left the tree on emergence. This is a case in which the actual number of 'particles' involved in the process can be counted, although not followed over time.

(a) Develop and fit a compartment model for these data.
(b) According to this model, what is the mean sojourn time in each stage?
(c) Does the mortality rate change over the various stages?

13.7 In studies of digesta flow through the gastrointestinal tract of ruminants, attributes of variation in this flow are collected along with variation in intake and digestive use of forages. Such data are different from many experiments in which compartment models are applicable in that initial mixing of a new meal with existing gastrointestinal contents is not instantaneous. As well, the digesta typically are heterogeneous.

In a first study, grazing Brahman–Jersey-cross cows were pulsed with Yb^{169}-marked coastal Bermuda grass. The proportion of marked material excreted over time was recorded, as shown in Table 13.10.

(a) How many compartments do you expect to require for these data?
(b) What stochastic assumptions do you require in order to obtain a reasonable fit?
(c) How does your model allow for the lack of perfect mixing?

Table 13.10. *Passage data of marked coastal Bermuda grass through grazing Brahman–Jersey-cross cows. (Hughes and Matis, 1984, from Lascano)*

Hours	0	4	8	16	20	24	28
Proportion	0.0000	0.0000	0.0201	0.0421	0.0649	0.0929	0.2011
Hours	40	48	64	72	88	96	112
Proportion	0.3563	0.5165	0.6559	0.7538	0.8261	0.8734	0.9034
Hours	120	136	144				
Proportion	0.9382	0.9711	1.0000				

13.8 A second digestion experiment was undertaken to examine the difference in effect of two diets, chopped straw and finely ground, pelleted straw, both alkali-treated. Each of four cows received both treatments in a paired design. Part of each kind of straw was stained with magenta. The fæces were supposed to be collected, mixed, and weighed every four hours during 360 h. Samples were taken from which stained particles were counted and recorded per gram for many, but not all, four-hour periods. The data for one of the cows are given in Table 13.11.

(a) What assumptions do you need to make in order to model these data?
(b) How many compartments do you require for a satisfactory model?
(c) Is there a difference between the two treatments?

13.9 In Section 13.3, we looked at data for which measurements were available in two compartments. Occasionally, information may be available on even more compartments. For example, concentration in the kidney may be measured indirectly, yielding a renogram. In one study, ortho-iodo hippuran tagged with I^{131} was injected intravenously into a dog. Radioactivity was measured over the kidney as well as in the blood and urine, as shown in Table 13.12.

(a) Develop and fit a compartment model for these data.
(b) According to this model, what is the mean sojourn time in each compartment?

Table 13.11. *Concentrations of stained particles in fæces of a cow for chopped and ground straw. (Matis* et al., *1989, from Ferreiro* et al.*)*

Hours	Chopped	Ground
2	0	0
6	0	0
10	0	6
14	0	0
18	60	0
22	57	303
26	211	316
30	374	444
34	321	491
38	603	636
42	531	651
46	659	746
50	791	729
54	763	638
58	902	494
65	778	456
77	879	—
89	746	248
101	856	—
113	520	141
125	376	—
137	267	115
149	149	—
161	194	50
173	94	—
185	116	20
197	82	—
209	16	—
221	14	—
233	12	8

Table 13.12. *Concentration of injected ortho-iodo hippuran tagged with iodine-131 in a dog. (Turner* et al., *1963)*

Minutes	Blood	Renogram	Urine
0.50	24.3	—	—
0.75	22.9	—	—
1.00	14.3	—	—
1.25	13.2	—	—
1.50	12.9	—	—
1.75	11.2	—	—
2.00	11.4	—	20.80
2.25	10.6	—	16.80
2.50	9.5	32.5	21.50
2.75	7.1	—	19.10
3.00	6.5	27.5	18.00
3.50	5.2	23.8	14.00
4.00	5.0	22.3	11.82
4.50	5.1	19.2	9.68
5.00	4.7	17.0	7.87
5.50	4.2	14.9	7.05
6.00	3.9	13.4	6.36
6.50	3.9	12.4	5.24
7.00	4.3	11.4	5.07
7.50	2.8	10.6	3.51
8.00	3.1	10.2	3.72
8.50	2.8	9.7	3.63
9.00	2.9	9.0	3.33
9.50	2.3	8.4	3.24
10.00	2.5	7.3	2.86
10.50	2.5	6.9	2.79
11.00	2.2	6.6	2.46
11.50	2.2	6.3	2.35
12.00	2.0	6.2	2.34
12.50	2.0	6.0	2.14
13.00	2.0	5.8	2.03
13.50	1.8	4.9	2.03
14.00	1.8	4.7	1.98
14.50	1.7	4.3	1.87
15.00	1.9	4.5	1.69
15.50	—	4.4	—

14

Repeated measurements

Up until this point, when several series were available, I have assumed that they arose from the same stochastic process (unless some covariates were available to distinguish among them). The one exception was Section 7.3 in which I introduced frailty models for recurrent events. It is now time to look at some standard methods for handling differences among series when adequate covariate information to describe those differences is not available.

14.1 Random effects

Suppose that we observe a number of series having the same basic characteristics. If we examine several responses from one of these series, we can expect them to be related more closely, independently of their distance apart in time, than if we take one response from each of the series. Indeed, generally we assume the responses from separate series to be independent, whereas those on the same series usually are not. If inadequate information is available from covariates in order to be able to model the differences among the series, then other methods must be used. The usual approach is to assume that one or more parameters of the distribution describing the stochastic process differ randomly (because we do not have appropriate covariates to describe systematic change) among the series. This type of construction yields a mixture distribution. However, in contrast to the dynamic models of Chapters 7 and 11, here the random parameters will be static, only varying among series and not over time.

14.1.1 Mixture models

A regression function serves to describe how characteristics of a stochastic process vary *systematically*, in observable ways, both in subsets of the population (Chapter 2) and as conditions change over time. If adequate covariates are available, distinguishing among the series and describing how conditions change over time, the set of states within a series should be independent, conditional on these covariates. However, all such conditions may not be observable and, indeed, may not be relevant for a given problem at hand. Here, I particularly shall be concerned about (the lack of) time-constant covariates that describe differences among the series.

Suppose, then, that some parameter in the distribution of states of the process is suspected to vary in unknown ways among the series. The ideal situation is to be able to model changes in this parameter as a regression function, conditioning on available series-specific covariates. However, if such covariates are unavailable, let us assume, instead, that this parameter varies randomly across the series, with a distribution independent of the distribution of the states. (In more complex cases, there may be more than one such parameter.) The hope is that allowing this parameter to vary randomly may help to account for the heterogeneity among the series. In recurrent event studies, this unobserved variable has come to be known as the subject's *frailty* (Section 7.3).

Construction of a mixture distribution

For the moment, let us simplify by ignoring dependencies over time within a series. Then, let $f(y_{it}|\psi_i)$ be the conditional probability density of the states in series i at time t, given the value of the random parameter ψ_i that we assume to vary across series but to be fixed (and unknown) for each given series. Next, let $p(\psi_i)$ be the *mixing distribution* describing the variation of this parameter across series. Then, the (marginal) distribution of the observed states can be obtained by integration to remove the unobservable ψ_i. As we have seen (Sections 4.3.2 and 7.3.1), the resulting construction is called a *mixture distribution*. Thus, the joint distribution of the observed responses in a series is obtained from

$$f(y_{i1}, \ldots, y_{iR}) = \int \prod_{t=1}^{R} f(y_{it}|u)p(u)\mathrm{d}u \tag{14.1}$$

In this way, we obtain a joint multivariate distribution of all observed states of a series. Except in special cases, this distribution cannot be written in closed form, so that numerical integration must be used.

Choice of random parameters

Let us now look a little more closely at ways in which to choose an appropriate random parameter. Random variation may be introduced in at least two ways:

(i) We may start with some appropriate probability distribution and, before introducing any regression function(s), choose a parameter to make random. Then, once the resulting mixture distribution is obtained, the regression function(s) may be introduced into it. This is a *population-averaged* model (Zeger *et al.*, 1988), because time-varying covariates, describing differences among the states within a series, cannot be used.

(ii) Appropriate regression function(s) may first be introduced into the probability distribution, including covariates describing differences among the states of a series. Then, some parameter in that function can be given a distribution. This is a *series-specific* model.

In either case, the result is a joint distribution of the observed states and the unobserved random parameter from which the latter is eliminated by integration.

Often, the most useful approach is to allow an *intercept* parameter to have the random distribution, so that the height of the regression curves varies among series. More occasionally, one may suspect that a *slope* parameter varies among series. For mechanistic models, such as those in Chapters 12 and 13, most parameters have physical meaning and any one of them might be interpretable when allowed to vary randomly among series.

14.1.2 Choice of mixing distribution

The next question is the choice of the mixing distribution for the parameter. Hopefully, this might be selected on theoretical grounds, in the context of a specific scientific study. Often, this is not possible. When the parameter can take on only a restricted range of values, this is another factor that must be taken into account.

Conjugate distributions

Let us, first, look at cases in which we can obtain a closed form for the multivariate distribution. A mixing distribution that yields such a result is said to be *conjugate* to the conditional probability density of the states. For members of the exponential (dispersion) family (Section 8.2.1), such a mixing distribution can be written down in general (see Diaconis and Ylvisaker, 1979; Morris, 1983).

In this family, giving a distribution to the mean ψ, when it is constant for all responses on a series (a population-averaged model), is equivalent to giving the distribution to the canonical parameter θ. From Equation (8.20), members of the exponential dispersion family have the form

$$f(y;\phi|\theta) = \exp\{\phi[y\theta - b(\theta)] + c(\phi, y)\} \tag{14.2}$$

where θ is related to the mean by the canonical link function $\theta = g(\psi)$ and ψ, or equivalently θ, is now a random parameter.

Then, the conjugate distribution will have the form

$$p(\theta;\kappa,\upsilon) = \exp[\kappa\theta - \upsilon b(\theta) + d(\kappa,\upsilon)] \tag{14.3}$$

where $d(\kappa, \upsilon)$ is a term not involving θ. The resulting closed-form marginal distribution, for one state, is

$$f(y;\kappa,\upsilon,\phi) = \exp\left\{d(\kappa,\upsilon) + c(\phi,y) - d\left[\kappa + y\phi, \frac{1}{\phi+\upsilon}\right]\right\} \tag{14.4}$$

The expected value of ψ can be found from the conjugate distribution, as a function of κ and υ. Then, a function of this, most often the same link function as would be used for the original exponential family distribution, can be used to set up a regression function in the usual way.

Other mixtures

If we first construct a regression function for the states and then give a distribution to one of its parameters, we can incorporate changing conditions for the states of a series within the regression function. Thus, we can add some parameter to the regression function that indexes the series explicitly. As mentioned above, a useful case is when the intercept varies randomly:

$$\mu_{it} = \beta_0 + h(\mathbf{x}_{it}, \boldsymbol{\beta}) + \psi_i \tag{14.5}$$

The parameter ψ_i will be constant for a given series, but will vary randomly across the different series, with mean zero. Thus, here the height of the response curve $h(\cdot)$ will vary randomly among the series around the value of β_0. Some of the variables in the vector $\mathbf{x}_{it}$ may describe changing conditions of the series, whereas others refer to conditions common to groups of series. In more complex models, one or more parameters in $\boldsymbol{\beta}$ may be random.

When we integrate to obtain the marginal distribution of the observed states, we shall again obtain a multivariate distribution for all of the observed states of each given series. Unfortunately, here no simple closed form generally will be available, no matter what distribution is chosen for the random parameter, so that estimation must rely on numerical methods.

The common models based on mixture distributions, as just described, often yield equal interdependence among all states of a series, although this will not generally be the case if a nonlinear parameter is made random. An important characteristic of all of these models is that they are *static*: the chosen difference (say a slope or intercept) among series remains constant over time. In other words, they may be thought to allow for missing interseries/time-constant covariates that describe inherent constant characteristics of each series. This contrasts with the dynamic random parameters in Chapters 7 and 11. The latter allow for missing time-varying covariates.

14.2 Normal random intercepts

In the simplest, and most easily interpretable, models, only one parameter, the intercept, is taken to vary randomly among time series. This means that the regression function is shifted randomly by different amounts for each series, as in Equation (14.5).

14.2.1 Collections of time series

A random intercept may provide a useful supplement to many stochastic models when there is unexplained variability among the series. Let us first consider one simple case that we have already modelled in a wide variety of ways in previous chapters, before going on to more complex cases.

Luteinising hormone For the luteinising hormone data in Chapter 11, I looked at a dynamic model involving continuous-time AR(1) dependence (Section 11.1.2) which did not fit as well as hidden Markov models (Section 11.2). From the profile plots in Figure 10.9, we know that there was a great deal of variability among the cows. Let us now see whether or not the problem with the poorer fit of the AR(1)

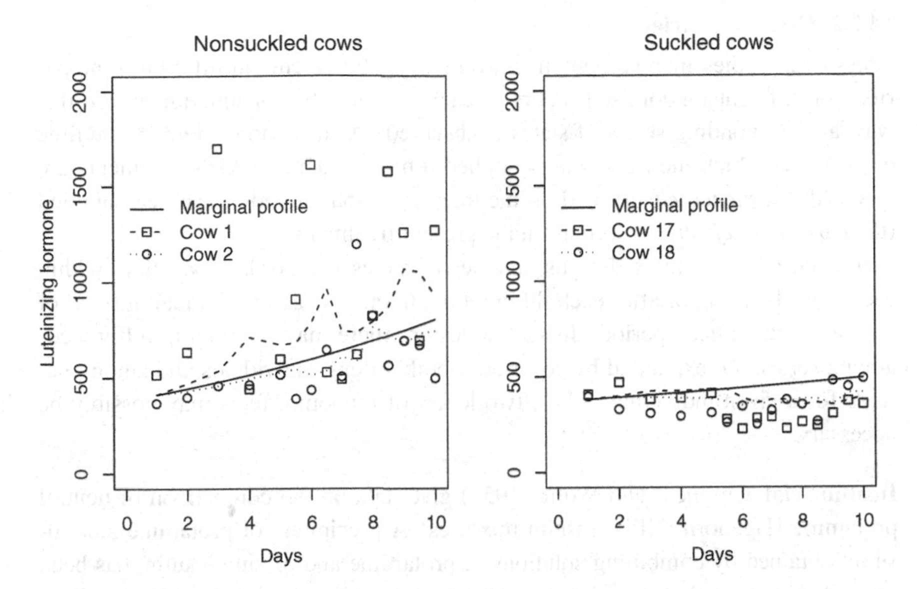

Fig. 14.1. The underlying profiles for the random intercept model with serial dependence for the two groups of luteinising hormone concentrations in cows, with the individual recursive predictions for the first two cows in each group.

model can be remedied by allowing for this heterogeneity among cows using a random intercept.

Recall from Table 11.1 that the AR(1) model with a log normal distribution and an interaction between time and suckling had an AIC of 3207.4 compared with 3239.4 for the independence model. The corresponding model only with the random intercept (without an AR(1)) has 3201.2, showing that much of the dependence is due to the static, systematic differences among cows. However, when the AR(1) is included, as well as the random intercept, the AIC is reduced further to 3193.6, so that both types of dependence are necessary.

The serial correlation is estimated to be $\hat{\rho} = 0.05$ and the intracow correlation to be 0.22. The regression functions are estimated to be

$$\begin{aligned}\mu_{1t} &= 6.40 + 0.074t \\ \mu_{2t} &= 6.09 + 0.029t\end{aligned}$$

respectively for nonsuckled and suckled cows, where μ_{jt} refers to the log level of luteinising hormone. The underlying profiles for the same cows as in Figures 10.10, 11.1, 11.2, and 11.4 are given in Figure 14.1.

Although this is an improvement on the AR(1) model, the fit is not nearly as good as the hidden Markov model (AIC 3161.3) or the ARCH model (3186.9), both also with a log normal distribution (of course, the Burr hidden Markov model fits even better: 3150.3).

14.2.2 Cross-over trials

Cross-over studies involve applying two or more treatments in different time periods in different randomised orders to each of a number of individuals. In this way, a corresponding series of states is observed. A state can depend on the time or *period* at which the treatment is applied, on the order in which treatments are given (the *sequence effect*), and on the treatments that already have been applied (the *carry-over effect*), as well as on the present treatment.

In even more complex designs, a series of states is recorded over time within each period, during or after each different treatment on each individual, instead of just one state in each period. In such a design, there may be random differences among series, not explained by covariates, both among individuals and among periods for a given individual. Thus, two levels of random effects may possibly be necessary.

Insulin trial Ciminera and Wolfe (1953) give data on the comparison of neutral protamine Hagedorn (NPH) insulin mixtures. A precipitate of protamine zinc insulin, obtained by combining solutions of protamine and of zinc insulin, has been used in treating diabetes. It has the advantage over regular insulin of controlling the patient's blood sugar level over a relatively long period of time. The isophane ratio of insulin to protamine results in equivalent amounts of insulin and protamine staying in the supernatant as tested by nephelometric methods. NPH insulin is a suspension of protamine zinc insulin whereby the protamine is never less than in this ratio and never exceeds it by more than 10 per cent. The goal of the experiment was to determine whether a preparation 5 per cent less than the isophane ratio produced a biological difference.

Two mixtures of insulin, the standard (A) and one containing 5 per cent less protamine (B), each with insulin equivalent to 40 units/ml, were tested on rabbits. Two groups of 11 female rabbits weighing between 2.5 and 3.5 kg were injected subcutaneously with 0.051 ml of the suspension at weekly intervals in the randomly assigned orders ABAB (sequence one) and BABA (sequence two). For each treatment, blood sugar level was measured at injection and at four post-injection times over six hours. The resulting response profiles are plotted in Figure 14.2, separately for each treatment, distinguishing among periods. Blood sugar level goes down and then climbs back close to the pretreatment level. The state of rabbit three in period one at three hours seems to be an outlier.

Many different analyses could be applied to these data. Here, I shall concentrate on the differences in profile of the states under the two treatments, because this should be the main point of interest in the experiment. Note that, if the blood sugar level at time zero is included in the analysis as a state, the average differences in treatment will be attenuated, because the treatment will not yet have had an effect. If overall differences between treatments were of prime interest, it might be preferable instead to use it as a baseline covariate. However, here, for profiles, I include it as a state.

I shall allow for two levels of random intercepts, among rabbits and among periods within rabbits, as well as for autocorrelation over the six hours. The ex-

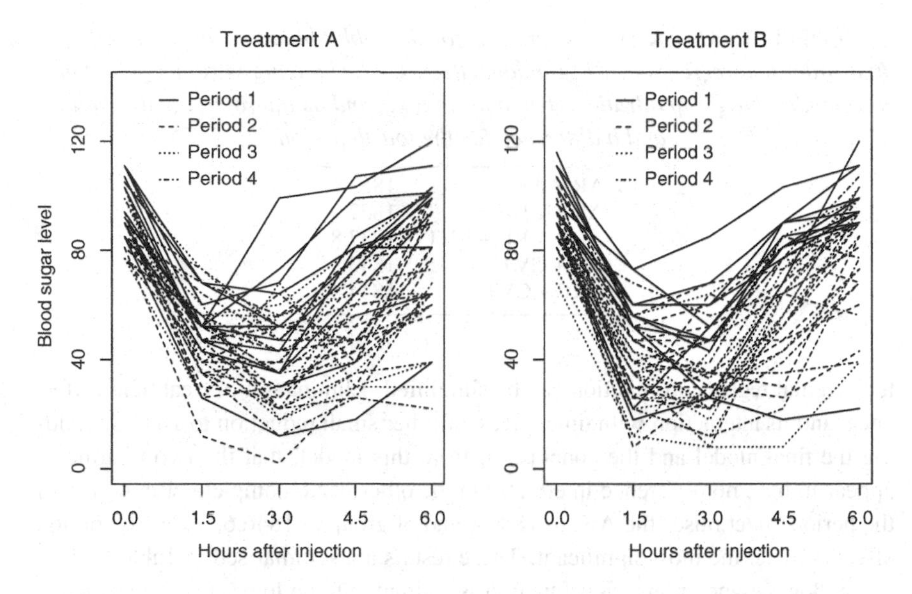

Fig. 14.2. Blood sugar level plotted against time (h) for the two treatments of rabbits.

planatory variables are the period, the treatment, and the group. There are only two treatments that alternate, so that the only sequence effect comes from which treatment is administered in the first period; this is covered by the group effect.

I shall begin by assuming a normal distribution with independence among responses, while I develop an appropriate regression function for the mean. The saturated regression function in time, with a different value at each time point, is equivalent to a quartic polynomial. However, here a quadratic proves to be sufficient. The other covariates will be differences in level for each treatment, period, and group, but without interactions among them (in any case, they are not all identifiable).

This model, with eight location parameters, has an AIC of 1899.0 under independence. Adding second-order interactions between time and these variables reduces the AIC to 1894.8. This can be simplified further by taking only the contrast of the fourth period with the other three, and using only the linear interaction of time with this new period effect and with treatment, yielding an AIC of 1891.7.

The next step is to introduce dependencies, producing a multivariate normal distribution of the states over time. Adding two levels of random effects and a first-order autoregression to the above model lowers the AIC to 1853.8 without changing the variables required in the model. However, not surprisingly, only the series-level random effect (among rabbits), and not the period level one, is required, with an AIC of 1852.8; the fourth period is also in the model as a fixed effect.

Another problem is that the variance, in Figure 14.2, does not appear to remain constant over time. Letting the log variance depend linearly on time reduces the AIC to 1838.2.

After including these improved distributional assumptions, we now can check if

Table 14.1. *AICs for various models for the rabbit blood sugar data. AR, first-order autoregression; RE, random effect; CVT, changing variance with time. All models have a quadratic polynomial in time, and its interaction with group, and a difference for the fourth period.*

AR + 2 RE	1853.9
AR + 1 RE	1852.9
AR + 1 RE + CVT	1837.8
AR + CVT	1843.9
1 RE + CVT	1844.4

the original regression function can be simplified. Elimination of treatment difference, and its interaction with time, gives a further small reduction to 1837.8, yielding the final model and the conclusion, from this model, that the two treatments appear to have no difference in effect. On the other hand, complete elimination of the period effect raises the AIC to 1880.4 and of group to 1846.6. Thus, the period effect is by far the most significant. These results are summarised in Table 14.1.

The log variance regression function is estimated to be $\log(\widehat{\xi_t}) = 5.45 + 0.24t$, the autocorrelation as $\hat{\rho} = 0.37$, and the random effect as $\hat{\delta} = 32.5$, so that the intrarabbit correlation varies from 0.12 at time zero to 0.033 at six hours.

The profile equations for the two groups and four periods, with time centred at three minutes, are

$$\begin{aligned} \mu_{i1kt} &= 37.90 - 1.33(t-3) + 5.10(t-3)^2 \\ \mu_{i2kt} &= 30.58 + 0.50(t-3) + 6.32(t-3)^2 \end{aligned} \qquad k = 1,2,3$$

$$\mu_{i14t} = 51.59 - 1.33(t-3) + 5.10(t-3)^2$$
$$\mu_{i24t} = 44.27 + 0.50(t-3) + 6.32(t-3)^2$$

for both treatments. The profiles are plotted in Figure 14.3. In the second group, the blood insulin level increases more quickly at the end of the six-hour period.

Lindsey (1999b) performs a more sophisticated analysis of these data, showing that, with more appropriate distributional assumptions, a treatment effect may be detected.

14.3 Normal random coefficients

A more complex extension of the previous models is obtained by allowing regression coefficients to be random, and not just the intercept. When the regression function is nonlinear, with parameters that have a physical interpretation, such a model can make sense. Some parameter, such as the absorption rate of a medication by patients (Chapter 13), may vary among subjects in some way that cannot adequately be predicted by the available covariates. However, in many cases, those that concern us here, the procedure is applied in a much more *ad hoc* way, usually with no reasonable scientific interpretation.

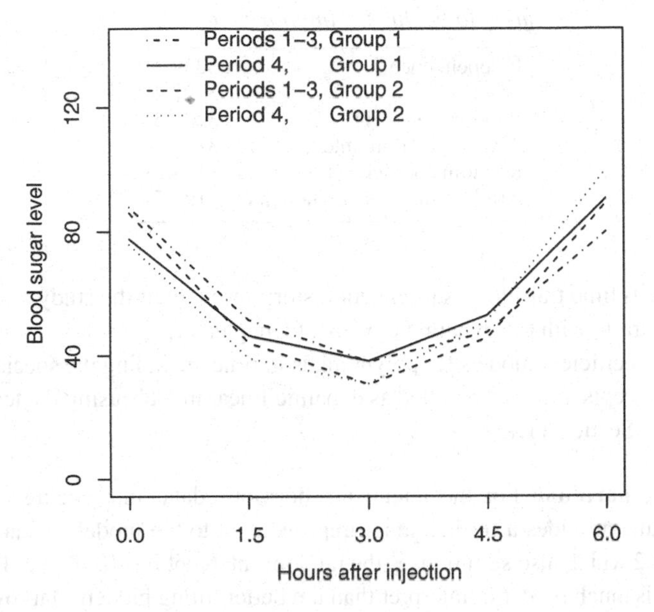

Fig. 14.3. The response profiles of blood sugar for the two groups of rabbits in the four periods for the cross-over study.

14.3.1 Random coefficients in time

Random time coefficient models originally were proposed by Elston and Grizzle (1962); they are now sometimes, rather unjustly, called the Laird and Ware (1982) model. In a simple case with only a linear trend in time, suppose that the states follow a regression function of the form

$$\mu_{it} = \beta_0 + \beta_1 t + h(\mathbf{x}_i, \boldsymbol{\beta}) + \psi_{0i} + \psi_{1i} t \tag{14.6}$$

where, in the same way that β_0 is the average intercept and ψ_{0i} is the difference from this for series i, β_1 is the average slope of the time trend and ψ_{1i} is the corresponding difference. More complex cases will include a higher order polynomial in time. Recall, however, that such polynomials rarely provide an interpretable description of time trends (Chapter 12).

The innovation of this model, with respect to the random intercept model, is that the parameter(s) describing the change in state over time for each series now can have random distributions. However, the results may depend on where zero time is located, because the slopes will be varying randomly around it. Thus, a random coefficient polynomial model can be quite different when times are centred at their mean rather than at the true time origin. As Elston (1964) argues, this makes this kind of model generally unrealistic.

Recall also that random coefficient models carry an assumption that is often not appropriate for stochastic processes: series differ randomly only by their value of the random coefficients, but these differences remain constant over time. Once a

Table 14.2. *AICs for several types of dependency using a log normal distribution, fitted to the luteinising hormone data.*

Independence	3239.4
AR(1)	3207.4
Random intercept	3201.2
AR(1) + random intercept	3193.6
Random coefficients	3194.3
AR(1) + random coefficients	3191.5

series is on its time trajectory, subsequent history throughout the study cannot alter it. This contrasts with the dynamic models of Chapter 11.

Random coefficient models for polynomials in time, including the special case of random intercepts, also can be fitted as dynamic linear models using the techniques described in Section 11.1.

Luteinising hormone For the luteinising hormone data, introducing a random slope for time provides a further small improvement to the model, as can be seen in Table 14.2 which also summarises the random intercept results above. However, this model is much harder to interpret than the better fitting hidden Markov models of Section 11.2.

14.4 Gamma random effects

In Section 11.3, I extended the dynamic models for recurrent events of Section 7.3.1 to ordinary time series. A gamma random effect for repeated measurements, instead of the normal random effects above, can be obtained by a similar modification to the frailty update in Equation (7.29):

$$\begin{aligned} \alpha_i &= \alpha_{i-1} + 1 \\ \beta_i &= \beta_{i-1} + \Lambda(y_i) \end{aligned} \tag{14.7}$$

As we saw in Section 7.3, in the analysis of recurrent events, this is called a frailty model. In the present context, this update also can be combined with serial dependence as defined in Section 1.2.4 to allow simultaneously for longitudinal dependence and heterogeneity among series.

Luteinising hormone For the luteinising hormone data, with a log normal distribution, a regression function containing the interaction between time and suckled, a gamma frailty distribution, and serial dependence, the AIC is 3189.5. This is a better fit than any of the models with normal random effects and AR(1) in Table 14.2, although still not as good as the hidden Markov models of Section 11.2.

The estimated regression functions are

$$\begin{aligned} \mu_{1t} &= 6.21 + 0.062t \\ \mu_{2t} &= 6.11 + 0.027t \end{aligned}$$

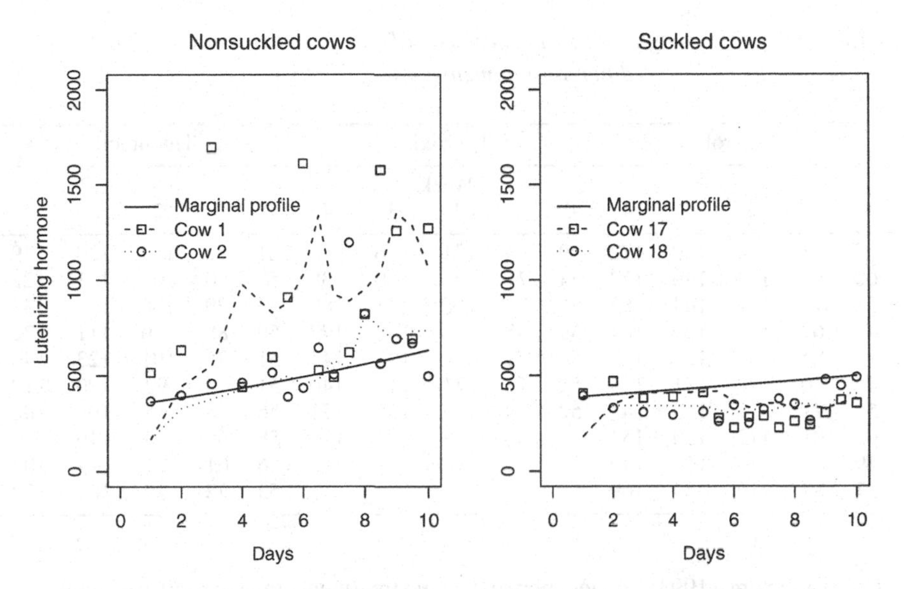

Fig. 14.4. The underlying profiles for the frailty model with serial dependence for the two groups of luteinising hormone concentrations in cows, with the individual recursive predictions for the first two cows in each group.

respectively for nonsuckled and suckled cows, where μ_{jt} is the marginal mean for the log level of luteinising hormone. The dependence parameter is estimated to be 0.09 and the frailty 0.27, the latter indicating an intracow correlation of about 0.16. Thus, the serial dependence is smaller than in the model of Section 11.3, compensated by the constant frailty dependence on each cow. On the other hand, the regression function is similar.

The profiles for the same cows as in Figures 10.10, 11.1, 11.2, 11.4, and 14.1 are given in Figure 14.4. These show a more pronounced attempt to follow the individual responses than any of the previous models except perhaps the ARCH model of Figure 10.10.

The results from this study should not be used to draw general conclusions. Hidden Markov models describe one particular type of dependence, spells, and other types often are necessary, as we have seen throughout this text. On the other hand, random coefficients of a polynomial in time are an *ad hoc* remedy that should be avoided if possible.

Further reading

For an introduction to repeated measurements, see Hand and Crowder (1996). Lindsey (1999a) provides a more advanced development of these types of models. Jones (1993) shows how to fit the random coefficient models as dynamic linear models.

Cross-over trials are a rather controversial area of statistics. Jones and Kenward

Table 14.3. *The weights (units unspecified) of rats over five weeks under three different treatments. (Box, 1950)*

Control					Thyroxin					Thiouracil				
							Week							
0	1	2	3	4	0	1	2	3	4	0	1	2	3	4
57	86	114	139	172	59	85	121	146	181	61	86	109	120	129
60	93	123	146	177	54	71	90	110	138	59	80	101	111	122
52	77	111	144	185	56	75	108	151	189	53	79	100	106	133
49	67	100	129	164	59	85	116	148	177	59	88	100	111	122
56	81	104	121	151	57	72	97	120	144	51	75	101	123	140
46	70	102	131	153	52	73	97	116	140	51	75	92	100	119
51	71	94	110	141	52	70	105	138	171	56	78	95	103	108
63	91	112	130	154						58	69	93	116	140
49	67	90	112	140						46	61	78	90	107
57	82	110	139	169						53	72	89	104	122

(1989) and Senn (1993) provide more extensive treatments than in Lindsey (1999a), although not from the point of view of stochastic process.

Exercises

14.1 Box (1950) was the first to develop methods for repeated measurements over time. His example involved thirty rats which were assigned randomly to one of three groups and kept in individual cages. The first group was a control, whereas the second received thyroxin and the third thiouracil in their drinking water. Due to an accident at the beginning of the experiment, three rats in the second group were lost. The initial weights, and those over the first four weeks, were recorded, as shown in Table 14.3.

(a) In what ways do the treatments affect the growth of the rats?
(b) How homogeneous a group were the rats used in this experiment?

14.2 In their classical paper introducing random time coefficient models, Elston and Grizzle (1962) analysed data from a dental study, reproduced in Table 14.4. These data come from a study of the ramus height measured in a cohort of 20 boys four times at half-year intervals.

(a) The purpose of the study was to try to determine what would be the shape of the growth curve under normal conditions, for use by orthodontists. Can you develop such a curve?
(b) Are there differences among the children?

14.3 In another one of the classical papers on repeated measurements, Potthoff and Roy (1964) provided data on the distance (mm) from the centre of the pituitary to the pterygomaxillary fissure for 16 boys and 11 girls at four different ages, as shown in Table 14.5. These measurements can occasion-

Table 14.4. *Ramus heights (mm) for 20 boys measured at four time points. (Elston and Grizzle, 1962)*

Age			
8	8.5	9	9.5
47.8	48.8	49.0	49.7
46.4	47.3	47.7	48.4
46.3	46.8	47.8	48.5
45.1	45.3	46.1	47.2
47.6	48.5	48.9	49.3
52.5	53.2	53.3	53.7
51.2	53.0	54.3	54.5
49.8	50.0	50.3	52.7
48.1	50.8	52.3	54.4
45.0	47.0	47.3	48.3
51.2	51.4	51.6	51.9
48.5	49.2	53.0	55.5
52.1	52.8	53.7	55.0
48.2	48.9	49.3	49.8
49.6	50.4	51.2	51.8
50.7	51.7	52.7	53.3
47.2	47.7	48.4	49.5
53.3	54.6	55.1	55.3
46.2	47.5	48.1	48.4
46.3	47.6	51.3	51.8

Table 14.5. *Distances (mm) from the centre of the pituitary to the pterygomaxillary fissure in girls and boys aged 8, 10, 12, and 14. (Potthoff and Roy, 1964)*

Girls				Boys			
Age							
8	10	12	14	8	10	12	14
21	20	21.5	23	26	25	29	31
21	21.5	24	25.5	21.5	22.5	23	26.5
20.5	24	24.5	26	23	22.5	24	27.5
23.5	24.5	25	26.5	25.5	27.5	26.5	27
21.5	23	22.5	23.5	20	23.5	22.5	26
20	21	21	22.5	24.5	25.5	27	28.5
21.5	22.5	23	25	22	22	24.5	26.5
23	23	23.5	24	24	21.5	24.5	25.5
20	21	22	21.5	23	20.5	31	26
16.5	19	19	19.5	27.5	28	31	31.5
24.5	25	28	28	23	23	23.5	25
				21.5	23.5	24	28
				17	24.5	26	29.5
				22.5	25.5	25.5	26
				23	24.5	26	30
				22	21.5	23.5	25

Table 14.6. *The heights (cm) of 20 girls in the London growth study, classified by their mother's height. (Tan* et al., *2001)*

Girl's age				
6	7	8	9	10
Short mother				
111.0	116.4	121.7	126.3	130.5
110.0	115.8	121.5	126.6	131.4
113.7	119.7	125.3	130.1	136.0
114.0	118.9	124.6	129.0	134.0
114.5	112.0	126.4	131.2	135.0
112.0	117.3	124.4	129.2	135.2
Medium mother				
116.0	122.0	126.6	132.6	137.6
117.6	123.2	129.3	134.5	138.9
121.0	127.3	134.5	139.9	145.4
114.5	119.0	124.0	130.0	135.1
117.4	123.2	129.5	134.5	140.0
113.7	119.7	125.3	130.1	135.9
113.6	119.1	124.8	130.8	136.3
Tall mother				
120.4	125.0	132.0	136.6	140.7
120.2	128.5	134.6	141.0	146.5
118.9	125.6	132.1	139.1	144.0
120.7	126.7	133.8	140.7	146.0
121.0	128.1	134.3	140.3	144.0
115.9	121.3	127.4	135.1	141.1
125.1	131.8	141.3	146.8	152.3

ally decrease with age because the distance represents the relative position of two points. No further information was provided.

(a) Develop a model to describe the change over time of these measurements that allows for differences between the two sexes.
(b) Are there systematic differences among the individuals participating in this study?

14.4 The London growth study had a number of objectives involving the growth patterns of children. One part involved boys and girls in a children's home outside London between 1948 and 1972, called the Harpendon Growth Study. Children were measured twice a year until puberty, then every three months until the end of their growth spurt, and finally every year until they were 20 years old. The results for a selection of 20 girls at ages 6, 7, 8, 9, and 10 years are shown in Table 14.6, classified by the mother's height.

(a) Develop an appropriate model to describe these data.
(b) How does growth depend on the mother's height?
(c) Are there any differences among the girls not explained by their mother's height?

References

Aalen, O.O. (1978) Nonparametric inference for a family of counting processes. *Annals of Statistics* **6**, 701–726.

Aalen, O.O. and Husebye, E. (1991) Statistical analysis of repeated events forming renewal processes. *Statistics in Medicine* **10**, 1227–1240.

Adler, R.J., Feldman, R.E., and Taqqu, M.S. (1998) *A Practical Guide to Heavy Tails. Statistical Techniques and Applications.* Basel: Birkhäuser.

Agresti, A. (1990) *Categorical Data Analysis*. New York: John Wiley.

Aickin, M. (1983) *Linear Statistical Analysis of Discrete Data.* New York: John Wiley.

Aitkin, M., Anderson, D., Francis, B., and Hinde, J. (1989) *Statistical Modelling in GLIM.* Oxford: Oxford University Press.

Akaike, H. (1973) Information theory and an extension of the maximum likelihood principle. In B.N. Petrov and F. Csàki (eds.) *Second International Symposium on Inference Theory*, Budapest: Akadémiai Kiadó, pp. 267–281.

Albert, P.S. (1994) A Markov model for sequences of ordinal data from a relapsing-remitting disease. *Biometrics* **50**, 51–60.

Altham, P.M.E. (1978) Two generalizations of the binomial distribution. *Applied Statistics* **27**, 162–167.

Andersen, P.K., Borgan, O., Gill, R.D., and Keiding, N. (1992) *Statistical Models Based on Counting Processes.* Berlin: Springer-Verlag.

Anderson, R.M. and May, R.M. (1991) *Infectious Diseases of Humans. Dynamics and Control.* Oxford: Oxford University Press.

Anderson, W.J. (1991) *Continous-time Markov Chains. An Applications-oriented Approach.* Berlin: Springer-Verlag.

Andrews, D.F. and Herzberg, A.M. (1985) *Data. A Collection of Problems from Many Fields for the Student and Research Worker.* Berlin: Springer-Verlag.

Anscombe, F.J. (1981) *Computing in Statistical Science through APL.* Berlin: Springer-Verlag.

Bailey, N.T.J. (1964) *The Elements of Stochastic Processes with Applications to the Natural Sciences.* New York: John Wiley.

Banks, R.B. (1994) *Growth and Diffusion Phenomena. Mathematical Frameworks and Applications.* Berlin: Springer-Verlag.

Bartholomew, D.J. (1973) *Stochastic Models for Social Processes.* New York: John Wiley.

Bartlett, M.S. (1955) *An Introduction to Stochastic Processes with Special Reference to Methods and Applications.* Cambridge: Cambridge University Press.

Bartlett, M.S. (1963) The spectral analysis of point processes. *Journal of the Royal Statistical Society* **B25**, 264–296.

Baskerville, J.C., Toogood, J.H., Mazza, J., and Jennings, B. (1984) Clinical trials designed to evaluate therapeutic preferences. *Statistics in Medicine* **3**, 45–55.

Baum, L.E. and Petrie, T.A. (1966) Statistical inference for probabilistic functions of finite state Markov chains. *Annals of Mathematical Statistics* **37**, 1554–1563.

Berg, H.C. (1993) *Random Walks in Biology.* Princeton: Princeton University Press.

Bishop, Y.M.M., Fienberg, S.E., and Holland, P.W. (1975) *Discrete Multivariate Analysis: Theory and Practice.* Cambridge, Mass.: MIT Press.

Bloomfield, P. (1973) An exponential model for the spectrum of a scalar time series. *Biometrika* **60**, 217–226.

Bloomfield, P. (1976) *Fourier Analysis of Time Series: An Introduction.* New York: John Wiley.

Box, G.E.P. (1950) Problems in the analysis of growth and wear curves. *Biometrics* **6**, 362–389.

Brauer, F. and Castillo-Chavez, C. (2001) *Mathematical Models in Population Biology and Epidemiology.* Berlin: Springer-Verlag.

Bremaud, P. (1999) *Markov Chains. Gibbs Fields, Monte Carlo Simulation, and Queues.* Berlin: Springer-Verlag.

Brockwell, P.J. and Davis, R.A. (1996) *Introduction to Time Series and Forecasting.* Berlin: Springer-Verlag.

Brown, J.H. and West, G.B. (2000) *Scaling in Biology.* Oxford: Oxford University Press.

Buckle, D.J. (1995) Bayesian inference for stable distributions. *Journal of the American Statistical Association*, **90**, 605–613.

Burnham, K.P. and Anderson, D.R. (1998) *Model Selection and Inference: A Practical Information-Theoretic Approach.* Berlin: Springer-Verlag.

Burr, I.W. and Cislak, P.J. (1968) On a general system of distributions. I. Its curve shaped characteristics. *Journal of the American Statistical Association* **63**, 627–635.

Bye, B.V. and Schechter, E.S. (1986) A latent Markov model approach to the estimation of response errors in multiwave panel data. *Journal of the American Statistical Association*, **81**, 375–380.

Cameron, M.A. and Turner, T.R. (1987) Fitting models to spectra using regression packages. *Applied Statistics* **36**, 47–57.

Carstensen, B. (1996) Regression models for interval censored survival data: application to HIV infection in Danish homosexual men. *Statistics in Medicine* **15**, 2177–2189.

Chatfield, C. (1989) *The Analysis of Time Series: An Introduction.* London: Chapman and Hall.

Chiang, C.L. (1968) *Introduction to Stochastic Processes in Biostatistics.* New York: John Wiley.

Churchill, G.A. (1989) Stochastic models for heterogeneous DNA sequences. *Bulletin of Mathematical Biology* **51**, 79–94.

Ciminera, J.L. and Wolfe, E.K. (1953) An example of the use of extended cross-over designs in the comparison of NPH insulin mixtures. *Biometrics* **9**, 431–446.

Clayton, D.G. (1988) The analysis of event history data: a review of progress and outstanding problems. *Statistics in Medicine* **7**, 819–841.

Cohen, J.E. (1968) On estimating the equilibrium and transition probabilities of a finite-state Markov chain from the same data. *Biometrics* **26**, 185–187.

Collett, D. (1994) *Modelling Survival Data in Medical Research.* London: Chapman and Hall.

Conaway, M.R. (1989) Analysis of repeated categorical measurements with conditional likelihood methods. *Journal of the American Statistical Association* **84**, 53–62.

Cox, D.R. (1955) Some statistical methods connected with series of events. *Journal of the Royal Statistical Society* **B17**, 129–164.

Cox, D.R. (1962) *Renewal Theory.* London: Chapman and Hall.

Cox, D.R. (1972) Regression models and life-tables. *Journal of the Royal Statistical Society* **B34**, 187–220.

Cox, D.R. and Isham, V. (1980) *Point Processes.* London: Chapman and Hall.
Cox, D.R. and Lewis, P.A.W. (1966) *The Statistical Analysis of Series of Events.* London: Chapman and Hall.
Cox, D.R. and Miller, H.D. (1965) *The Theory of Stochastic Processes.* London: Methuen.
Cox, D.R. and Oakes, D. (1984) *Analysis of Survival Data.* London: Chapman and Hall.
Crouchley, R., Davies, R.B., and Pickles, A.R. (1982) Identification of some recurrent choice processes. *Journal of Mathematical Sociology* **9**, 63–73.
Crowder, M.J. (1985) A distributional model for repeated failure time measurements. *Journal of the Royal Statistical Society* **B47**, 447–452.
Crowder, M.J. (1998) A multivariate model for repeated failure time measurements. *Scandinavian Journal of Statistics* **25**, 53–67.
Daley, D.J. and Vere-Jones, D. (1988) *An Introduction to the Theory of Point Processes.* Berlin: Springer-Verlag.
Davidian, M. and Giltinan, D.M. (1995) *Nonlinear Models for Repeated Measurement Data.* London: Chapman and Hall.
Davison, A.C. and Ramesh, N.I. (1996) Some models for discretized series of events. Journal of the American Statistical Association **91**, 601–609.
de Angelis, D. and Gilks, W.R. (1994) Estimating acquired immune deficiency syndrome incidence accounting for reporting delay. *Journal of the Royal Statistical Society* **A157**, 31–40.
Dellaportas, P., Smith, A.F.M., and Stavropoulos, P. (2001) Bayesian analysis of mortality data. *Journal of the Royal Statistical Society* **A164**, 275–291.
Derman, C., Gleser, L.J., and Olkin, I. (1973) *A Guide to Probability Theory and Applications.* New York: Holt, Rinehart, and Winston.
de Stavola, B.L. (1988) Testing departures from time homogeneity in multistate Markov processes. *Applied Statistics* **37**, 242–250.
Diaconis, P. and Ylvisaker, D. (1979) Conjugate priors for exponential families. *Annals of Statistics* **7**, 269–281.
Diggle, P.J. (1990) *Time Series. A Biostatistical Introduction.* Oxford: Oxford University Press.
Dinse, G.E. (1982) Nonparametric estimation for partially-complete time and type of failure data. *Biometrics* **38**, 417–431.
Dobson, A.J. (2002) *An Introduction to Generalized Linear Models.* 2nd edn. London: Chapman and Hall.
Doob, J.L. (1953) *Stochastic Processes.* New York: John Wiley.
Duncan, O.D. (1979) How destination depends on origin in the occupational mobility table. *American Journal of Sociology* **84**, 793–803.
Durbin, J. and Koopman, S.J. (2001) *Time Series Analysis by State Space Methods.* Oxford: Oxford University Press.
Durbin, R., Eddy, S., Krogh, A., and Mitchison, G. (1998) *Biological Sequence Analysis. Probabilistic Models of Proteins and Nucleic Acids.* Cambridge: Cambridge University Press.
Efron, B. (1986) Double exponential families and their use in generalized linear regression. *Journal of the American Statistical Association* **81**, 709–721.
Elliott, R.J., Aggoun, L., and Moore, J.B. (1995) *Hidden Markov Models. Estimation and Control.* Berlin: Springer-Verlag.
Elston, R.C. (1964) On estimating time–response curves. *Biometrics* **20**, 643–647.
Elston, R.C. and Grizzle, J.F. (1962) Estimation of time response curves and their confidence bands. *Biometrics* **18**, 148–159.
Engle, R.F. (1982) Autoregressive conditional heteroscedasticity with estimates of the variance of the United Kingdom inflation. *Econometrica* **50**, 987–1007.

Ewens, W.J. and Grant, G.R. (2001) *Statistical Methods in Bioinformatics. An Introduction.* Berlin: Springer-Verlag.

Faddy, M.J. (1993) A structured compartmental model for drug kinetics. *Biometrics* **49**, 243–248.

Faddy, M.J. (1997) On extending the negative binomial distribution, and the number of weekly winners of the UK national lottery. *Mathematical Scientist* **22**, 77–82.

Faddy, M.J. and Fenlon, J.S. (1999) Stochastic modelling of the invasion process of nematodes in fly larvae. *Applied Statistics* **48**, 31–37.

Fahrmeir, L. and Tutz, G. (1994) *Multivariate Statistical Modelling Based on Generalized Linear Models.* Berlin: Springer-Verlag.

Feigin, P.D. (1981) Conditional exponential families and a representation theorem for asymptotic inference. *Annals of Statistics* **9**, 597–603.

Feigl, P. and Zelen, M. (1965) Estimation of exponential survival probabilities with concomitant information. *Biometrics* **21**, 826–838.

Feller, W. (1950) *An Introduction to Probability Theory and its Applications.* Volume I. New York: John Wiley.

Finch, P.D. (1982) Difficulties of interpretation with Markov chain models. *Australian Journal of Statistics* **24**, 343–349.

Fingleton, B. (1984) *Models of Category Counts.* Cambridge: Cambridge University Press.

Finkelstein, D.M. and Wolfe, R.A. (1985) A semiparametric model for regression analysis of interval-censored failure time data. *Biometrics* **41**, 933–945.

Fitzmaurice, G.M. and Laird, N.M. (1993) A likelihood-based method for analysing longitudinal binary responses. *Biometrika* **80**, 141–151.

Gamerman, D. (1991) Dynamic Bayesian models for survival data. *Journal of the Royal Statistical Society* **C40**, 63–79.

Gardiner, C.W. (1985) *Handbook of Stochastic Methods for Physics, Chemistry, and the Natural Sciences.* Berlin: Springer-Verlag.

Gehan, E.A. (1965) A generalized Wilcoxon test for comparing arbitrarily singly-censored samples. *Biometrika* **52**, 203–223.

Gibaldi, M. and Perrier, D. (1982) *Pharmacokinetics.* Basle: Marcel Dekker.

Goodman, L.A. (1962) Statistical methods for analyzing processes of change. *American Journal of Sociology* **68**, 57–78.

Grandell, J. (1997) *Mixed Poisson Processes.* London: Chapman and Hall.

Grimmett, G.R. and Stirzaker, D.R. (1992) *Probability and Random Processes.* 2nd edn. Oxford: Oxford University Press.

Gutfreund, H. (1995) *Kinetics for the Life Sciences. Receptors, Transmitters, and Catalysts.* Cambridge: Cambridge University Press.

Guttorp, P. (1995) *Stochastic Modeling of Scientific Data.* London: Chapman and Hall.

Hamilton, J.D. (1990) Analysis of time series subject to changes in regime. *Journal of Econometrics* **45**, 39–70.

Hand, D.J. and Crowder, M.J. (1996) *Practical Longitudinal Data Analysis.* London: Chapman and Hall.

Hand, D.J., Daly, F., Lunn, A.D., McConway, K.J., and Ostrowski, E. (1994) *A Handbook of Small Data Sets.* London: Chapman and Hall.

Harvey, A.C. (1989) *Forecasting, Structural Time Series Models and the Kalman Filter.* Cambridge: Cambridge University Press.

Harvey, A.C. (1993) *Time Series Models.* London: Harvester Press.

Hay, J.W. and Wolak, F.A. (1994) A procedure for estimating the unconditional cumulative incidence curve and its variability for the human immunodeficiency virus. *Journal of the Royal Statistical Society* **C43**, 599–624.

Healy, M.J.R. and Tillett, H.E. (1988) Short-term extrapolation of the AIDS epidemic. *Journal of the Royal Statistical Society* **A151**, 50–61.

Heckman, J.J. and Willis, R.J. (1977) A beta-logistic model for the analysis of sequential labor force participation by married women. *Journal of Political Economy* **85**, 27–58.

Henderson, R. and Matthews, J.N.S. (1993) An investigation of changepoints in the annual number of cases of hæmolytic uræmic syndrome. *Applied Statistics* **42**, 461–471.

Hoffmann-Jørgensen, J. (1994) *Probability with a View Toward Statistics.* 2 Vols. London: Chapman and Hall.

Hoppensteadt, F.C. and Peskin, C.S. (2002) *Modeling and Simulation in Medicine and the Life Sciences.* Berlin: Springer-Verlag.

Hsu, D.A. (1979) Detecting shifts of parameter in gamma sequences with applications to stock price and air traffic flow analysis. *Journal of the American Statistical Association* **74**, 31–40.

Hughes, T.H. and Matis, J.H. (1984) An irreversible two-compartment model with age-dependent turnover rates. *Biometrics* **40**, 501–505.

Iosifescu, M. (1980) *Finite Markov Processes and their Applications.* New York: John Wiley.

Janacek, G. and Swift, L. (1993) *Time Series. Forecasting, Simulation, Applications.* London: Ellis Horwood.

Jarrett, R.G. (1979) A note on the intervals between coal-mining disasters. *Biometrika* **66**, 191–193.

Joe, H. (1997) *Multivariate Models and Dependence Concepts.* London: Chapman and Hall.

Jones, B. and Kenward, M.G. (1989) *Design and Analysis of Cross-over Trials.* London: Chapman and Hall.

Jones, P.W. and Smith, P. (2001) *Stochastic Processes. An Introduction.* London: Edward Arnold.

Jones, R.H. (1984) Fitting multivariate models to unequally spaced data. In E. Parzen (ed.) *Time Series Analysis of Irregularly Observed Data.* Berlin: Springer-Verlag, pp. 158–118.

Jones, R.H. (1993) *Longitudinal Data Analysis with Serial Correlation: A State-space Approach.* London: Chapman and Hall.

Jones, R.H. and Boadi-Boateng, F. (1991) Unequally spaced longitudinal data with AR(1) serial correlation. *Biometrics* **47**, 161–175.

Jørgensen, B. (1986) Some properties of exponential dispersion models. *Scandinavian Journal of Statistics* **13**, 187–198.

Jørgensen, B. (1987) Exponential dispersion models. *Journal of the Royal Statistical Society* **B49**, 127–162.

Jørgensen, B. (1992) Exponential dispersion models and extensions: a review. *International Statistical Review* **60**, 5–20.

Jørgensen, M., Keiding, N., and Sakkebak, N.E. (1991) Estimation of spermarche from longitudinal spermaturia data. *Biometrics* **47**, 177–193.

Juang, B.H. and Rabiner, L.R. (1991) Hidden Markov models for speech recognition. *Technometrics* **33**, 251–272.

Kalbfleisch, J.D. and Lawless, J.F. (1985) The analysis of panel data under a Markov assumption. *Journal of the American Statistical Association* **80**, 863–871.

Kalbfleisch, J.D., Lawless, J.F., and Vollmer, W.M. (1983) Estimation in Markov models from aggregate data. *Biometrics* **39**, 907–919.

Kalbfleisch, J.D. and Prentice, R.L. (1980) *The Statistical Analysis of Failure Time Data.* New York: John Wiley.

Kalbfleisch, J.G. (1985) *Probability and Statistical Inference.* Vol. 2. Berlin: Springer-Verlag.

Kaplan, E.L. and Meier, P. (1958) Nonparametric estimation from incomplete observations. *Journal of the American Statistical Association* **53**, 457–481.

Karlin, S. and Taylor, H.M. (1975) *A First Course in Stochastic Processes.* New York: Academic Press.

Karlin, S. and Taylor, H.M. (1981) *A Second Course in Stochastic Processes.* New York: Academic Press.

Karr, A.F. (1991) *Point Processes and their Statistical Inference.* Basel: Marcel Dekker.

Keene, O.N. (1995) The log transformation is special. *Statistics in Medicine* **14**, 811–819.

Kein, S., Timmer, J., and Honerkamp, J. (1997) Analysis of multichannel patch clamp recordings by hidden Markov models. *Biometrics* **53**, 870–884.

Kendall, M.G. and Ord, J.K. (1990) *Time Series.* London: Edward Arnold.

Kingman, J.F.C. (1992) *Poisson Processes.* Oxford: Oxford University Press.

Klein, J.P., Klotz, J.H., and Grever, M.R. (1984) A biological marker model for predicting disease transition. *Biometrics* **40**, 927–936.

Klein, J.P. and Moeschberger, M.L. (1997) *Survival Analysis. Techniques for Censored and Truncated Data.* Berlin: Springer-Verlag.

Kitagawa, G. (1987) Non-Gaussian state-space modeling of nonstationary time series. *Journal of the American Statistical Association* **82**, 1032–1063.

Klotz, J.H. (1973) Statistical inference in Bernoulli trials with dependence. *Annals of Statistics* **1**, 373–379.

Kodell, R.L. and Matis, J.H. (1976) Estimating the rate constants in a two-compartmental stochastic model. *Biometrics* **32**, 377–400.

Kotz, S., Kozubowski, T.J., and Podgórski, K. (2001) *The Laplace Distribution and Generalizations. A Revisit with Applications to Communications, Economics, Engineering, and Finance.* Basel: Birkhäuser.

Kramer, W. and Sonnberger, H. (1986) *The Linear Regression Model Under Test.* Heidelberg: Physica-Verlag.

Küchler, U. and Sørensen, M. (1997) *Exponential Families of Stochastic Processes.* Berlin: Springer-Verlag.

Laird, N.M. and Ware, J.H. (1982) Random-effects models for longitudinal data. *Biometrics* **38**, 963–974.

Lawal, H.B. and Upton, G.J.G. (1990) Alternative interaction structures in square contingency tables having ordered classificatory variables. *Quality and Quantity* **24**, 107–127.

Lawless, J.F. (1982) *Statistical Models and Methods for Lifetime Data.* New York: John Wiley.

Lawless, J.F. and Nadeau, C. (1995) Some simple robust methods for the analysis of recurrent events. *Technometrics* 37, 158–168.

Lawrance, A.J. and Lewis, P.A.W. (1980) The exponential autoregressive moving EARMA(p,q) process. *Journal of the Royal Statistical Society* **B42**, 150–161.

Leroux, B.G. and Puterman, M.L. (1992) Maximum-penalized-likelihood estimation for independent and Markov-dependent mixture models. *Biometrics* **48**, 545–558.

Lindsey, J.K. (1992) *The Analysis of Stochastic Processes Using GLIM.* Berlin: Springer-Verlag.

Lindsey, J.K. (1995a) *Modelling Frequency and Count Data.* Oxford: Oxford University Press.

Lindsey, J.K. (1995b) Fitting parametric counting processes by using log-linear models. *Applied Statistics* **44**, 201–212.

Lindsey, J.K. (1997) *Applying Generalized Linear Models.* Berlin: Springer-Verlag.

Lindsey, J.K. (1998) A study of interval censoring in parametric regression models. *Lifetime Data Analysis* **4**, 329–354.

Lindsey, J.K. (1999a) *Models for Repeated Measurements.* 2nd edn. Oxford: Oxford University Press.

Lindsey, J.K. (1999b) Multivariate elliptically-contoured distributions for repeated measurements. *Biometrics* **55**, 1277–1280.

Lindsey, J.K. (2000) A family of models for uniform and serial dependence in repeated measurements studies. *Journal of the Royal Statistical Society* **C49**, 343–357.

Lindsey, J.K. (2001) *Nonlinear Models for Medical Statistics.* Oxford: Oxford University Press.

Lindsey, J.K. (2004) *Introduction to Applied Statistics. A Modelling Approach.* 2nd edn. Oxford: Oxford University Press.

MacDonald, I.L. and Zucchini, W. (1997) *Hidden Markov and Other Models for Discrete-Valued Time Series.* London: Chapman and Hall.

Maguire, B.A., Pearson, E.S., and Wynn, A.H.A. (1952) The time intervals between industrial accidents. *Biometrika* **39**, 168–180.

Marshall, A.W. and Proschan, F. (1972) Classes of distributions applicable in replacement with renewal theory implications. *Proceedings of the Sixth Berkeley Symposium* **1**, 395–415.

Marubini, E. and Valsecchi, M.G. (1994) *Analyzing Survival Data from Clinical Trials and Observational Studies.* New York: John Wiley.

Matis, J.H. (1972) Gamma time-dependency in Blaxter's compartmental model. *Biometrics* **28**, 597–602.

Matis, J.H. and Kiffe, T.R. (2000) *Stochastic Population Models. A Compartmental Perspective.* Berlin: Springer-Verlag.

Matis, J.H., Wehrly, T.E., and Ellis, W.C. (1989) Some generalized stochastic compartment models for digesta flow. *Biometrics* **45**, 703–720.

Matthews, D.E. (1988) Likelihood-based confidence intervals for functions of many parameters. *Biometrika* **75**, 139–144.

McCullagh, P. and Nelder, J.A. (1989) *Generalized Linear Models.* 2nd edn. London: Chapman and Hall.

McGilchrist, C.A. and Aisbett, C.W. (1991) Regression with frailty in survival analysis. *Biometrics* **47**, 461–466.

Meyn, S.P. and Tweedie, R.L. (1993) *Markov Chains and Stochastic Stability.* Berlin: Springer-Verlag.

Morris, C.N. (1983) Natural exponential families with quadratic variance functions: statistical theory. *Annals of Statistics* **11**, 515–529.

Nelson, W. (1995) Confidence limits for recurrence data—applied to cost or number of product repairs. *Technometrics* **37**, 147–157.

Notari, R.E. (1987) *Biopharmaceutics and Clinical Pharmacokinetics. An Introduction.* Basle: Marcel Dekker.

Nowak, M.A. and May, R.M. (2000) *Virus Dyanamics. Mathematical Principles of Immunology and Virology.* Oxford: Oxford University Press.

Oliver, F.R. 1970) Estimating the exponential growth function by direct least squares. *Applied Statistics* **19**, 92–100.

Parzen, E. (1979) Nonparametric statistical data modelling. *Journal of the American Statistical Association* **74**, 105–131.

Pierce, D.A., Stewart, W.H., and Kopecky, K.J. (1979) Distribution free regression analysis of grouped survival data. *Biometrics* **35**, 785–793.

Pollock, K.H., Winterstein, S.R., and Conroy, M.J. (1989) Estimation and analysis of survival distributions for radio-tagged animals. *Biometrics* **45**, 99–109.

Potthoff, R.F. and Roy, S.N. (1964) A generalized multivariate analysis of variance model useful especially for growth curve problems. *Biometrika* **51**, 313–326.

Priestley, M.B. (1981) *Spectral Analysis and Time Series.* San Diego: Academic Press.

Proschan, F. (1963) Theoretical explanation of observed decreasing failure rate. *Technometrics* **5**, 375–383.

Rabiner, L.R. (1989) A tutorial on hidden Markov models and selected applications in speech recognition. *Proceedings of the IEEE* **77**, 257–286.

Raz, J. (1989) Analysis of repeated measurements using nonparametric smoothers and randomization tests. *Biometrics* **45**, 851–871.

Rogers, L.C.G. and Williams, D. (1987) *Diffusions, Markov Processes, and Martingales. II. Ito Calculus.* New York: John Wiley.

Rogers, L.C.G. and Williams, D. (1994) *Diffusions, Markov Processes, and Martingales. I. Foundations.* New York: John Wiley.

Ross, S.M. (1989) *Introduction to Probability Models.* New York: Academic Press.

Royston, P. and Thompson, S.G. (1995) Comparing non-nested regression models. *Biometrics* **51**, 114–127.

Samorodnitsky, G. and Taqqu, M.S. (1994) *Stable Non-Gaussian Random Processes: Stochastic Models with Infinite Variance.* London: Chapman and Hall.

Sandland, R.L. and McGilchrist, C.A. (1979) Stochastic growth curve analysis. *Biometrics* **35**, 255–271.

Scallon, C.V. (1985) Fitting autoregressive processes in GLIM. *GLIM Newsletter* **9**, 17–22.

Schmidt, P. and Witte, A.O. (1988) *Predicting Recidivism using Survival Models.* Berlin: Springer-Verlag.

Senn, S. (1993) *Cross-over Trials in Clinical Research.* New York: John Wiley.

Seshradi, V. (1999) *The Inverse Gaussian Distribution. Statistical Theory and Applications.* Berlin: Springer-Verlag.

Shephard, N. (1996) Statistical aspects of ARCH and stochastic volatility. In D.R. Cox, D.V. Hinkley, and O.E. Barndorff-Nielsen (eds.) *Time Series Models in Econometrics, Finance, and other Fields.* London: Chapman and Hall.

Smith, R.L. and Miller, J.E. (1986) A non-Gaussian state space model and application to prediction of records. *Journal of the Royal Statistical Society* **B48**, 79–88.

Smith, W.L. (1958) Renewal theory and its ramifications. *Journal of the Royal Statistical Society* **B20**, 243–302.

Snyder, D.L. and Miller, M.I. (1991) *Random Point Processes in Time and Space.* Berlin: Springer-Verlag.

Sullivan, T.J., Still, R.G., Sakmar, E., Blair, D.C., and Wagner, J.G. (1974) *In vitro* and *in vivo* availability of some commercial prednisolone tablets. *Journal of Pharmacokinetics and Biopharmaceutics* **2**, 29–41.

Tan, F.E.S., Ouwens, M.J.N., and Berger, M.P.F. (2001) Detection of influential observations in longitudinal mixed effects regression models. *Statistician* **50**, 271–284.

Taylor, J.M.G., Cumberland, W.G., and Sy, J.P. (1994) A stochastic model for analysis of longitudinal AIDS data. *Journal of the American Statistical Association* **89**, 727–736.

Taylor, J.M.G. and Law, N. (1998) Does the covariance structure matter in longitudinal modelling for the prediction of future CD4 counts? *Statistics in Medicine* **17**, 2381–2394.

Thompson, W.A. (1988) *Point Process Models with Applications to Safety and Reliability.* London: Chapman and Hall.

Tijms, H.C. (1994) *Stochastic Models. An Algorithmic Approach.* New York: John Wiley.

Turner, M.E., Monroe, R.J., and Homer, L.D. (1963) Generalized kinetic regression analysis: hypergeometric kinetics. *Biometrics* **19**, 406–428.

Upton, G.J.G. and Fingleton, B. (1989) *Spatial Data Analysis by Example.* Vol. 2. *Categorical and Directional Data.* New York: John Wiley.

Wei, L.J. and Lachin, J.M. (1984) Two-sample asymptotically distribution-free tests for incomplete multivariate observations. *Journal of the American Statistical Association* **79**, 653–661.

Weng, T.S. (1994) Toward a dynamic analysis of disease-state transition monitored by serial clinical laboratory tests. *Communication in Statistics* **A23**, 643–660.

West, M. and Harrison, J. (1989) *Bayesian Forecasting and Dynamic Models.* Berlin: Springer-Verlag.

West, M., Harrison, P.J., and Migon, H.S. (1985) Dynamic generalized linear models and Bayesian forecasting. *Journal of the American Statistical Association* **80**, 73–97.

Whitmore, G.A. (1995) Estimating degradation by a Wiener diffusion process subject to measurement error. *Lifetime Data Analysis* **1**, 307–319.

Wolff, R.W. (1989) *Stochastic Modelling and the Theory of Queues.* Englewood Cliffs: Prentice-Hall.

Yafune, A. (1999) Application to pharmacokinetic analysis. In H. Akaike and G. Kitagawa (eds.) *The Practice of Time Series Analysis.* Berlin: Springer-Verlag, pp. 153–162.

Zeger, S.L. (1988) A regression model for time series of counts. *Biometrika* **75**, 621–629.

Zeger, S.L., Liang, K.Y., and Albert, P.S. (1988) Models for longitudinal data: a generalized estimating equation approach. *Biometrics* **44**, 1049–1060.

Author index

Subject index

Printed in the United States
By Bookmasters